The Molecular Basis of Gene Expression

4/XI

The Molecular Basis of Gene Expression

Benjamin M. Lewin
School of Biological Sciences, University of Sussex, England

WILEY-INTERSCIENCE
a division of John Wiley & Sons Ltd.
LONDON NEW YORK SYDNEY TORONTO

Library of Congress Catalog card No. 70-116654
ISBN 0 471 53160 X Cloth bound
ISBN 0 471 53161 8 Paper bound

Printed in Great Britain by
Dawson & Goodall Ltd.,
The Mendip Press, Bath.

Preface

This book has been written in the belief that our knowledge of the macromolecular systems involved in gene function has now reached a stage where it would be useful to integrate the ideas that have so far evolved. The purpose of the book, then, is to collate recent advances in the understanding of the mechanisms of nucleic acid and protein synthesis, and of the control of these processes. In short, its theme concerns the topics often popularly referred to as 'molecular biology'. Since this is a rapidly advancing field, some aspects of the discussion will, of course, be overtaken fairly soon by further research. However, although important problems remain to be solved, it seems likely that much of our present conceptual framework will retain its validity, and future research result, by and large, in additions and in changes of detail rather than in basic revision. It is my hope, therefore, that this book will provide such a framework, and prove of use both to students and to those who are interested in applying concepts developed in this field to related topics.

As the meeting point of several related disciplines, molecular biology derives from experiments which differ widely both in technical design and in conceptual rationale, encompassing a range of approaches from classical genetics, the biochemistry of nucleic acids, and microbial genetics. In light of this, a preliminary section provides a background from first principles in each of these three topics. Subsequent sections discuss protein biosynthesis and the components of the synthetic apparatus, the types of control system used for the regulation of gene expression, and the reproduction of DNA. Discussion is developed in what are essentially conceptual terms. Of course, as with any subject depending upon experiments, when lines of deductive reasoning interrelate experimental result and conceptual advance, this demands an appreciation of the techniques involved. However, whilst I have tried to meet this demand by discussing the underlying principles of experimental approaches, in general, emphasis is placed upon considering concepts rather than providing experimental details.

Almost all of this work has been performed in bacterial systems, in which our understanding of the processes involved in protein synthesis appears now to be soundly based; and we have clear ideas as to how gene function is controlled. Somewhat surprisingly, our knowledge of the behaviour of DNA itself is not so well advanced, and many of the enzymic processes which operate on DNA remain to be characterised. Although present evidence suggests that the mechanisms of protein synthesis are likely to be very similar in all organisms, our understanding of the behaviour of chromosomes and of the control of gene action in higher organisms is at a rather more preliminary stage—data is being gathered and concepts developed, but this aspect of the field is in a high degree of flux, and there is as yet no established conceptual framework within which to place and evaluate further discoveries. For this reason, discussion has been restricted largely to bacterial systems; and the analogous control mechanisms of higher organisms, such as those concerned with differentiation and development, must at present fall outside the scope of this study.

The references cited are intended to be sufficient to permit recourse to original papers, and especially to reviews more detailed in nature; however, no attempt has been made to provide an exhaustive survey of the literature, which might tend to obscure the main lines of research and turn the text into little more than a list of references. In many instances, therefore, there is further evidence to support the conclusions drawn, and where possible, this has been indicated by reference to reviews.

I gratefully acknowledge permission from many authors and publishers to reproduce diagrams from scientific journals; these are referred to individually. I should like to thank my publishers, John Wiley and Sons, for their unfailing courtesy and assistance. I am indebted to many of my colleagues at the University of Sussex for their assistance, and am especially grateful to Drs. J. G. Little and S. Shall for their helpful comments. I am deeply indebted also to Drs. B. J. Smith and M. Waring, who read the entire manuscript and suggested many improvements. Any errors remaining are, of course, the sole responsibility of the author. Finally, it is a pleasure to acknowledge the immeasurable debt I owe to my father, Dr. S. Lewin, whose enthusiasm first interested me in this subject, for innumerable stimulating discussions and invaluable advice and encouragement.

October 1969 B.M.L.

Contents

I INTRODUCTION

9 Suppression

III THE CONTROL OF PROTEIN SYNTHESIS

10 The operon

11 Systems of regulation

12 Translational control mechanisms

IV THE REPRODUCTION OF DNA

13 Replication

14 Repair of DNA

15 Recombination

I

Introduction

CHAPTER 1

The gene

Mendelian inheritance

The basis of modern genetics is the work performed on plant hybridization by Mendel in 1865. This was carried out using the garden pea, in which under normal conditions pollen from the flower falls on its own stigma to effect self-fertilization. However, it is possible to open the flower bud and remove the stamens before the pollen is shed; pollen from another plant can then be added to effect a *hybridization*. The second hybrid generation following the cross can be observed by allowing self-fertilization of the first hybrid generation.

This work succeeded, in contrast to the failure of earlier attempts to understand the mechanism of heredity, for two reasons. Whereas previously, all the traits in which the hybrids differed from their parents and each other were studied, Mendel confined his attention to one character at a time. The work was extended to following the inheritance of two or more characters together only after the pattern of behaviour of a single trait had been established. Further, his experiments were placed on a quantitative basis by counting the numbers of each progeny type resulting from the hybridization.

The procedure was to cross two plants differing in some contrasted character, such as flower colour, and to follow the behaviour of this character through subsequent generations. The parent generation is termed P_1, the first hybrid generation F_1, and subsequent generations F_2, F_3, etc. It is important that the parental plants should be *true-breeding* for the character under consideration, that is, they should be the result of a line of plants showing only the one particular form of the characteristic. For example, if red and white flower colours are used, the parent plants should be derived from lines showing only red and white flowers respectively.

When plants breeding true for red and white flowers were crossed, all the progeny had red flowers. Thus the red characteristic was termed *dominant*, and its white alternative *recessive*. All of the seven characters studied behaved in this way, including a round form of seed which was dominant over the recessive wrinkled form, and a yellow colour of cotyledon which was dominant over green. However, in contrast to the uniformity of the F_1 generation, the F_2 generation (obtained by self-fertilization of the F_1) showed a segregation of plants into the two original parental types. Quantitatively, there were three individuals of the dominant characteristic to every one with the recessive.

This led to the hypothesis that the characteristics studied are determined by particulate 'factors' of inheritance that are passed from parent to offspring in the *gametes* (sex cells). These factors are now known as *genes*, and the alternative forms of a gene which result in the expression of a different characteristic (such as red and white flower colour) are termed *alleles*. Mendel's law of independent segregation states that these alleles do not affect each other when present in a hybrid plant, but segregate unchanged by passing into different gametes when this hybrid gives rise to the next generation.

According to this theory, each true-breeding parent possesses two identical copies of its allele for the character under consideration, and one of these copies passes into the gamete. When the gametes of two parents unite, the *zygote* formed thus gains two alleles for this character, but these are different when the two parents are of different genotypes. Each parent is said to be *homozygous* for its particular allele, and the F_1 hybrid *heterozygous*. When the hybrid forms gametes, each gamete gains one of the alternative alleles, and it is purely random as to which gamete receives which allele. At union of the gametes from the F_1, each of the F_2 zygotes will receive one allele from each parent, but in contrast to the previous generation where each parent produces only one type of allele, in this generation each parent may produce either type. The F_2 zygotes are then formed by a random association of these gametes.

It is conventional to represent dominant genes by capital letters and their recessive alleles by the corresponding small letters. Thus the true-breeding parental types for red and white flower colour can be represented as *RR* and *rr*, where *R* is the dominant allele responsible for the production of red coloration. The two types of gamete, *R* and *r*, unite to form the hybrid F_1 plants which can be represented as *Rr*. Any *Rr* plant can in turn form both *R* and *r* gametes, and these then unite in the zygotes of the F_2 to give the types *RR*, *Rr* and *rr*, as illustrated in Figure 1.1.

If association of the gametes produced by the F_1 is random, there being no preference for any particular type of zygote, plants should be produced in the proportions 1 *RR* : 2 *Rr* : 1 *rr*. Since *R* is dominant over *r*, the *Rr* plants appear red as are the *RR*, so that the ratio observed

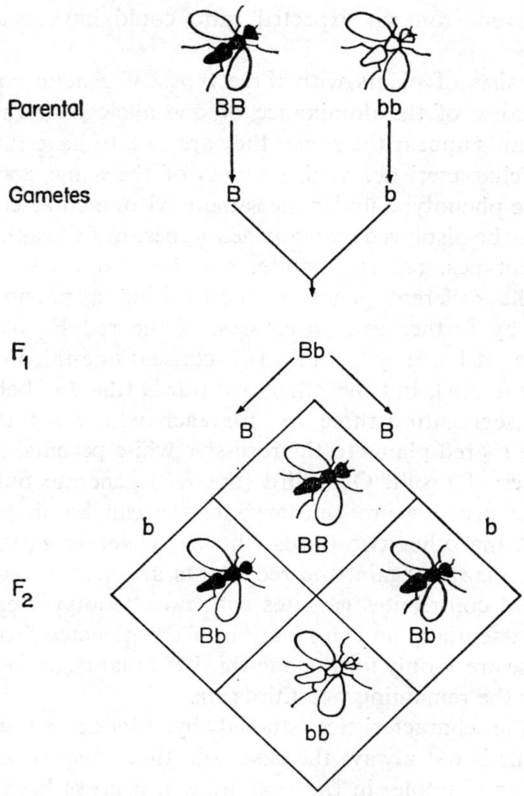

Figure 1.1. Independent segregation of alleles for body colour in *Drosophila*. Each F_1 fly receives one *B* allele (dominant for dark body) and one *b* allele (recessive for light body) to yield the hybrid *Bb*. At gamete formation, each F_1 forms equal numbers of gametes bearing *B* or *b*, and these unite randomly at zygote formation to generate the F_2 of *BB* : 2*Bb* : *bb*. Because *B* is dominant to *b*, the genotypes *BB* and *Bb* exhibit the dark phenotype, so that the observed segregation is 3 dark : 1 light

becomes three red to each white. This is a statistical probability and not an exact result; in any experiment the ratios of progeny will approximate to this ratio, and the more individuals bred, the closer the approximation will be. Statistical tests may be used to estimate whether the deviation actually observed from an expected ratio could have occurred merely by chance.

The F_2 consists of plants with three types of genetic constitution, or *genotype*. Because of the dominance of one allele over the other, both *RR* and *Rr* plants appear the same; they are said to have the same *phenotype* for this characteristic. Whilst plants of the same genotype always show the same phenotype under the same environmental conditions, one phenotype may be displayed by many genotypes. In this case, there are two possible phenotypes, red and white, and the red class comprises two genotypes. The different genotypes comprising a phenotype may be distinguished by further genetic crosses. If the red F_2 plants are bred further to give an F_3, they fall into two classes; one-third of them breed true for red (the *RR*), but the other two-thirds (the *Rr*) behave similarly to the F_1 in segregating three red for each white. Another test is to *back-cross* the F_2 red plants to the recessive white parents, and again this gives two types of result. One-third (the *RR*) generate only red plants, although these progeny are heterozygous, as can be shown by further crosses, whilst the other two-thirds (the *Rr*) generate equal numbers of red and white plants. Again, the red plants are heterozygous. Since the recessive parent contributes gametes with no phenotypic expression, the back-cross is essentially an examination of the gametes produced by the F_2 plant. These are *R* only for the one-third *RR* plants, and equal numbers of *R* and *r* for the remaining two-thirds *Rr*.

Although the characteristics studied by Mendel showed complete dominance, this is not always the case, and there may be alleles without dominance. For example, in the snapdragon, a cross between red (*RR*) and white (*rr*) produces pink (*Rr*) progeny. In this instance, the amount of red colour depends upon the number of alleles present which carry the red specification. Although there is no dominance, the same rule that the F_1 is uniform is observed, and all are pink. In the F_2, the same genotypes as before are produced, only these are now 1 red (*RR*) : 2 pink (*Rr*) : 1 white (*rr*). Further genetic crosses show exactly the same pattern of inheritance as in the fully dominant situation, except that the heterozygote *Rr* has a phenotype different from that of either parent homozygote. Alleles can show any degree of dominance, ranging from the fully dominant, through the partially dominant where the heterozygote resembles one parent more than the other, to the non-dominant where the hybrid is exactly intermediate between the homozygotes.

In the *dihybrid cross* the inheritance of two different types of character is followed, as opposed to only one in the *monohybrid cross* discussed above. One such cross was between a plant with round and yellow seeds and one with wrinkled and green seeds. The F_1 were uniform as round and yellow, since these are the dominant characters. In the F_2 each single character showed the usual segregation, with three round per wrinkled and three yellow per green. However, an examination of both characters together showed that the F_2 contained not only the round and yellow or wrinkled and green parental types of plant, but also two completely new types which were round and green, and wrinkled and yellow. These are termed the *recombinant* types. The proportions found were:

9 round+yellow: 3 round+green: 3 wrinkled+yellow: 1 wrinkled+green

Thus 9/16 show both dominant characters, 3/16 show one combination of dominant and recessive and another 3/16 show the other (*reciprocal*) combination, whilst only 1/16 shows the double recessive parental combination.

If the gene for round seeds is R and its recessive allele r (wrinkled phenotype), and the yellow gene is Y with a recessive allele y (green phenotype), the original true-breeding parental types can be represented as $RRYY$ and $rryy$. These produce gametes RY and ry which unite to give the F_1 $RrYy$. When this species gives gametes, each gamete obtains one of the alleles for seed shape (that is R or r) and one of the alleles for seed colour (that is Y or y). Mendel's law of independent assortment states that it is purely random which of the two alleles for the one characteristic combines with which of the two alleles for the other. Thus all four possible types of gamete, RY, Ry, rR and ry, are equally likely to be formed. When two F_1 plants are crossed, the association between these types of gamete is purely random, so that zygotes are formed as shown in Figure 1.2. Exactly the same F_1 and F_2 would result from the *reciprocal cross* of true-breeding round and green plants ($RRyy$) with true-breeding wrinkled and yellow ($rrYY$).

Because of the dominance relationships the four genotypes $RRYY$, $RrYY$, $RRYy$ and $RrYy$ all have the round and yellow phenotype, giving this a total fraction of 9/16. The round and green recombinant phenotype comprises one $RRyy$ and two $Rryy$ genotypes, and similarly, the wrinkled and yellow 1 $rrYY$: 2 $rrYy$, so that each recombinant class represents 3/16 of the total. The double homozygous recessive parental type class $rryy$ comprises only the remaining 1/16. These interpretations can be confirmed by back-crosses to the double recessive parent in a manner similar to the monohybrid F_2 back-cross.

The number of characteristics in the phenotype directly determined by single genes is rather small, most characters being influenced by the interactions of many genes. The phenotypic proportions obtained in

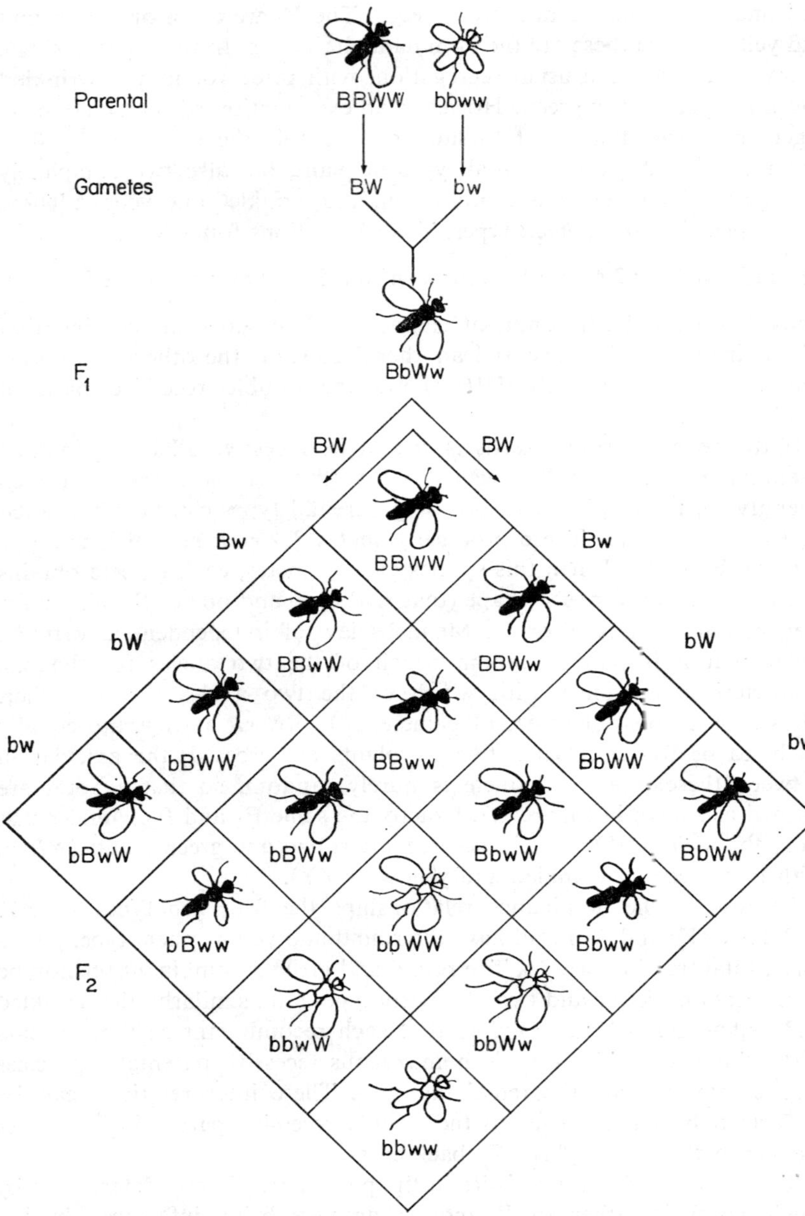
Parental
BBWW
bbww
Gametes
BW
bw
F1
BbWw
BW
BW
Bw
Bw
bW
bW
bw
bw
BBWW
BBwW
BBWw
bBWW
BBww
BbWW
bBwW
bBWw
BbwW
BbWw
bBww
bbWW
Bbww
F2
bbwW
bbWw
bbww

breeding experiments vary according to the number of genes and type of interaction, but can always be accounted for on the basis of Mendelian inheritance. For example, if the presence of two dominant genes is required for expression of some character, a 9:7 ratio is obtained in the F_2—only the '*RY*' phenotypic class of Figure 1.2 would show the character, and it would be absent in the '*Ry*', '*rY*' and '*ry*' classes.

Although many characteristics are determined by interactions between several genes, the converse is also true in that a single gene may influence more than one phenotypic characteristic. A gene which causes changes in unrelated characters is said to be *pleiotropic*. Thus although a gene governs the synthesis of only a single product at the molecular level, this product may influence several characteristics of the phenotype.

Not all genetic traits are concerned with clear-cut qualitative differences, but many show as quantitative variations. These may be accounted for on the basis that many genes each contribute a small effect to the characteristic. A comparatively simple example is the production of red colour in wheat kernels, which can be explained by the hypothesis that two genes are each concerned with colour production, and at each locus the presence of a dominant allele contributes a certain amount of coloration. If the genes are R_1 and R_2, with recessive alleles r_1 and r_2, a cross between red and white parents can be represented as:

$$R_1 R_1 R_2 R_2 \times r_1 r_1 r_2 r_2 \longrightarrow R_1 r_1 R_2 r_2$$

Thus the F_1 is intermediate in colour between the two parents, that is half red. The F_2 is essentially the same as that of Figure 1.2, and if each dominant gene present produces the same amount of colour, irrespective of its locus, the total outcome in terms of coloration produced can be represented as shown in Table 1.1.

Thus the pattern of colours in the F_2 follows a binomial distribution, 1 : 4 : 6 : 4 : 1. In other similar crosses it has been found that only 1/64 of the F_2 is white, suggesting the segregation of three genes, as was confirmed by finding the phenotypic distribution 1 : 6 : 15 : 20 : 15 : 6 : 1

Figure 1.2. Independent assortment of two genes in a dihybrid cross. Each gamete receives one of the alleles at the *B*/*b* locus and one from the *W*/*w* locus (winged *versus* wingless). Which allele is received at one locus does not affect that donated from the other locus, so that the F_1 parents generate equal numbers of each of the four classes, *BW*, *Bw*, *bW* and *bw*. Their random association produces the F_2 genotypes shown; because of the dominance relationships, the phenotypic proportions observed are 9 '*BW*' : 3 '*Bw*' : 3 '*bW*': 1 '*bw*'. (Note: phenotypes may be indicated by the same nomenclature used for genotypes, but with the letters placed in inverted commas.)

Table 1.1. The quantitative inheritance of kernel colour in wheat

Genotype	Phenotype	Number	Total
$R_1R_1R_2R_2$	4 dominants = full red	1	1
$R_1R_1R_2r_2$	3 dominants = $\frac{3}{4}$ red	2	4
$R_1r_1R_2R_2$		2	
$R_1R_1r_2r_2$	2 dominants = $\frac{1}{2}$ red	1	6
$r_1r_1R_2R_2$		1	
$R_1r_1R_2r_2$		4	
$r_1r_1r_2R_2$	1 dominant = $\frac{1}{4}$ red	2	4
$R_1r_1r_2r_2$		2	
$r_1r_1r_2r_2$	0 dominants = white	1	1
			16

for colours from fully red to white. As the number of genes concerned in the quantitative determination of a characteristic increases, so does the width of the distribution. Eventually, this becomes a normal (Gaussian) distribution, such as is found in the population for continuously varying factors such as intelligence, height, etc. Of course, in these instances there is also an environmental effect imposed upon the genetic distribution.

The chromosome

The cell theory established in the middle of the last century proposed that all organisms are composed of cells, and that these can arise only from pre-existing cells. A typical cell consists of a dense nucleus separated by a membrane from the less dense surrounding cytoplasm. During the process of cell division, chromosomes appear as chemically stainable threads inside the nucleus, any particular species of organism having a defined number of chromosomes (the chromosome *complement*) of differing but characteristic sizes. When a cell divides, each chromosome appears to be longitudinally split to give rise to two copies, and each of the daughter cells of this *mitosis* receives one of these new chromosomes, ensuring the constancy of the chromosomal complement in the various cell lines. Replication of DNA has already occurred prior to the mitosis, so that although this is not visible, the chromosome then actually contains the two double helices of DNA produced by this replication; the observed chromosome splitting represents their separation into two identical daughter chromosomes (sometimes termed *chromatids* at this stage of the process).

In sexually reproducing organisms the cell has two copies (*homologues*) of each particular chromosome. Heredity is determined by the germ cells produced by the reproductive tissues of the parent, and in the production of gametes the chromosomes undergo a reductive division (*meiosis*) as a result of which each gamete receives one chromosome from each homologous pair. The gamete cells are referred to as *haploid*, compared with the normal *diploid* cells containing both homologues; possession of a haploid set of chromosomes is a feature of both egg and sperm. When the gametes unite, the resultant zygote obtains one homologue of each chromosome pair from each parent and is restored to the diploid state.

The behaviour of chromosomes closely parallels the requirements for Mendel's particulate factors of inheritance. Genes should occur in allelic pairs, one member of each pair having been contributed by the gamete from each parent; the diploid set of chromosomes results from the contribution of a haploid set by each parent. Further, the assortment of the non-allelic genes into gametes should be independent of the parent from which they have come, the only proviso being that each gamete obtains a complete haploid set. It is found that non-homologous maternal and paternal chromosomes undergo segregation independently of each other.

Direct evidence for the role of chromosomes in heredity came from identification of an association between a specific gene and a particular chromosome. In many bisexual organisms, there is an exception to the rule that chromosomes occur in homologous pairs whose disjunction at meiosis results in the production of two identical haploid sets. Male and female cells may differ visibly in their chromosome constitutions, the most common form of difference being the absence in one sex of one of the homologues of a certain pair and its replacement by a different chromosome. These chromosomes are referred to as the *sex chromosomes*, and the remaining normal homologous pairs as the *autosomes*. If the haploid set of autosomes is denoted by A, the sex chromosome present in both sexes as X and the other Y, the sex containing both X chromosomes is the *homogametic* (chromosome complement 2A+XX), and the sex with a single X and one Y the *heterogametic* (2A+XY). These form gametes of A+X only, and A+X or A+Y equally, respectively, so that their random union gives equal numbers of zygotes of each sex. Either sex may be the homogametic; in the insect *Drosophila* and in mammals (including man) it is the female, whilst in birds it is the male.

According to Mendelian prediction, the results of a genetic cross between certain characteristics should be the same irrespective of which parent introduces which characteristic. However, in *Drosophila* the cross between white eyes (recessive) and red eyes (dominant) gives different results depending upon which sex carries the mutant. The cross of white male × red female gives the expected entirely red F_1, but all the white-eyed flies reappearing as one-quarter of the F_2 are males. In the reciprocal cross of red male × white female, all the F_1 males are white eyed and all the females red eyed. Crossing these gives an F_2 with equal proportions of white and red eyes in each sex. These results can be explained on the assumption that the alleles for red and white eyes are carried on the X chromosome, with no such gene locus for eye colour present at all on the Y chromosome. This means that the eye colour phenotype of a male fly is determined only by the allele on its single X chromosome. This pattern of sex inheritance is shown in Figure 1.3.

The number of genetic factors greatly exceeds the number of chromosomes, and this led to the proposal that the genes are contained in the chromosomes. Whilst it is clear that genes in different chromosomes will obey Mendel's laws, there remains the problem of the inheritance of genes within the same chromosome. In the first decade of this century the phenomenon of *linkage* was observed; some characters do not assort independently, but tend to remain in the parental arrangement. If two genes *A* and *B* come from the same parent, as in the cross *AABB* × *aabb*, they tend to remain together in subsequent gamete formation, whilst if they come from different parents, as in the cross *AAbb* × *aaBB*, they tend to remain apart. The first was termed *coupling*, the second *repulsion*. Examining the gametes produced by the F_1 of each cross through a back-cross to the double recessive parent showed that although both F_1 hybrids have the same genotype, they produce different proportions of the various classes of gamete. Writing the back-crosses so as to show how each F_1 arrangement of alleles was derived from its parents, they may be represented as:

Coupling	$\frac{AB}{ab} \times \frac{ab}{ab}$		parental	>	recombinant	
		$\longrightarrow$	'*AB*' '*ab*'		'*Ab*' '*aB*'	
Repulsion	$\frac{Ab}{aB} \times \frac{ab}{ab}$		recombinant	<	parental	

In these crosses, the parental and recombinant genotypes have been reversed through introducing the alleles in opposing combinations in the parental generation, and in each there is an excess of the parental arrangement over the recombinant type.

When meiosis commences, homologous chromosomes approach each other and become tightly paired (*synapsed*), so that they are arranged as the haploid number of chromosome pairs. At this point the chromosomes

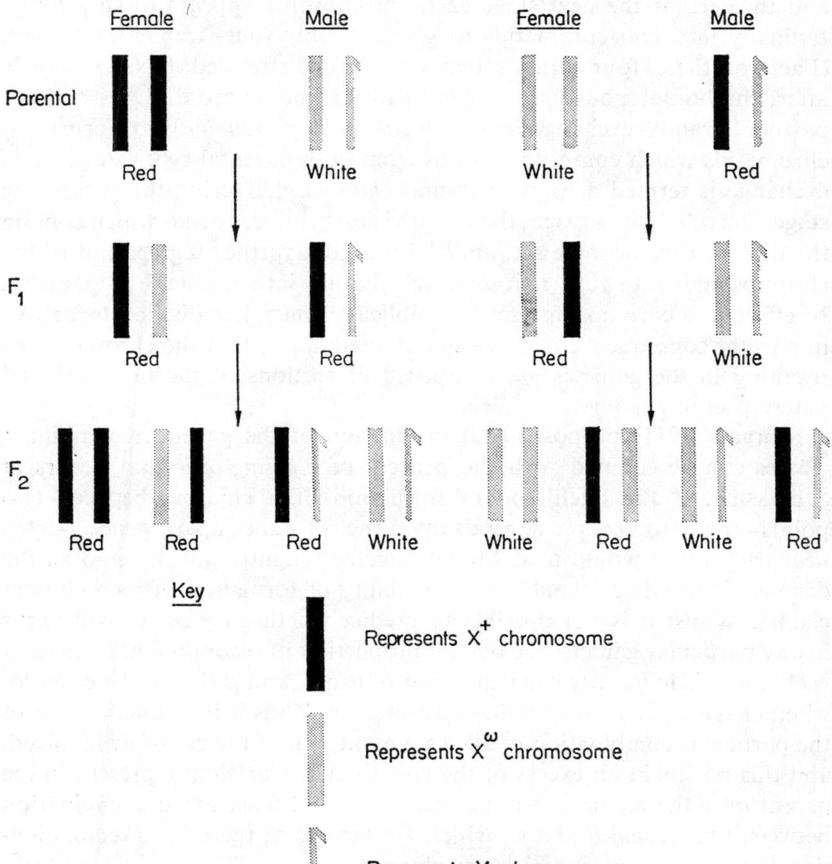

Figure 1.3. Sex-linked inheritance of eye colour in *Drosophila*. The X^+ chromosome carries the dominant red allele for eye colour and the X^ω chromosome bears its recessive white allele. The Y chromosome carries no allele. Thus the eye colour of a male fly depends only upon the X chromosome received from its mother, whilst the phenotype of a female is determined by whether it receives a red allele from either parent. This generates the 'criss-cross' pattern of inheritance as above

are long slender threads, but as meiosis proceeds they become shorter and thicker. At the next stage each chromosome appears to split longitudinally into two chromatids to give a visibly four-stranded structure. (Each of these four strands contains double-stranded DNA.) Shortly after, the homologous chromatids break in one or more places and the partner strands exchange parts (*breakage and reunion*) to form new chromatids which comprise sections from both parental types. A place of exchange is termed a *chiasma* (plural *chiasmata*). During the succeeding stages the cell divides twice; the first division produces *dyads* which contain the diploid chromosome set, and the second a further segregation of the chromosomes into four *tetrads* which mature into the haploid gametes. In effect, the chromosome set has replicated once but divided twice. An important consequence of breakage and reunion is that the chromosomes resulting in the gametes are composed of sections of the maternal and paternal chromosomes.

Morgan (1911) proposed that production of the genetic recombinant classes can be equated with the process of *crossing-over* that occurs at a chiasma. If the likelihood of formation of a chiasma between two points on a chromosome depends upon their distance apart, genes located near each other would tend to stay together (tightly linked), and as the distance increases so would the probability of formation of recombinant classes. Whilst it is not possible to predict whether crossover will occur in any particular gamete, the overall proportion of recombinant formation is characteristic for any particular pair of genes, and is the same no matter whether the cross is in coupling or repulsion. Thus it is a function not of the particular combination of alleles present, but of the *gene-loci* involved, and this results in an excess of the combinations originally present in the parent over the recombinant classes. Figure 1.4 illustrates recombination between two genes, *B* and *C*, which are far apart; there is no recombination between *A* and *B* which are closer together.

The final chromosomes are distributed to different gametes. In the illustration, one chiasma has formed between two of the chromatids so that their gametes show recombination between *B* and *C*, whilst the other two receive unchanged combinations of genes. The two classes *ABc* and *abC* resulting from the recombination event are termed the *reciprocal* recombinants. There may be any number of chiasmata formed involving up to all of the chromatids, and it is possible for one chromatid to have more than one chiasma—these may not necessarily be with the same second chromatid.

Direct evidence that crossing-over corresponds to genetic recombination has been provided by correlation between cytological observation and the

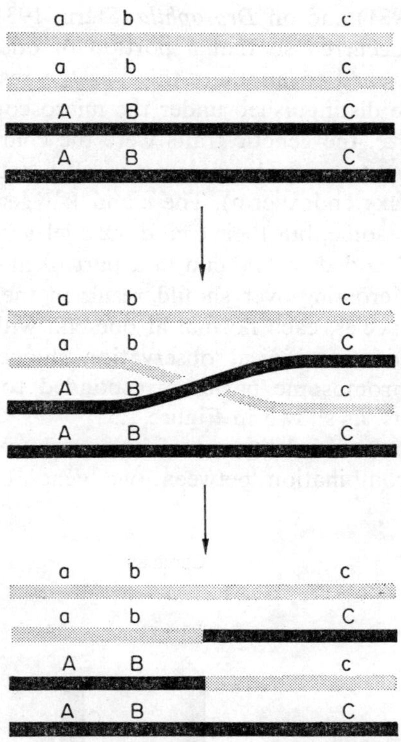

Figure 1.4. Chiasma formation at meiosis. A chiasma has formed between the *B*/*b* and *C*/*c* loci, resulting in recombination between them; *A*/*a* and *B*/*b* remain in their parental arrangement, but *C*/*c* has recombined with respect to both of them. Because recombination occurs on only two of the four strands at one chiasma, two of the resultant chromosomes bear the reciprocal recombinants *ABc* and *abC*, whilst the remaining two carry the parental arrangements *ABC* and *abc*. This means that recombination cannot rise above 50 per cent in frequency

pattern of genetic inheritance. Work was performed on maize (Creighton and McClintock, 1931) and on *Drosophila* (Stern, 1931). In both cases, a *translocation* had occurred so that a portion of one chromosome had broken off and attached to another; this has the result that this latter chromosome can be distinguished under the microscope from its normal homologue. In maize, the genetic traits were the characters *C* (coloured aleurone) dominant to *c* (colourless), and *Wx* (starchy endosperm) dominant to *wx* (waxy endosperm). The *c* and *Wx* genes were associated on a normal chromosome, but their *C* and *wx* alleles were present on one which had a 'knob' added to one end and part of another chromosome to the other. Thus crossing-over should result in the formation of two new chromosome species, each normal at one end with an addition only at the other. Indeed, cytological observation showed that a physical exchange of the chromosome parts corresponded to the formation of genetic recombinants, as shown in Figure 1.5.

Sturtevant (1913) extended the use of linkage data by proposing that the amount of recombination between two genes can be taken as a

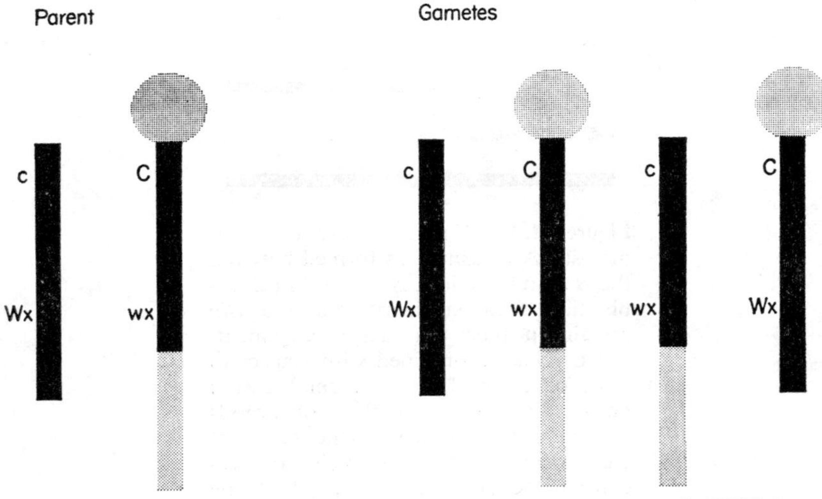

Figure 1.5. Correlation between cytological and genetic observations in maize. The light-coloured segments represent additions to the normal chromosome. In the heterozygote, whenever recombination takes place between the genetic loci *C*/*c* and *Wx*/*wx* a corresponding physical rearrangement of the chromosome is observed

measure of their distance apart. This *map distance* is usually expressed as the per cent of recombination, that is recombinants × 100/total progeny. When the inheritance of three instead of only two characters was followed, this concept led to the finding that map distances are additive. For example, after the cross

$$AABBCC \times aabbcc \longrightarrow ABC/abc$$

the back-cross of the F_1 to the recessive parent might give the phenotypes:

Parental	'*ABC*'	575	
	abc	615	1190
a–bc crossover	*aBC*	140	
	Abc	160	300
ab–c crossover	*ABc*	270	
	abC	230	500
ac–b crossover	*AbC*	6	
	aBc	4	10
		2000	

This gives the crossover values for linkage of:

$$a\text{–}b \text{ crossover} = \frac{300+10}{2000} = 15{\cdot}5 \text{ per cent.}$$

$$b\text{–}c \text{ crossover} = \frac{500+10}{2000} = 25{\cdot}5 \text{ per cent.}$$

$$a\text{–}c \text{ crossover} = \frac{500+300}{2000} = 40{\cdot}0 \text{ per cent.}$$

The total of the map distances *a–b*+*b–c* is 41 per cent., compared with the recombination frequency actually found of 40 per cent. Thus the genes can be assigned in the linear order *a–b–c*, with distances of 15·5 units from *a* to *b*, and 25·5 units from *b* to *c*. This is an example of a general phenomenon; if distances *a–b* and *b–c* are mapped for genes *a*, *b* and *c*, the distance *a–c* will equal either their sum or difference, depending upon whether *c* is beyond *a* and *b* or in between them. In other words, *a*, *b* and *c* lie on a straight line, and this is the basis of the concept that chromosomes consist of a linear array of genes.

In principle, this procedure may be extended to map the position of any new mutant discovered by measuring its linkage with two known genes on the chromosome. In practice, there is a difficulty due to the phenomenon of *double crossover*. In the example above, a crossover between *b* and *c* would generate the *aBC* and *Abc* classes, and a crossover between *b* and *c* would generate *ABc* and *abC*. In order to generate a class with the parental combinations of the *A* and *C* loci, but showing recombination to the middle marker, *b*, it is necessary for two crossovers to occur. One must be between *a* and *b* to generate *aBC* and *Abc*, and then a second must occur between *b* and *c* to give *aBc* and *AbC*, so that the parental *A–C* and *a–c* combinations, separated by the first crossover, are restored. The consequence of two such crossovers is to increase the number of progeny with the parental combinations of the far apart markers, so that the apparent distance between them is lowered. This is why the sum of the *a–b* and *b–c* distances (41 per cent.) is greater than that actually found for *a–c* crossover (40 per cent.). Further, it is not possible for a recombination greater than 50 per cent. to occur—this is the independent segregation predicted by Mendel's law. When two loci are sufficiently far apart on the same chromosome the probability that a crossover will form between them is very large, but because a chiasma involves only two of the four strands in meiosis, only half of the gametes contain such recombinants, that is at most 50 per cent. This means that genetic maps cannot be extended directly, but the recombination distances between two genes which are far apart can be calculated by addition of the distances from intermediate genes. A chromosome could thus, for example, be 120 recombination units long, but linkage is only shown between loci closer to each other than 50 units.

There is a correlation between the behaviour of chiasmata and the formation of double crossovers which affords further support to the concept of the latter as sites of genetic recombination. In the example above, if there is 15·5 per cent. and 25·5 per cent. recombination in the single crossover classes, their coincidence to give double crossovers should occur in 15·5 per cent. × 25·5 per cent. × 100 per cent. of the progeny. This gives almost 4 per cent., yet the double recombinant class actually observed totals only 10/2000 = 0·5 per cent. This is due to *positive interference*; the occurrence of one crossover greatly diminishes the probability that another will occur nearby, lowering the number of double crossovers. Studies on the cytology of chiasma formation have shown that there is a definite pattern in their formation, with only a low probability of two occurring near each other. A further correlation has emerged from studies on *Drosophila*; in the male of this species there is no chiasma formation at meiosis, and correspondingly, it is found that genes on the same chromosome show absolute linkage, there being recombination only between genes on different chromosomes.

In general, the calculation of *linkage maps* in an organism correlates with the properties of its chromosomes. As linkage relationships are extended, mutants are found to fall into one of a number of groups, genes showing linkage with others in their group, but demonstrating independent segregation with regard to genes located in other groups; there are the same number of linkage groups as chromosomes. For example, *Drosophila* has three autosomes, two large and one small, in addition to the (large) sex chromosomes. Four linkage groups, independent of each other, have been extended, and their relative sizes correspond roughly to the relative sizes of the chromosomes; of the three large groups one is sex linked, and the fourth group is much smaller. The pattern of Mendelian inheritance developed the concept of a gene as some discrete genetic factor, and these studies validate the concept of the chromosome as a linear unit divided into many genes, each of which is much the same order of size.

The cistron

The mapping of mutations affecting certain characteristics has shown that it is possible for there to be more than just alternative states of a gene at its locus. That is, as well as the wild type, there may be a series of mutants available at the site. These are termed *multiple alleles*; for example, several different mutants affecting eye colour in *Drosophila* all map at the same site. It is thus possible to have a heterozygote for two mutant alleles.

When two mutants affecting the same characteristic lie close together on the genetic map, the precision of the recombination technique means that it is not always possible to tell whether these are in the same gene (that is alleles), or in adjacent genes affecting the same character. A basic difference is that mutants in different genes show *complementation* when their homozygotes are crossed to give the heterozygote; those in the same gene do not. In the latter case the progeny are mutant for

$$\frac{a}{a} \times \frac{b}{b} \longrightarrow \frac{a}{b} \text{ heterozygous mutant}$$

whereas in the former they are wild type for

$$\frac{a\,+}{a\,+} \times \frac{+\,b}{+\,b} \longrightarrow \frac{a\,+}{+\,b} \text{ complementation}$$

with a wild-type allele at each locus. This led to the definition of the gene as the unit defined by the complementation test; that is, alleles do not show complementation, whilst mutants in different genes do so.

In contrast to an early idea representing the gene as the unit of recombination, Lewis (1945) demonstrated recombination between alleles affecting eye size in *Drosophila*. The mutant Star (*S*) is a dominant which causes a reduction in eye size and a roughening of its texture, and the mutant asteroid (*ast*) is a recessive which causes an extreme reduction in eye size. *S* and *ast* are alleles since they do not show complementation and the heterozygous *S/ast* causes an even more extreme reduction in eye size. However, when the heterozygotes give progeny, some 0·02 per cent. are almost wild type phenotypically. This is explained if the two alleles are due to mutation at different sites in the same gene and recombination has occurred between them. Thus:

$$\frac{S \quad +}{+ \quad ast} \quad \text{no complementation—heterozygous mutant}$$

$$\frac{+ \quad +}{S \quad ast} \quad \text{wild type}$$

The only difference between these genotypes is the arrangement of the mutant sites; in the mutant phenotype they are on different chromosomes (*trans*), but in the wild type they are on the same chromosome (*cis*). This has led to the definition of the gene as a unit which can only function when all its parts are together in a single unit, its function being impaired by mutation anywhere along its length. In the *trans* arrangement, each allele is mutant. In the *cis*, one allele carries both mutants, but recombination has also produced the reciprocal chromosomes carrying an allele which is wild type at both sites of mutation. The term *cistron* is often used for the unit defined by the *cis/trans* complementation test, and this is equivalent to the older term *gene*. When two mutant sites are in different genes they show complementation in either *trans* or *cis* positioning, since both the genotypes $a+/+b$ and $ab/++$ are wild type, each possessing a functional copy of each gene. When the two sites are in the same gene they show complementation in only the *cis* arrangement. This is illustrated in Figure 1.6.

Many examples are now known of mutants which map at very closely linked but separable sites, and show the *cis/trans* position effect, such as the classical series of *w* alleles affecting eye colour in *Drosophila*. These fall into three groups mapping at very close but different sites, and every pair-wise combination between these sites shows the *cis/trans* effect.

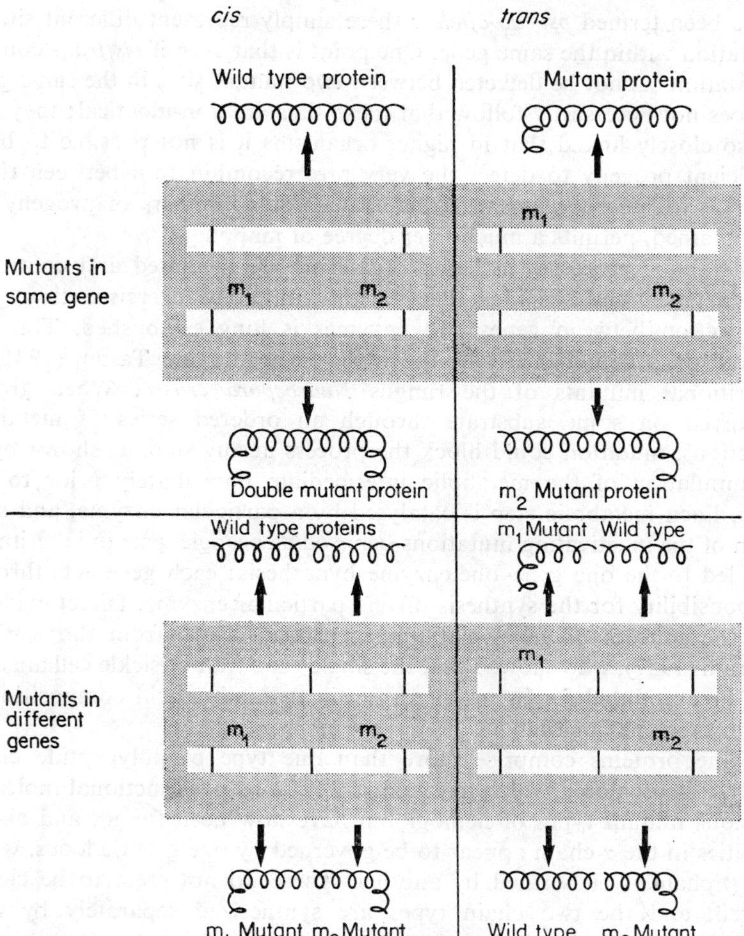

Figure 1.6. The *cis/trans* test for the gene. If the mutant sites m_1 and m_2 are located in the same cistron, in the *trans* arrangement both alleles produce a mutant protein, but in the *cis* arrangement one allele synthesizes a double mutant and the other a normal protein. If the mutant sites lie in different genes, equal proportions of normal : mutant protein are produced in either arrangement, so that both are phenotypically wild type

Mutants which are non-complementary *trans* but map at different sites have been termed *psuedo-alleles;* these simply represent different sites of mutation within the same gene. One point is that even if *cis/trans* complementation cannot be detected between two mutant sites in the same gene, it does not necessarily follow that these sites are non-identical; they may be so closely linked that in higher organisms it is not possible to breed sufficient progeny to detect the very rare recombination between them. Work with microorganisms, where far greater numbers of progeny can be obtained, permits a much finer degree of mapping.

Metabolic processes in living organisms are mediated and controlled by enzymes, and the idea that heredity might be exercised through a connection between genes and enzymes is long established. This was placed on a definite basis by the work of Beadle and Tatum (1941) on nutritional mutants of the fungus *Neurospora crassa*. When growth occurred on some substrate through an ordered series of metabolic reactions, mutation could block the process at any step, as shown by an accumulation of the metabolic intermediate immediately prior to this step. Each metabolic step is catalysed by a particular enzyme, and since each of the inactivating mutations behaved as a single gene in inheritance, this led to the one gene–one enzyme hypothesis; each gene acts through responsibility for the synthesis of one particular enzyme. Direct evidence for the concept of genes elaborating proteins came from the work of Ingram (1957), who showed that the single gene trait of sickle cell anaemia can be accounted for by an alteration in the amino acid composition of the protein hemoglobin.

Some proteins comprise more than one type of polypeptide chain, such as hemoglobin which has 2α and 2β chains per functional molecule. Various mutant types of hemoglobin have now been found, and abnormalities in the α chain appear to be governed by one genetic locus, whilst the β chain is elaborated by another. These do not seem to be closely linked, and the two chain types are synthesized separately by their respective genes and then undergo random association to give the completed molecule. This has led to the more precise definition of one cistron–one polypeptide chain, which has been amply confirmed by detailed work at the molecular level.

There is an apparent exception to the rule that alleles show only *cis* complementation and never *trans*; this is the phenomenon of inter-allelic complementation. A number of mutants may be isolated which have the same phenotype, map in the same genetic segment, and give negative complementation tests in all combinations; thus they belong to the same cistron. However, as larger numbers of such mutants are isolated, a few pair-wise combinations of these may show *trans* complementation, although most remain non-complementary. The most obvious explanation is for each of two adjacent cistrons to be concerned in producing one of two different polypeptide chains which comprise a single enzyme. Mutants in the different cistrons complement because each wild-type cistron supplies a normal chain, resulting in at least a proportion of normal enzyme. This explanation holds for tryptophan synthetase of *Escherichia coli* (*E. coli*) which is a heteromultimeric protein.

However, there is also strong evidence for complementation between different mutants in homomultimeric enzymes (that is, the protein consists of several identical polypeptide chains). A comprehensive review is given by Fincham (1966). It has been suggested that one of the stages in the formation of such an enzyme would be the random aggregation of its component chains, accompanied by their folding into a specific tertiary structure. The mutations result in chains which fold incorrectly, but in inter-allelic complementation the different mutations would somehow correct each other's tendencies to faulty folding, giving an active molecule. This is illustrated in Figure 1.7.

It is reasonable to suppose that although *some* degree of complementation may occur between different mutant derivatives of the same gene, such complementation is not complete in the sense of resulting in a gene product identical to the wild type. Thus:

$\frac{m_1 \; +}{+ \; m_2}$ *trans*—produces partially functional, although non wild-type product

$\frac{m_1 \; m_2}{+ \; +}$ *cis*—produces at least proportion of true wild type

Thus the definition of the cistron by the *cis/trans* test still holds, although the *trans* is more active than either mutant by itself. Indeed, it has been shown experimentally in some cases that the enzyme formed as the consequence of such inter-allelic complementation is qualitatively different from the wild type. For example, the alkaline phosphatase of *E. coli* resulting from this situation is less thermostable than the true wild-type enzyme, the degree of instability depending upon the particular complementing mutants involved.

The nature of the genetic material

Chromosomes are composed of both deoxyribonucleic acid (DNA), ribonucleic acid (RNA), and protein, and many early suggestions proposed

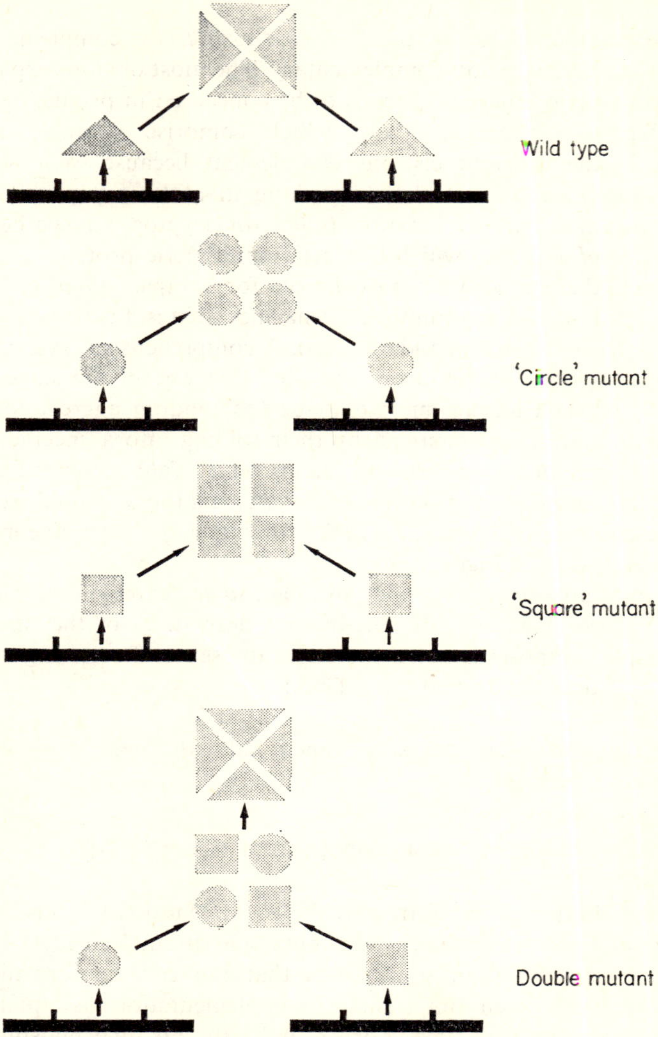

Figure 1.7. Inter-allelic complementation. The enzyme consists of an aggregate of four identical polypeptide subunits. The wild-type subunits have a conformation represented by a triangle, which forms the functional enzyme. The 'circle' and 'square' mutants synthesize subunits with an altered conformation which therefore lack enzyme activity. The heterozygote of these two mutants forms all possible tetramer aggregates of 'squares' and 'circles'; amongst these is one in which the two mutants interact with each other to produce a pseudo-wild protein with at least some enzymic function

that genetic specificity must reside in the protein component since nucleic acid was thought to have too little variety to be able to carry the information required for synthesis of the enormous numbers of different proteins. However, DNA is now known to comprise the chemical basis of heredity, and much of the detailed mechanism of how it specifies protein is established. Indirect evidence supports its role in the heredity of higher organisms; the DNA content of the nuclei of all diploid somatic cells of an organism is roughly constant, as would be expected from their common chromosome complement, and further, the gametic content is half of that found in the somatic cells. The respective behaviour of the protein and DNA components of the chromosomes in replication for meiosis is consistent with the role of the latter as genetic material.

Direct evidence comes from work on microorganisms; the initial discovery leading to this finding was that of *transformation* by Griffith (1928). Mice were infected with either of two morphologically distinguishable strains of *Pneumococcus;* encapsulated bacteria are virulent and cause death, whilst a non-encapsulated mutant strain is harmless. When virulent bacteria were killed by heat treatment and then injected they proved harmless, but if live bacteria of the harmless strain were injected together with them, the mice died as the result of production of live bacteria of the virulent form. The inference is that although dead, the virulent strain contains genetic information which can be transferred in some manner to use the live organization of the other bacterial cell type. Studies on active cell-free preparations by Avery, Macleod and McCarty (1944) identified the nature of this transferable factor. Chemical, physical and serological tests all confirmed the conclusion that DNA is the carrier of genetic information from one strain to the other. Further, proteases (enzymes which degrade proteins) had no effect on the factor, whilst deoxyribonuclease (which destroys DNA) abolished its biological activity completely. (For a more detailed discussion, see pages 43–47.)

The final confirmation that DNA comprises the genetic material came from work on bacteriophages by Hershey and Chase (1951,1952). These are small viruses with bacteria as their hosts, and they consist of a protein coat surrounding nucleic acid (there are roughly equal proportions of each component). They act by attaching themselves to a bacterium and injecting genetic material into it; the rest of the phage remains adsorbed to the outside of the bacterial cell. The injected information then takes over the cell and destroys it in the process of producing more bacteriophage particles.

T2 phage particles, which have *E. coli* as host, were grown in media containing radioactive isotopes. By using S^{35} the radioactive label was incorporated into only the protein component of the phage, or alternatively, by using P^{32} only the nucleic acid was labelled. After the phage had been allowed to infect bacteria, unadsorbed virus and the external particles remaining after the injection of genetic information were removed. The S^{35} label could then be recovered from the phage removed (that is, the isotope had not been injected into the bacteria), but the P^{32} label could not be retrieved. Virtually no radioactive sulphur appeared in the progeny phages, whereas by contrast there was a high degree of incorporation of P^{32} into progeny particles, implying its entry into the bacterium and subsequent role as genetic information in the reproduction of the bacteriophage.

CHAPTER 2

Nucleic acids

Polynucleotide structure

Nucleic acids consist of a chemically linked sequence of their fundamental components, *nucleotides*, analogous to the sequence of linked amino acids comprising a protein. Nucleotides comprise three elements: a heterocyclic ring containing nitrogen (the *nitrogenous base*), a five-carbon sugar (*pentose*) in ring form, and a phosphate group. The bases fall into the two classes of *pyrimidines* and *purines*; the former have a six-membered ring and the latter consist of two fused six- and five-membered rings. Each nucleic acid commonly contains four different types of base, two of each ring form, although the two types of nucleic acid, deoxyribonucleic acid (DNA) and ribonucleic acid (RNA), differ in one of the pyrimidines present. Both DNA and RNA contain the same two purines, adenine and guanine, but whilst DNA possesses the pyrimidines cytosine and thymine, in RNA the thymine is replaced by uracil. The bases are shown in Figure 2.1, and it can be seen that they differ according to their amino or keto substituents. The two pyrimidines thymine and uracil differ only in the presence in the former of a C_3 methyl substituent, which does not affect the chemical properties of the other substituents.

It is possible to write alternative tautomeric structures for all these bases in which the keto groups exist as hydroxyls and the amino as imino. Infrared spectroscopy has suggested that the usual forms are those shown above—this is important for their interactions in nucleic acid structure.

The nitrogenous base is linked to the sugar by a glycosidic bond from N_1 of pyrimidines or N_9 of purines to the 1 position of the pentose. In order to avoid ambiguity between the numbering systems of the heterocyclic rings and the sugar, positions on the latter are usually given a prime. Thus the pyrimidine glycosidic bond can be written as $N_1—C_1'$. As indicated by their names, DNA and RNA differ in the sugar present; DNA contains 2′-deoxyribose whilst RNA contains ribose. Species comprising a base linked to sugar are termed *nucleosides*. Nucleotides are

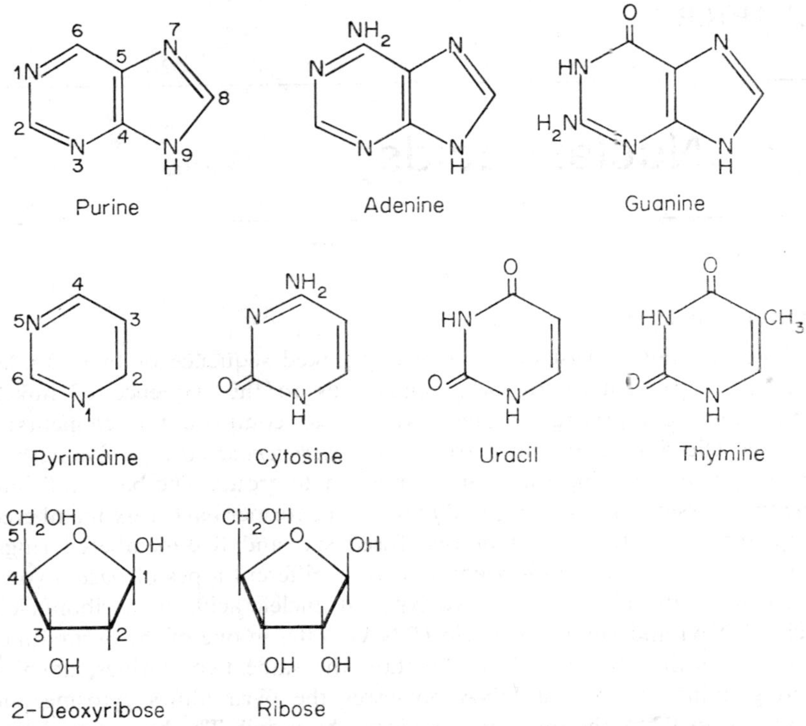

Figure 2.1. The chemical components of nucleic acids. Unless otherwise indicated, ring junctions are carbon atoms. The ring numbering systems are as shown; positions on the sugar are usually given a prime (′) to distinguish them from substituents on the heterocyclic ring. (Note: the terms pyrimidine and purine apply only to the type of structure and do not represent actual compounds.)

esters of orthophosphoric acid (H_3PO_4), with one of the hydroxyls of the nucleoside pentose linked to the phosphate group. This may be either the 3′ or 5′ hydroxyl (termed nucleoside-3′-monophosphates and nucleoside-5′-monophosphates). These are shown in Figure 2.2.

These nucleosides and nucleotides may exist in either the deoxy or hydroxy form; if in the former they are written with a prefix *d*, for example *d*GMP. All the nucleotides can exist in a form possessing more than one phosphate moiety when there is a sequence of two or three phosphate groups linked to the 5′ position; these are known as nucleoside diphosphates (NDP) and nucleoside triphosphates (NTP). The bonds between the first (α) and second (β), and second (β) and third (γ)

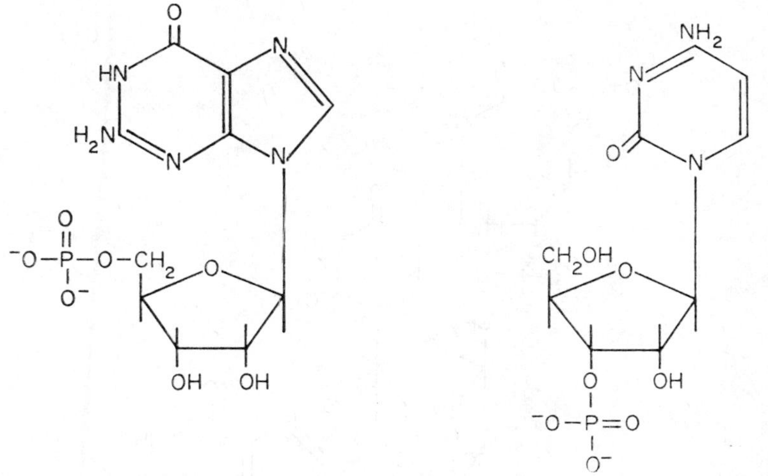

Figure 2.2. The 5′ nucleotide of guanine and the 3′ nucleotide of cytosine

phosphate groups are rich in energy, and these forms are used to provide an energy source for various cellular activities. The triphosphates are the forms from which nucleic acids are synthesized.

Nucleotide phosphate groups can undergo ionization; at acid pH they carry a single charge, and at alkaline (above pH 6) they are doubly charged. Titration of nucleic acids shows that their phosphate groups do not undergo secondary ionization, indicating that two of the three functional groups are involved in chemical linkage. In fact, the phosphate moieties act as bridges between the nucleotide subunits of the nucleic acid polymer by forming diester linkages. A nucleic acid thus comprises a long polynucleotide chain connected by from 3′ to 5′ phosphate linkages, the bases 'sticking out' from the resultant sugar–phosphate backbone. Figure 2.3 shows the primary structure of a polydeoxynucleotide and the abbreviated nomenclature used to represent it.

Table 2.1. The nomenclature of nucleic acid components

Base	Nucleoside	Nucleotide
Adenine	Adenosine	Adenylic acid (AMP)
Guanine	Guanosine	Guanylic acid (GMP)
Cytosine	Cytidine	Cytidylic acid (CMP)
Uracil	Uridine	Uridylic acid (UMP)
Thymine	Thymidine	Thymidylic acid (TMP)

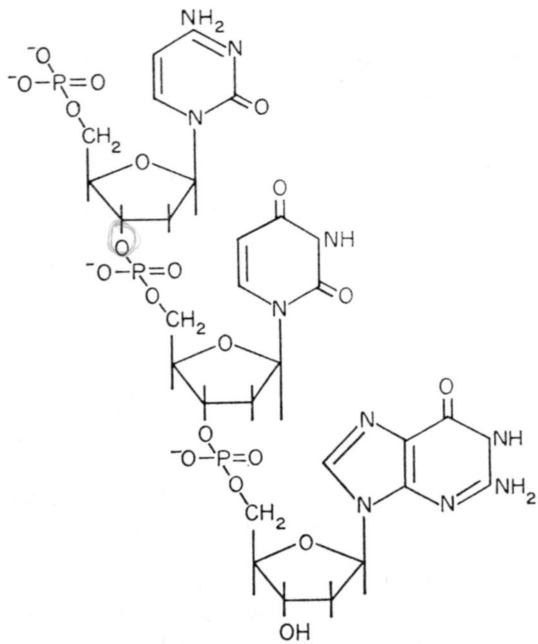

Figure 2.3. The structure of a polynucleotide chain. 3′–5′ phosphodiester linkages run from the free 5′-phosphate residue (top) to the 3′-hydroxyl terminus (bottom). Nucleotide sequences are generally written from 5′ to 3′, so this trinucleotide may be represented by *p*C*p*U*p*G

The secondary structure of DNA

In general, DNA consists of two very long polynucleotide chains wound around each other so as to give a double helix (for review see Wilkins, 1963); results from X-ray crystallography (Wilkins, Stokes and Wilson, 1953) are consistent with such a structure. This requires the duplex to have a constant width, and Watson and Crick (1953a, 1953b) suggested that this could be accounted for if a purine from one polynucleotide chain always partners a pyrimidine from the other chain; both purine–purine (too wide) and pyrimidine–pyrimidine (too narrow) interactions would be prohibited. They suggested that the two strands would be held together by hydrogen bonding, and would not be covalently linked. A feature of these double-stranded species is that the content of adenine is usually the same as that of thymine, and that of guanine equals that of cytosine. The building of molecular models explained this by suggesting that hydrogen bonding could occur between the nitrogenous bases such that the purine guanine could bond only to the pyrimidine cytosine, and not to thymine. Similarly, adenine and thymine would form hydrogen bonds only with each other. Thus a cytosine of one chain would always face a guanine in the other, and a thymine would always oppose an adenine. As illustrated in Figure 2.4, the adenine–thymine (A–T) base pair has two hydrogen bonds, and the guanine–cytosine (G–C) pair has three. The two bases comprising each pair are said to be *complementary*.

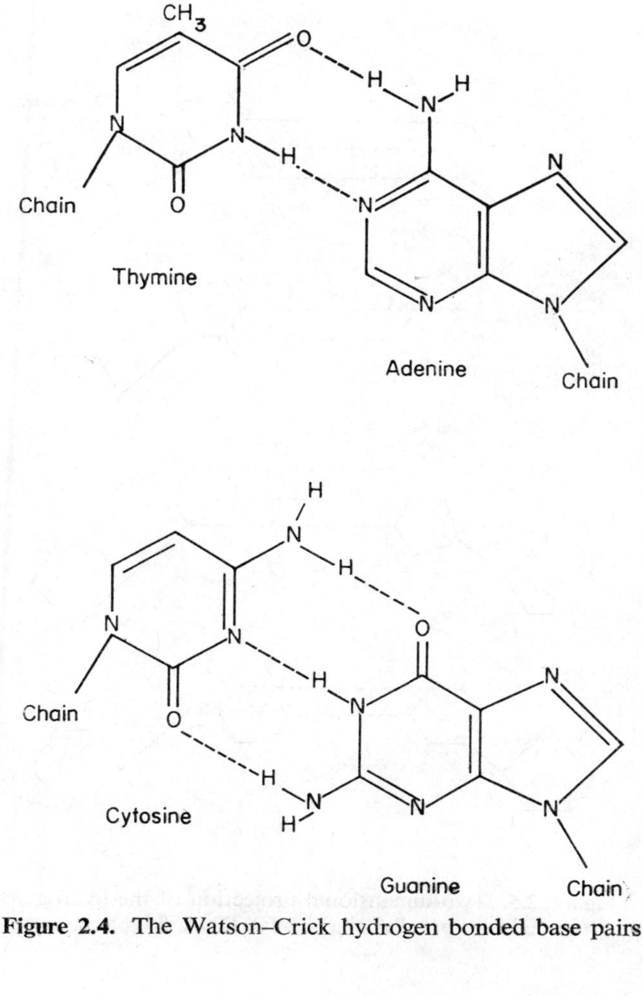

Figure 2.4. The Watson–Crick hydrogen bonded base pairs

This model requires the two polynucleotide chains to run in opposite directions (*antiparallel*), as shown in Figure 2.5. The arrangement has been confirmed chemically by the technique of nearest neighbour analysis (Josse, Kaiser and Kornberg, 1961). In these studies, DNA was synthesized using a radioactive precursor, for example *d*ATP labelled with P^{32} in the 5′α-phosphate position. This then forms a bond to the 3′ position of the next nucleotide in the chain. Cleavage of the DNA with enzymes which specifically break the 3′–5′ phosphate link on its 5′ side gives a mixture of nucleoside-3′-monophosphates, so that the net result

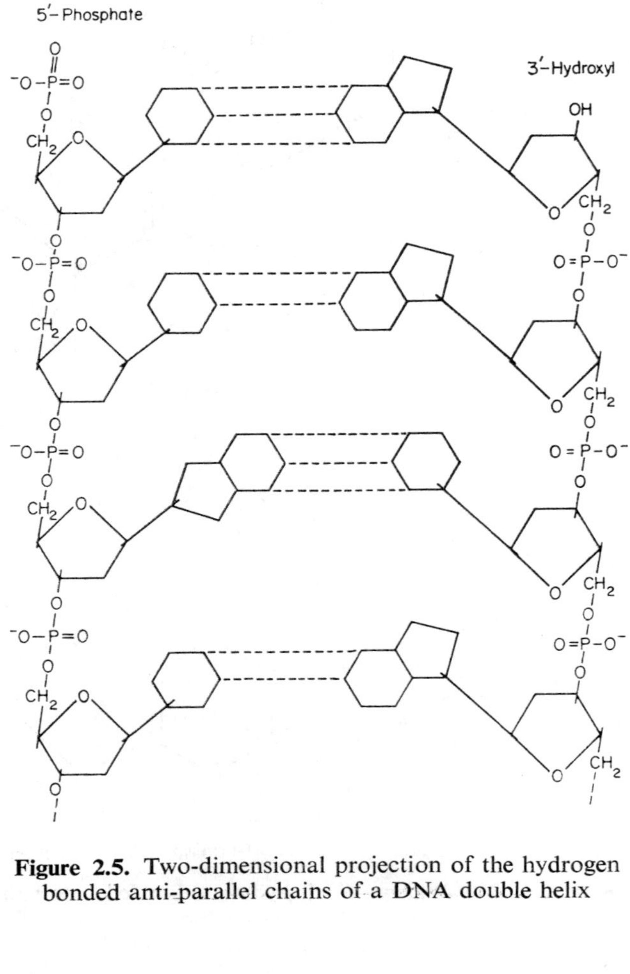

Figure 2.5. Two-dimensional projection of the hydrogen bonded anti-parallel chains of a DNA double helix

is a transfer of the radioactive phosphate from the base with which it was originally associated to its 5′ neighbour in the chain. This is illustrated in Figure 2.6.

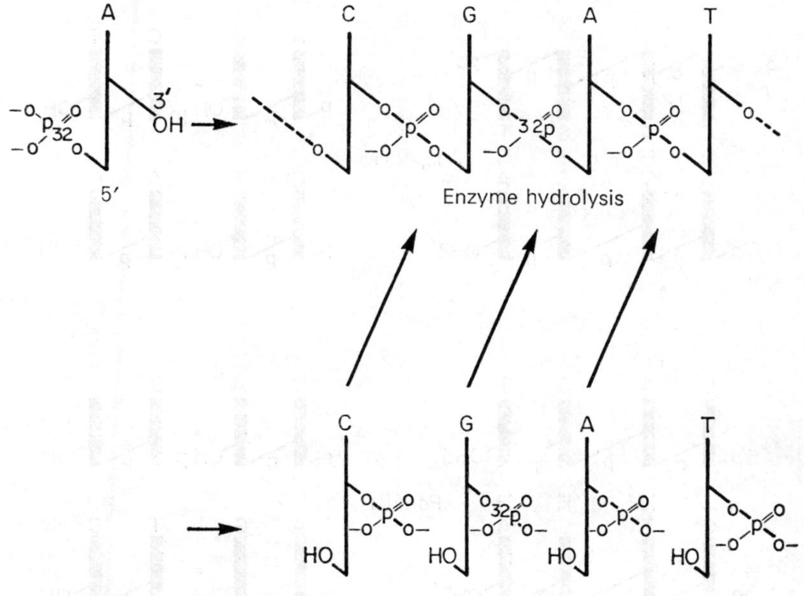

Figure 2.6. Nearest neighbour frequency analysis. A radioactive label in the 5′ position of an adenine nucleotide is incorporated into a nucleic acid chain. Hydrolysis by an enzyme which specifically splits the phosphodiester bonds on their 5′ sides generates nucleoside-3′-phosphates. This results in transfer of the radioactive isotope from its original nucleotide to its 5′ neighbour

Determination of the proportions of nucleotides of each type containing the radioactive label gives a measure of the frequency with which adenine has each of them for a neighbour on its 5′ side. Use, in turn, of each of the four possible labelled nucleotides can give the frequencies of each of the sixteen possible dinucleotide sequences. Certain equalities can be predicted in these sequences, and the predictions are different for the antiparallel and parallel arrangements, as shown in Figure 2.7. The actual data obtained are consistent only with the antiparallel arrangement.

The planar purine and pyrimidine bases in the interior of the duplex

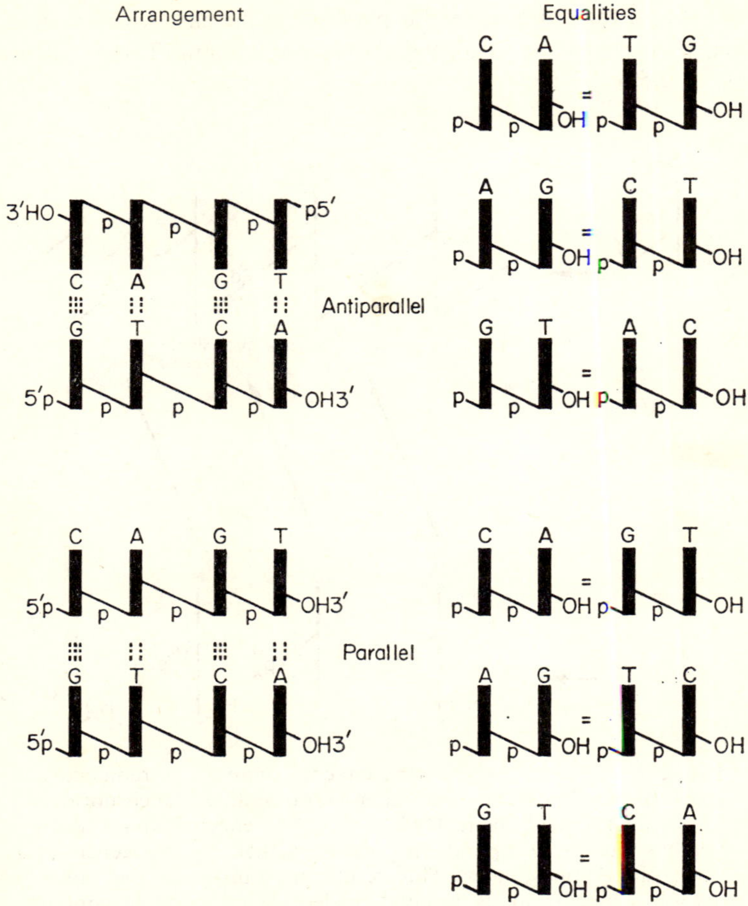

Figure 2.7. Some of the dinucleotide equalities expected according to either antiparallel or parallel arrangements of the opposing chains of a DNA duplex. As customary, the dinucleotide sequences are all written from the 5′-phosphate to 3′-hydroxyl terminus

are aligned parallel to each other and have the planes of the heterocyclic ring roughly perpendicular to the axis of the helix. They are thus said to be 'stacked' above each other in this parallel array. There is a distance of 3·4Å between each base pair, and ten base pairs for each turn of the helix (34Å). The duplex has a shallow groove (about 12Å across) and a wide groove (about 22Å). Figure 2.8 (Plate I) is a photograph of a scale model of the DNA double helix.

The phosphodiester backbone of the helix is highly negatively charged (one charge per phosphate group), and this is neutralized by ionic interactions at pH values above 2. The neutralization may be due to counterions such as Na^+, although for the DNA of chromosomes of higher organisms, it has been attributed to association with basic proteins (histones). It is now clear (for review see Marmur, Rownd and Schildkraut, 1963; Lewin, 1967a, 1967b) that other forces besides hydrogen bonding contribute to the stability of the double-helical structure. It has been suggested that energy arising from base-stacking may play a considerable role through hydrophobic interactions between the non-polar heterocyclic rings. Inter-strand hydrophilic interactions are also thought to be important.

The joining of the two strands by only hydrogen bonds, and not by covalent links, has implications for the biological role of DNA. The specificity of the hydrogen bonding between the bases suggests that, if the strands could separate, each could associate with the appropriate nucleotides to form its *complementary* strand. This would give rise to two DNA duplexes, each identical to the original molecule, and would account for the reproduction of the genetic material. The lack of covalent bonds suggests an easy way for the two strands to separate; one possibility would be for them to start unwinding at one end, each of the free single strands then assembling its complement, as illustrated in Figure 2.9. The mode of synthesis in which each of the original strands of the parent DNA is conserved and handed down intact to the progeny has been termed *semi-conservative* replication.

RNA is usually encountered as a single-stranded polymer of ribonucleotides with a primary structure comparable to that of the individual polynucleotide strands of DNA. Its base content does not exhibit the pattern of A = T, G = C equalities characteristic of DNA, and its physical properties show that its secondary molecular organization is rather less complete than that of DNA. In solutions of low ionic strength RNA molecules tend to behave as typical single polyelectrolyte chains, but at higher ionic levels they appear to possess some degree of secondary structure. The data fit a model in which there are regions of double-helix formation, but these are separated by a length of single-strand structure. The duplex regions must arise from a doubling back on itself of some complementary length of the single strand, and not from interaction between two separate strands. There are various species of RNA in the cell, and these differ with regard to their characteristic content of duplex regions. All are constructed from the DNA of the genome by the same property of complementary assembly thought to be involved in replication of DNA, but in this case, however, only a comparatively short sequence

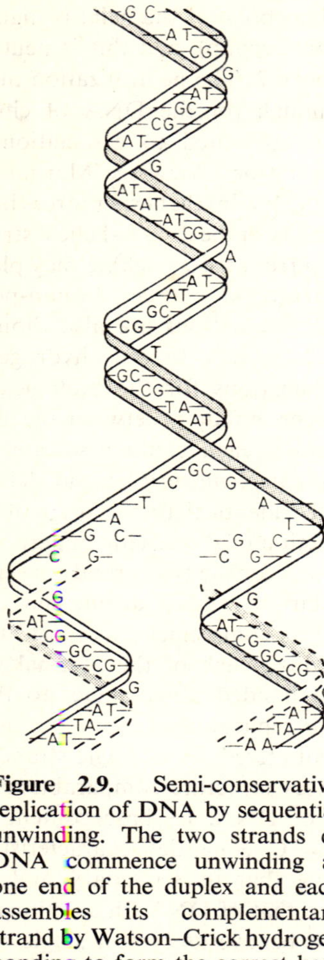

Figure 2.9. Semi-conservative replication of DNA by sequential unwinding. The two strands of DNA commence unwinding at one end of the duplex and each assembles its complementary strand by Watson–Crick hydrogen bonding to form the correct base pairs. This generates two daughter duplexes, each identical in nucleotide sequence to the parent

of the DNA is concerned, and over this sequence the new species is assembled complementary to only one of the DNA strands. The duplex probably partially unwinds to allow RNA synthesis from one of its strands, and after its completion the RNA species leaves the DNA and the double helix reforms.

Denaturation and renaturation

When a neutral solution of DNA is heated, very striking changes occur in many of its physical properties, such as viscosity, light scattering, and optical density. This occurs in a narrow temperature range which depends upon the pH and type of buffer (Lewin and Pepper, 1965); at physiological ionic strength it is usually somewhere in the region of 80–90°C, and has been interpreted as a disruption of the double-stranded helix to give single strands. The physical processes involved have been discussed by Marmur, Rownd and Schildkraut (1963). The phenomenon is known as the *melting* or *denaturation* of DNA, and is usually characterized by the temperature of the mid-point of transition (*Tm*), or *melting temperature*.

The heterocyclic rings of the nucleotides absorb ultraviolet light very strongly, with a maximum around 260mμ, depending upon the particular base, but the ultraviolet absorbance of DNA itself is some 40 per cent. less than would be expected from a mixture of free nucleotides of the same composition. This is referred to as the *hypochromic* effect, and although the cause is not completely understood, it appears to result from mutual interactions of the electron systems of the bases. It is thus at its greatest when they are stacked in the parallel array of the double helix, and the degree of hypochromicity is a sensitive measure of the physical state of DNA, since any departure from the ordered helical configuration is reflected by a loss of hypochromicity, that is increase in optical density. The denaturation reaction can readily be followed by this increase in ultraviolet absorbance as the temperature is increased, and curves obtained from a variety of DNA species are shown in Figure 2.10.

Since the denaturation is due to disruption of the hydrogen bonds which help maintain the double helix, and the G–C pair has three such bonds compared with the two of the A–T pair, DNA helices containing a greater proportion of G–C base pairs should be more resistant to denaturation and exhibit a higher melting temperature. This was established experimentally, as shown in Figure 2.11. (We should note that it has been calculated that increased stability with G–C content should result also from increased base-stacking interactions.)

Doty and coworkers (1960) reported that when DNA which had been denatured by heating was cooled, the ultraviolet absorbance decreased until it was about 12 per cent. above that of the original solution. This solution could then be reheated to see whether it would give the characteristic transition at the T_m. It was found that there was a difference in the characteristics of the increase in optical density during this reheating, according to the speed with which the denatured DNA had been cooled. After quick cooling, reheating resulted in a gradual increase of optical

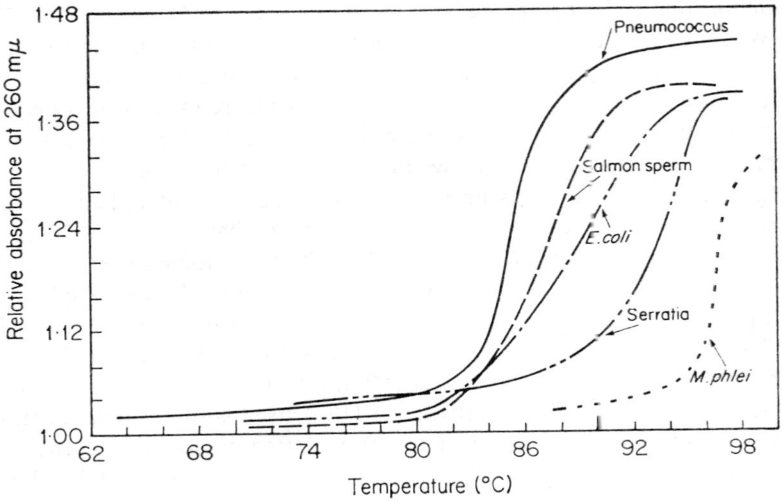

Figure 2.10. Melting curves of DNA from different species. (After Marmur and Doty, 1959)

density without a sudden transition. This was taken to indicate that the optical density decrease on cooling had been due to a random association of short regions which happened to be complementary. The various resulting double-helical regions would have different characteristics with regard to their subsequent denaturation, giving a wide range of transition temperatures, so that optical density would increase gradually on heating. However, if the solution was cooled slowly, there was a characteristic transition at the usual T_m on reheating. This suggests that substantial amounts of the original complementary regions have reformed to give a DNA very like the initial double-stranded species. This process has been termed *renaturation*.

Clearly, the likelihood of two complementary strands meeting and rejoining after their separation will depend upon their relative concentrations in the solution. If only one type of DNA molecule is present, the chances of renaturation occurring are much greater than if, say, there are ten different types present. In accord with this, Marmur and Doty (1961), studying the extent of renaturation in a variety of species, found that the greatest renaturation was obtained from a bacteriophage (that is, the smallest genome). As the complexity of the organism was increased through bacteria to mammalian cells, that is, as a greater quantity of DNA comprises the genome, the extent of renaturation declined.

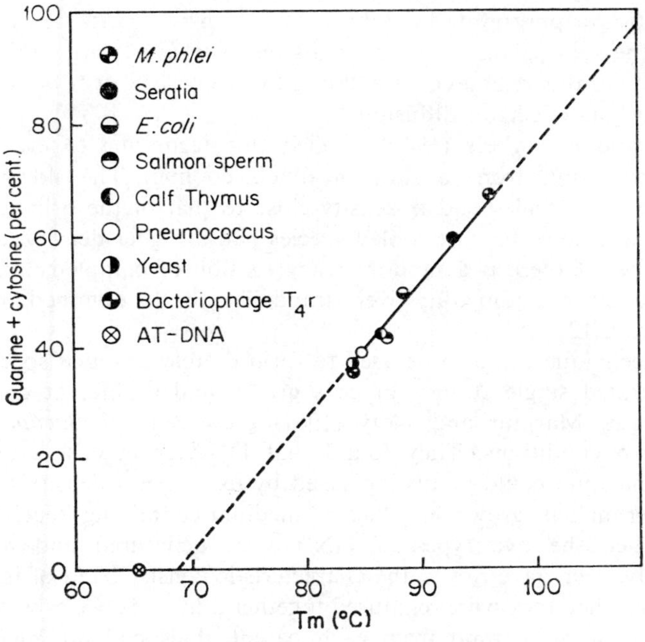

Figure 2.11. Dependence of melting temperature upon G–C content of DNA. Helices are more stable with increasing numbers of G–C over A–T base pairs. (After Marmur and Doty, 1959)

The reformation of double-stranded species from denatured single strands is quite specific for complementary DNA; strand reunion does not occur between DNA molecules from unrelated organisms. Indeed, the extent of renaturation of DNA from one species with that from another indicates the degree to which they have similar nucleotide sequences in their genomes, so that it can be taken as a criterion of their evolutionary relationship.

A technique which has been used for studying the properties of DNA is that of density gradient equilibrium centrifugation. Meselson, Stahl and Vinograd (1957) observed that when a solution of some dense low-molecular weight solute (such as sucrose or caesium chloride) is centrifuged at high speed, an equilibrium is reached between the opposing tendencies of sedimentation and diffusion. This results in a continuously increasing density of solution along the direction of centrifugal force. If a macromolecule, such as a nucleic acid, is placed in this solution, it moves to an appropriate equilibrium position where its density equals that of the solute; this is known as its *buoyant density*. It stabilizes in a band around this region corresponding to a Gaussian distribution due to the disturbing effects of diffusion.

Doty and coworkers (1960) applied this technique to examine the species resulting from a slow or quick cooling. The slow-cooled—renatured—molecules had a density close to that of the original DNA. By contrast, however, fast-cooled species had a higher density, as would be expected if there is a random reaggregation of complementary parts of various strands along one given strand. The results obtained are shown in Figure 2.12.

This technique can also be used to form double-stranded species from the separated single strands of cells grown under different conditions. Schildkraut, Marmur and Doty (1961) grew cells of *Bacillus subtilis* under two conditions. They found that DNA from cells grown in a normal medium could be distinguished by its buoyant density from that derived from cells grown in a 'heavy' medium containing deuterium and N^{15}. When the two types of DNA were denatured and renatured separately, each gave rise to the characteristic density band of its parent. However, when they were renatured together a hybrid species was formed consisting of one strand from each parent. This had an intermediate density. The resolution of the technique was greatly improved by the introduction of a specific phosphodiesterase enzyme which attacks only single-stranded DNA; when this was used to remove non-renatured regions it showed quite clearly that DNA double strands were formed both by renaturation of the parental types, and by formation of a hybrid with one strand from each. This confirmed the postulated mechanism for denaturation and renaturation; the experimental results are shown in Figure 2.13.

Hybridization

Hall and Spiegelman (1961) extended the technique of renaturation—also known as the *annealing reaction*—to an examination of RNA species. If strand complementarity is the requirement for annealing, denatured DNA should be able to anneal with the RNA which it has produced, to form an RNA–DNA hybrid duplex with one strand of each type of nucleic acid (*hybridization*). Experimentally, the RNA produced by phage T2 would anneal with the separated single strands of T2 DNA, showing the sequence homology expected, but would not anneal with heterologous DNA from another species, confirming the specificity of

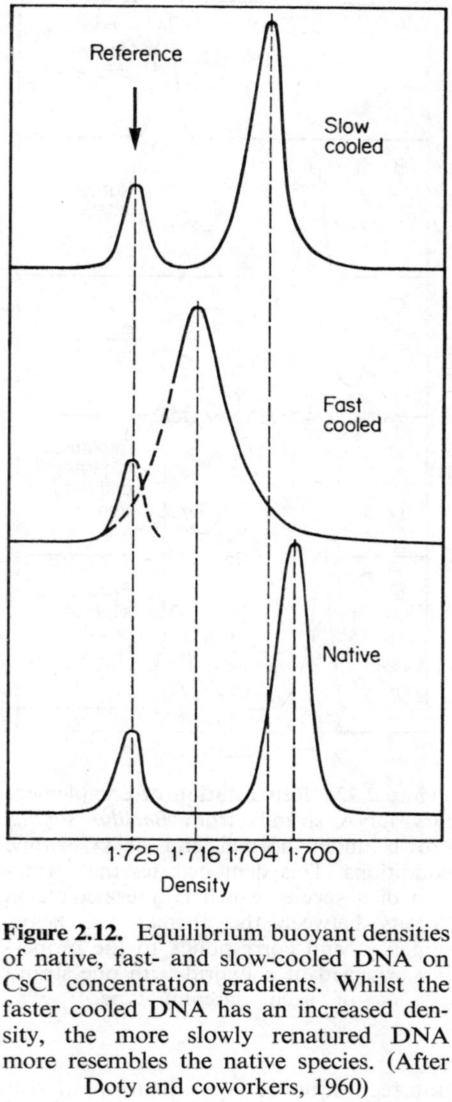

Figure 2.12. Equilibrium buoyant densities of native, fast- and slow-cooled DNA on CsCl concentration gradients. Whilst the faster cooled DNA has an increased density, the more slowly renatured DNA more resembles the native species. (After Doty and coworkers, 1960)

the technique. Hybridization can take place between complementary sequences as short as twenty nucleotides, although this minimum may have to be increased to some extent to allow for mismatching of short non-complementary regions (for review see Walker, 1969).

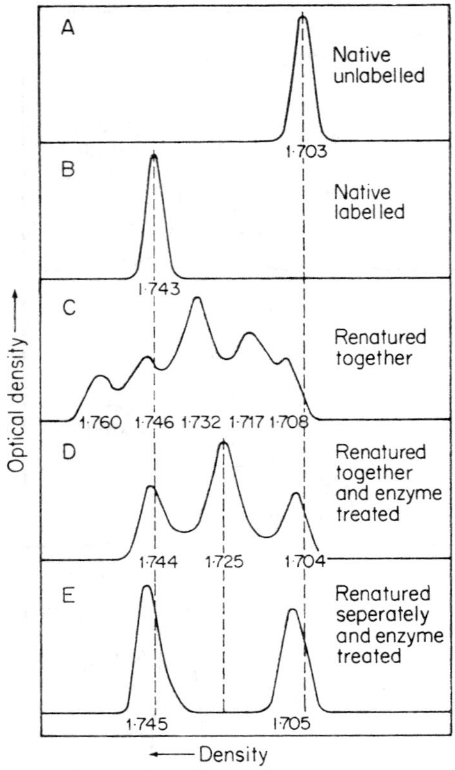

Figure 2.13. Renaturation of complementary DNA strands from *Bacillus subtilis* grown under normal and heavy-isotope conditions. This demonstrates the formation of a species which is intermediate in density between the normal and heavy duplexes, and corresponds to the properties expected of a hybrid with one strand of each parental type

The experimental technique of hybridization in solution is at a disadvantage in examining the formation of hybrids between DNA from different sources or between DNA and RNA, because hybrid formation must compete with renaturation of the original parent strands. This difficulty has been overcome by the development of a method for immobilizing the DNA single strands physically, so that although they cannot renature they can anneal with other species which may subsequently be added to them (Bolton and McCarthy, 1962). Denatured, single-stranded DNA is immobilized in an agar gel and the DNA-agar is incubated with the species to be hybridized, usually by using the DNA-agar as a column to which the molecules to be hybridized are added. Complementary species are retained on the column, whilst non-complementary molecules are washed through. If the species added for hybridization is radioactively labelled, the radioactivity which becomes associated with the DNA indicates the extent of hybridization.

The use of agar columns has now given way to a more convenient experimental method of immobilization of the DNA (Gillespie and Spiegelman, 1965; Nygaard and Hall, 1963, 1964) which provides a faster and more quantitative production of the hybrids. In moderately concentrated salt solution, nitrocellulose filters adsorb single strands of DNA; RNA species can then be added and those hybridizing with the DNA remain on the filter while the remainder pass through. If labelled RNA is used, it is only necessary to measure the radioactivity retained on the filter to obtain the efficiency of hybridization.

A further approach, that of competition for hybridization, has been developed when the extent of homology between two RNA species is to be investigated. A single preparation of immobilized denatured DNA is made, and this is then used in a series of incubations. Each incubate contains a constant amount of a radioactively labelled RNA (or melted DNA), but increasing amounts of a different non-labelled species. The influence of the unlabelled fragments on the extent of binding of the radioactive sample is a measure of their degree of homology. If they are unalike, they will hybridize with different regions of the DNA and not interfere with each other, so that increasing the concentration of unlabelled species has no effect upon the binding of the radioactive species. If they have similarities in base sequence, as the concentration of unlabelled species increases, it will compete with rising efficiency for the DNA, and less of the radioactive species will be hybridized; the greater the similarity, the more pronounced will be the displacement of the radioactive species by increasing concentrations of unlabelled species.

Whilst hybridization studies with bacterial systems have achieved a high degree of accuracy, the technique has proved rather less satisfactory with higher organisms (see Church and McCarthy, 1968; Walker, 1969). This arises from the greater complexity of the genome—mammals may contain of the order of a thousand-fold more DNA than bacteria, and many of the nucleotide sequences may be repeated. This provides binding sites at which RNA may undergo 'false' hybridization, in the sense that the DNA to which it is bound is not that responsible for its synthesis, but another sequence with sufficient similarity to allow annealing. Such hybridization considerably lowers the accuracy of the technique. Further, the repetition of sequences also results in a much increased tendency for reassociation to occur between denatured single strands of DNA, creating certain technical difficulties.

CHAPTER 3

Microbial genetics

A considerable proportion of the studies relating to the mechanisms and control of nucleic acid function has been performed in bacteria, and an extensive use has been made of techniques in microbial genetics. The various systems available for transferring genetic information from a donor bacterium to a recipient (transformation, transduction, conjugation, sexduction) have been used to establish very detailed genetic maps of bacterial chromosomes, on occasion extending even to the location of mutant sites within genes. In many instances, genetic identification of a mutant has been correlated with the biochemical function impaired. Further, it has proved possible to introduce genes representing particular functions into recipient bacteria, so that they are converted from their normal haploid state into a partial diploid (*heterogenote*). This allows investigation of the relationship between normal and mutant alleles in the same cell—an option not otherwise available with bacteria—and this has been of especial value in elucidating control mechanisms.

Transformation

As outlined in Chapter 1, if mice are injected simultaneously with heat-killed virulent bacteria and a live mutant strain, they die due to infection with bacteria characteristic of the heat-killed species. A *transforming factor* (DNA) carries genetic information from the virulent strain to use the live organization of the other cell type for its expression, and it is now possible to achieve transformation by the addition of DNA prepared from a donor strain to a recipient culture. Acceptance of the transforming DNA introduces genetic information from the donor into the recipient cell, which is then diploid for the genes introduced. These can express their characteristics in progeny cells if the transforming DNA undergoes genetic recombination with the recipient genome to replace the alleles previously present. One of the particular uses of transformation is that it is the only genetic system in which native DNA can be used directly as such, so that its structure and function can be correlated. For example, the DNA resulting from denaturation and renaturation experiments has been tested for biological activity by its ability to transform.

Transformation occurs only in certain bacterial species, and the exact conditions for its success vary with the strain. The ability of recipient bacteria to be transformed (*competence*) depends upon their physiological state, including conditions of culture. Competence also varies as a function of the growth cycle, with a rapid increase at some particular point (often just before the end of exponential growth). This lasts for around fifteen minutes, on average, in the individual bacterium, although the duration of competence in a culture may vary rather more. It develops independently of the addition of transforming DNA, being a function only of the culture. The supernatant of a competent culture is able to confer competence on an incompetent culture through the presence of a heat labile factor which appears to be protein; this would suggest that the development of competence in a few cells of a culture could induce competence in the remaining population through this secretion. There are two current theories accounting for the basis of competence (although they are not mutually exclusive). One suggestion is that it may be due to changes in cell wall properties related to the division cycle (Thomas, 1955), which would render the cell permeable to large molecules. Fox and Hotchkiss (1957) proposed that it might be determined by the synthesis of specific enzymically active receptor sites at the bacterial surface.

There are two steps in successful transformation. First, the transforming DNA must enter the competent cell (*penetration*). This transfer passes through two phases; first, there is a reversible fixation of the transforming molecule to the recipient, and then attachment becomes irreversible. Once in the recipient cell, the entering DNA must undergo recombination with the appropriate region of its chromosome; successful penetration does not necessarily mean that the inserted DNA will be able to integrate with the recipient genome. The transformation is completed by the segregation of transformed cells after cellular division, with phenotypic expression of donor characteristics in the daughter cells.

There are specific conditions for the penetration of transforming DNA into competent cells. The DNA can be damaged by various treatments and then tested for its ability to penetrate. Heat denaturation shows that single-stranded DNA penetrates very poorly and the system is specific for double-stranded species. Reduction of the size of the DNA by shear degradation or limited treatment with deoxyribonuclease (DNAase) shows that a minimum molecular weight of 10^6 (about 1500 base pairs) is required for penetration. There appears to be no restriction on the maximum size of the transforming DNA accepted, but generally, the procedure used for its extraction fragments it somewhat—to about 25×10^6 for *Bacillus subtilis* and 5×10^6 for *Pneumococcus* (their genomes have molecular weights around 10^9). Transformation is proportional to the numbers of donor DNA molecules added up to a saturation level, suggesting that each recipient can accept up to about ten transforming molecules. Although the system is quite specific for double strands above a minimum size, it is not specific with respect to source, and DNA from species different from the recipient can undergo penetration, even although it is not transforming. For example, calf thymus DNA competitively inhibits the uptake of pneumococcal-transforming DNA by competent bacteria.

It is possible to obtain hybrid double-stranded molecules of DNA carrying different genetic information on the two strands by annealing single strands obtained by denaturation of two strains with alternative genetic markers, such as *str-s* and *str-r* (streptomycin sensitivity and resistance). This shows that although two strands may be required for penetration, only one is needed for the subsequent genetic events. Fox (1960, 1962) and Lacks (1962) have followed the fate of irreversibly fixed DNA by re-extraction experiments. In principle, recipient cells are penetrated with donor DNA, and at some time after this the DNA within the recipient is extracted and examined. The donor is radioactively labelled to distinguish it from the recipient chromosome. For a short time (5–10 minutes) after fixation the inserted donor DNA passes through a transient inactive state (the *eclipse period*), after which it regains its normal biological transforming activity. Re-extraction of the radioactively labelled DNA a short time after penetration shows that it is in the form of a single strand; the other strand has been degraded to acid-soluble material, that is nucleotides. This initial conversion of the duplex to a single strand accounts for the eclipse period, and together with the necessity for double-stranded DNA for successful transformation, suggests that strand separation may be an essential part of the process.

It is probably random as to which of the two entering strands is degraded, and the single strand remaining intact is physically incorporated into the recipient chromosome almost immediately. Fox and Allen (1964) subjected the DNA of competent cells to density-gradient equilibrium centrifugation at various times after their exposure to donor DNA labelled with the heavy isotopes, deuterium and N^{15}; during the eclipse period, the state of the donor DNA corresponds to either single heavy strands or degraded fragments. After a longer period, when biological activity has been recovered, it is found in sections of light DNA containing stretches of hybrid possessing one heavy and one light strand.

The segments of donor DNA which are inserted in the recipient chromosome by replacing part of one of its strands are somewhat smaller than the lengths entering the cell, and are covalently bound within the recipient strands. Further work along these lines has been reviewed by Fox (1966) and reported by Gurney and Fox (1968). This has confirmed that after recombination the recipient DNA is genetically heterozygous as well as physically hybrid, by growing clones from the two single daughter bacteria of one division; this gives one with a recipient and one with a donor marker.

On average, each transforming DNA carries only a small fragment of the donor genome, of the order of from five to ten genes (*Pneumococcus*) or from thirty to fifty (*Bacillus subtilis*). Linkage may be shown between genetic markers when they lie close enough to be transformed together on the same DNA fragment. Thus transformation generally allows genetic mapping within only small regions, and cannot be used for the construction of complete genetic maps, although extraction of the transforming DNA under particularly mild conditions can give species rather larger than usual, with an extension of the size of region mapped.

Since each recipient cell can accept up to about ten transforming molecules, unlinked characters may be transformed into the same recipient cell. In principle, the probability of joint transformation of unlinked markers is the product of the probabilities of their independent transformations, whereas linked markers would demonstrate a higher double recombinant frequency. Unfortunately, the calculation is not this simple since the frequency of transformation should be expressed relative to the proportion of competent cells in the recipient population; this is characteristic of the cell type and ranges from 10 to 100 per cent. If the proportion is low, spurious linkage may be deduced. For example, if two markers are each transformed with 1 per cent. efficiency (typical values are from 1 to 10 per cent.), their joint transformation frequency should be 0·01 per cent. if they are unlinked. Thus a measurement above this would imply linkage. But if competent cells comprise only 20 per cent. of the culture, the single transformation frequencies become 5 per cent. each and the proportion of joint transformants should be 0·25 per cent. competent cells, that is 0·05 per cent. relative to the culture as a whole. Since this is somewhat higher than the spurious value, a measurement in between these levels would imply a linkage which does not exist.

Goodgal (1961) suggested a method for establishing linkage relationships by transformation. Below the saturation level the number of transformants is proportional to the number of transforming DNA molecules present. If two markers are linked, the proportions of single and joint transformants will bear the same linear relationship to the DNA concentration. However, if they are unlinked the production of double transformants depends upon the independent interactions of two transforming molecules with the recipient, rather than just one; thus if the DNA concentration is reduced by half, double transformation decreases by a quarter. The curve of transformants versus DNA concentration is therefore much steeper for joint transformation than for the single.

Bacteriophages

These are viruses which can reproduce themselves only by using the live organization of a bacterial cell. Their nucleic acid genomes may be either duplex DNA, single-stranded DNA, or RNA. Phages vary in size from the relatively large with duplex DNA genomes to the rather small with single-stranded DNA or RNA. The smaller phage types have a morphology similar to that of small plant and animal viruses, although the larger phages are somewhat different. Some of the relevant functions of the more common systems which have been employed may be summarized as shown in Table 3.1.

Table 3.1.

Phage	Host	Nucleic acid	M.W. genome	Number of genes
T2	*Escherichia coli*	duplex DNA	120×10^6	200
SP8	*Bacillus subtilis*	duplex DNA	110	200
λ	*Escherichia coli*	duplex DNA	33	40–50
P22	*Salmonella typhimurium*	duplex DNA	28	40–50
φX174	*Escherichia coli*	single DNA	1·6	7+
MS2	*Escherichia coli*	single RNA	1·0	3
f_2	*Escherichia coli*	single RNA	1·0	3

Bacteriophages have been classified on the basis of their interaction with host bacteria as *virulent* or *temperate*. After infection of a bacterium, the former lyse it in the process of reproduction to give more phage particles. Temperate phages may either enter this *lytic cycle*, or they may instead establish a state in which they exist in harmony with their host; this non-infectious state is termed *lysogeny*. The distinction between these phage types is not absolute since a temperate phage can be converted into a virulent by mutation. Environmental conditions, such as the physiological state of the bacterium, decide whether a temperate phage enters the lytic cycle or establishes lysogeny.

The best understood large virulent phages comprises a set of seven known as the T series, which infect *E. coli.* The T-even phages (T2, T4 and T6) are distinguished by possession of 5-hydroxylmethylcytosine in their DNA in place of the normal cytosine. Their general behaviour is closely similar, and their intra-cellular development is independent of the integrity of the bacterial chromosome, which is destroyed following infection. T5 also shows this independence, but differs in retaining normal cytosine in its DNA. Development of the T1, T3 and T7 phages requires the presence of an intact, bacterial chromosome, but otherwise their behaviour is quite different, both from the other T phages and from each other.

The process of infection by T2 and T4 is the best characterized. Phage T2 is illustrated in Figure 3.1, and has a *head* containing DNA surrounded by protein, followed by a tail consisting of a *sheath* surrounding a *core*, and terminating in a *base plate* with protruding *spikes* and *tail fibres*. The phage is about one-thousandth the size of its bacterial host.

Chemically, phage is almost entirely DNA and protein, approximately half of each. These can be dissociated by osmotic shock; on transfer from a high salt solution to water the DNA dissociates from the protein, although the 'ghost' phage particles remaining retain their characteristic appearance, except that the heads appear collapsed and empty. Dissociation of the protein components shows that the head protein is composed of a large number of identical subunits of molecular weight around 80,000; the sheath comprised of some 200 subunits of 50,000 molecular weight; and the tail fibres have a subunit of molecular weight greater than 100,000. The DNA is a single molecule, and as a continuous double helix should be around five hundred times the length of the head. The nature of its condensation is not understood, but may be assisted by the presence of an internal protein which is released from inside the head together with the DNA after osmotic shock. This may act by neutralising the acidity of the nucleic acid to provide some sort of matrix for its condensation.

The first step in infection is a random collision between phage and bacterium when its tail makes contact with the cell wall. Collision is succeeded by an unwinding of the tail fibres from the base plate (they are usually wound around the base of the tail and only unwind on infection), and these then ensure a more durable attachment of the phage tail. Some phages attach to the inner lipo-protein layer of the bacterial cell membrane and some to the rigid outer lipo-polysaccharide. Phage infectivity tends to be highly host specific, and depends upon the antigenic properties of the bacterial surface. The tails of T phages contain a lysozyme which bores a hole through the outer layer of the cell wall after attachment, and

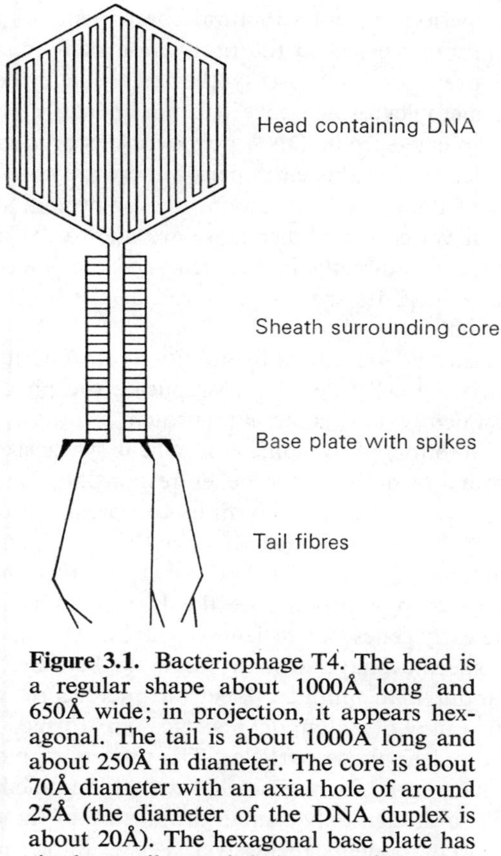

Figure 3.1. Bacteriophage T4. The head is a regular shape about 1000Å long and 650Å wide; in projection, it appears hexagonal. The tail is about 1000Å long and about 250Å in diameter. The core is about 70Å diameter with an axial hole of around 25Å (the diameter of the DNA duplex is about 20Å). The hexagonal base plate has six short spikes and six long tail fibres (not all drawn) projecting from it

some of the products of this lytic activity probably trigger the next step, in which the sheath contracts to about half its former length and the core penetrates the bacterium. The DNA is then passed as a linear thread through the core into the host cell, together with the internal protein (about 3 per cent. of the total protein). The nature of the force propelling the DNA into the recipient cell is not known.

After infection, protein synthesis continues in the host cell at about the same rate as formerly, but within three to four minutes different proteins are being made. The protein synthesis occurring after infection has been divided into the *early* and *late* periods. Proteins synthesized during the early period are not structural components of the phage, but rather are enzymes concerned in the production of its components. For example, the T-even phages must synthesize the enzymes required to make 5-hydroxymethylcytosine, since it is not normally produced in the host cell. Also, a phage-specific DNA polymerase is produced for replication of its genome. During this early period there are changes in some of the components of the apparatus responsible for protein synthesis in the host cell; it is not yet clear whether these are due to the introduction of new species coded for directly in the genes of the infecting phage, or whether the phage modifies species produced by the host (for review see Subak-Sharpe, 1968).

If early synthesis is prevented by addition of inhibitors of protein synthesis, the infected cell is unable to replicate the phage DNA. Also, it cannot subsequently undertake late protein synthesis (that is after removal of the inhibitor), since some component synthesized by the early system is necessary to activate the genes responsible for late synthesis. Normally, the early genes do not continue to operate through the entire period of infection, but are switched off when the late functions commence activity. In mutants defective for late functions this may not occur, suggesting that in turn a product of the late genes is responsible for switching off the early genes (for review see Hayes, 1968b).

Serologically specific phage proteins begin to be detected in infected bacteria after about nine minutes. Electron microscopy shows that the individual head, tail, fibre components, etc., are formed prior to their assembly into complete phage particles. The process of morphogenesis is fairly complex since some forty genes have been identified as controlling the process in T4, although only a small proportion of these actually code for the structural components themselves. For example, mutations have resulted in the production of very long tubes composed of normal head protein (*polyheads*); normally, assembly is halted at the appropriate stage of one head size. Another mutation leads to the production of heads which are too short. Edgar and Wood (1966) showed that when lysates of phages defective for different morphological components (such as heads and tails) were mixed, *in vitro* it is often possible for them to complement one another to produce active phage particles. This suggests that some of the completed protein components of the phage are capable of self-assembly.

Synthesis of phage genomes starts about seven to eight minutes after infection, and prior to this the phage is not infective—it is said to be *vegetative*. During replication, the physical form of the DNA of some phages alters. Phage λ is replicated through a circular form (for review see Sinsheimer, 1968), although this does not seem to happen with T phages. Single-stranded DNA phages are replicated through (circular) double-stranded species (the *replicative form*). After synthesis of T2-phage DNA has begun, its quantity rises sharply so that when the first infective particles appear at twelve minutes, the infected bacterium contains some fifty to eighty phage genomes. As soon as infective particles are produced, their number rises in a linear manner, and the rate of increase in the number of phage genomes levels off from an exponential to a linear curve. This is consistent with a model in which, once the DNA pool has reached a certain size, phage genomes are irreversibly withdrawn from it to form mature phage particles at a rate proportional to their synthesis. Virtually nothing is known about the mechanism of insertion of the DNA into its protein coat. The assembly of genomes and protein coats is random, as shown by *phenotypic mixing*. In mixed infection of one host with two phage types, progeny particles are generated which carry the nucleic acid genome of one type but which are inserted into the coat protein of the other.

The final step in phage reproduction is synthesis of a lysozyme which destroys the bacterial cell wall so that the mature phages are released. Generally, some hundred infective particles are released about twenty minutes after infection. There are mutants in which this lysis does not occur, and the phage cannot then escape from its host.

Bacteriophage properties can be altered by mutation. Two of the most common characteristics used are those of plaque size and appearance, and host specificity. The *r* (rapid lysis) mutants of phage T4 form larger plaques than normal on *E. coli B* because they are insensitive to lysis inhibition (this is a delay in lysis caused by reinfection). The *r* mutants occur at three well-separated loci so they are particularly useful for genetic mapping. Host range (*h*) mutants of a phage can attack bacteria which are resistant to the wild-type phage. If some *E. coli* strains are infected with a phage which attacks them, a few bacteria will survive; these are mutants resistant to attack by this phage because they have different cell wall properties. If these bacteria are grown and then infected with phage, a few will be susceptible to attack—this will be by host-range mutant phages. This process can then be repeated to give bacteria resistant to the mutants, and further phage mutants which will attack this bacterial class also, so that a whole series of host-range mutants are obtained.

Many of the mutants isolated in phages have been as conditional lethal mutants. These grow normally under one set of conditions (*permissive*), but show mutant behaviour under another (*restrictive*). They fall into the two classes of temperature-sensitive and nonsense mutants. Temperature-sensitive phages grow normally at 25–30°C, but are unable to develop normally at 42°C. Nonsense mutants (see Chapter 9) can grow in one strain of bacteria, but show as mutants on another strain. Conditional lethal mutation is not restricted to modifiable characters, such as plaque morphology and host specificity, but may occur in any function, including the essential. This enables mutations to be obtained in every single gene as a step to obtaining a complete genetic map of the organism (for review see Hayes, 1968a).

Genetic recombination can take place if a host is mixedly infected with phages carrying different genetic markers. If *E. coli B* are mixedly infected with the T2 mutant strains h^+r and hr^+, the progeny phages include both possible recombinant classes in about equal frequencies, as well as the parental types. As expected, a characteristic recombination value is found for any particular pair of markers. However, this is the result obtained as the average of a whole population of infected bacteria. To gain information on the events taking place in an individual bacterium *single burst* experiments are used; bacteria are multiply infected with two or more phage types, and the culture is then diluted so that there is a probability of only one bacterium present in the aliquot examined. This yields different results. When two markers are not closely linked, although both recombinant types are usually present, their frequencies are often unequal. For closely linked markers, many bursts contain only one recombinant type.

Visconti and Delbruck (1953) suggested that even taking phage progeny from a single bacterium is not comparable to a single meiosis, but the progeny are actually the result of a study in population genetics. They proposed that replication and recombination of phage DNA takes place in a single vegetative pool; whilst the pool material is replicating, some genomes are being withdrawn for insertion in mature phage particles, and the remainder will continue to undergo rounds of replication and recombination. Levinthal and Visconti (1953) found that in triparental crosses, using infection with abc^+, ab^+c, a^+bc, some $a^+b^+c^+$ recombinants may be obtained; these cannot be explained on the basis of recombination between any two of the parental types, and show that there must be at least two rounds of mating. In T-even phage, the average number of rounds of recombination appears to be between 0·5 and 2·0, with a pool size of some thirty genomes. This model explains the results of crossing closely linked markers; even if both recombinant types are present in the pool, because their frequency is low, the random withdrawal of genomes means that only one type may be inserted into the mature phage particles.

The genetic maps of T-even phages have two particular characteristics. There is a high degree of functional grouping; mutations affecting early synthesis are grouped together, as are regions controlling the structural components. The second feature of the T4 map is that it is circular. However, whereas the chromosomes of some bacteria, for example *E. coli*, are known to be physically circular, in T phages the genome is physically linear. (Stahl, 1967, has discussed the advantages of circular genetic maps). The chromosomes of a T-even phage population are now known to be circular permutations of one another; thus the linear genomes behave as though they were the result of breakage at some random point of an initially circular structure (Streisinger, Edgar and Denhardt, 1964), although no such physical circularity in their formation has been detected. This gives rise to the circular genetic map. A population of genomes can thus be represented as:

A B C .... *X Y Z*
H I J .... *E F G*
Q R S .... *N O P*
Z A B .... *W X Y*

The experimental support for circular permutation lies in findings on the annealing of denatured phage genomes. One such approach (Thomas, 1963) depends upon extracting the ends of fragmented chromosomes, denaturing them to single strands, and annealing with denatured single strands from an unfragmented population. If all genomes are identical, these ends will only anneal with the few ends available on the intact species. If the chromosome population is circularly permutated, end fragments will be able to anneal with any region of the intact molecules. These situations can be distinguished by comparing the binding of the ends with the binding obtained when all the fragments are used; for identical genomes there will be much less annealing in the former case (only with intact ends), whereas with circular permutation binding ability should be the same. The results showed that phage T2 is circularly permutated, but phage T5 is not.

Another approach (Thomas and MacHattie, 1964) is to denature and renature intact DNA molecules. If these are circularly permutated, the terminal regions of some strands will bind to the more central regions of others, to produce molecules with a greater length than the normal genome, which are double stranded in the centre but have single-stranded extremities. For example,

a b c d e f g h i j k
a b c d e f g h i j k

denature and → anneal

a b c d e f g h i j k
h i j k a b c d e f g

h i j k a b c d e f g
h i j k a b c d e f g

The extremities are complementary and so will hydrogen-bond with each other to yield a double-stranded circular structure, with one break in the covalent links along each strand, in a different location on each strand. Again, T2 gave the result expected from circularly permuted genomes, whilst T5 (and recently T3 and T7 also) renatured only to linear species.

Further, in all the phages with linear genomes so far examined, the sequence of bases at one end of the chromosome is repeated at the other end (*terminal redundancy*). Because the chromosomes are circularly permuted, the terminally redundant sequence (1–3 per cent. of the total genome in T2) is different in each phage particle. Thus the representation above can be amended to:

A B C D E .... *X Y Z A B C*
H I J K L .... *E F G H I J*
Q R S T U .... *N O P Q R S*
Z A B C D .... *W X Y Z A B*

Redundant ends have been demonstrated experimentally (MacHattie and coworkers, 1967; Ritchie and coworkers, 1967) in phages T2, T3 and T7 by subjecting their genomes to attack by the enzyme exonuclease III from *E. coli*. This 'nibbles away' one strand of a double-stranded structure from its free 3′ hydroxyl end; this occurs at both ends of a DNA duplex to give a species which is double stranded in the middle, but has opposing single strands protruding free at each end. If a molecule is terminally redundant these single-stranded sequences are complementary, and thus undergo hydrogen bonding with each other to yield a circle.

The presence of circular permutation and terminal redundancy suggests a mechanism for production of the phage genome (reviewed by Hayes, 1968a). Fixed lengths of DNA longer than a genome would be 'chopped off' from long chains comprising several genome units in sequence (*concatenates*); thus:

//A B C D E F G H I J A B//C D E F G H I J A B C D//E F G H I J A B...

If the chromosome consists of ten units and the genome actually inserted into the phage has twelve units, the result would be circularly permuted genomes with a terminal redundancy of two units. There is some experimental evidence suggesting the existence of concatenates, but as yet it is far from conclusive. One possible mechanism which has been suggested is that the phage head might act as a 'chopper' of the concatenate when it is full with the correct length of DNA. In support of the idea that the length of the chromosome corresponds to a 'headful' of DNA, Streisinger, Emrich and Stahl (1967) found that when part of the genome was deleted it was compensated by a corresponding lengthening in terminal redundancy.

Lysogeny

In the lysogenic state, temperate bacteriophages exist in harmony with their host cells in a non-infectious manner (see recent reviews by Luria and Darnell, 1968; Signer, 1968; Thomas, 1968). The stages of infection are the same up to the entry of DNA, but then, instead of entering the lytic cycle, the phage genome becomes specifically associated with the host cell chromosome in a form known as *prophage*, which then replicates only in synchrony with the host. Occasionally, release of the phage DNA from the prophage state into a lytic form is triggered (*induction*) either spontaneously or by ultraviolet irradiation, and this leads to multiplication of the phage and destruction of the host. The basis of this induction is not well understood. If the progeny phage then infect other bacteria, they do so in the temperate form, not the virulent.

When a bacterium is lysogenic, it is immune to vegetative attack by another of these phages. In the phage λ–*E. coli* system, three genes of the phage genome, the *C* cistrons, are responsible for the association between phage and host genomes. The lysogenic state is maintained through synthesis by the c_1 cistron of an immunity substance which prevents transcription of the phage genome and thus prevents expression of the functions necessary for the lytic cycle. This *repressor* is thus also responsible for the immunity to super-infection since it prevents the functioning of any further infecting λ phage. The exact functions of the c_2 and c_3 genes are not known, but they are involved with the transition of vegetative phage to the prophage state. Immunity does not extend to protection against vegetative attack by other species of phage, and, indeed, a bacterium may carry several different lysogenic phages.

Genetic crosses between lysogenic and non-lysogenic bacteria show that prophage is inherited as though occupying a specific locus on the bacterial chromosome. Campbell (1962) proposed what is now generally accepted as the model for prophage formation; this depends upon a physical insertion of the phage DNA into the bacterial chromosome at a specific site. Earlier suggestions that prophage might be a form of association with the bacterial chromosome, rather than insertion, have now been abandoned. The insertion model proposes that the phage has regions which are homologous with regions on the bacterial chromosome. Pairing takes place between these, just as between homologous chromosomes at meiosis, and is followed by a breakage and reunion event which results in physical incorporation of the phage into the bacterial genome. When phage is released from the prophage state, this process is reversed. This model is illustrated in Figure 3.2.

Indirect support for insertion rather than association is provided by the finding that deletions in lysogenic bacteria may on occasion extend along the host chromosome into the phage genome. A demonstration of physical insertion was provided by Inselberg (1968). DNA was extracted from bacteria either lysogenic or non-lysogenic for phage PI, fragmented into a range of sizes and denatured. After denaturation these fragments were tested for hybridization with PI-DNA-agar. This showed that *all* the size fractions contained approximately 2 per cent. of specific hybridizable material when the bacteria were lysogenic. This is the result expected if one prophage is inserted into each chromosome and the chromosome is then fragmented in a random manner during its preparation for denaturation. If the prophage was only associated with the host DNA and not integrated into it, the larger fractions (larger than the size of the prophage itself) would not show the same degree of hybridization.

The Campbell model makes various predictions which have been tested experimentally. The insertion of the phage into the host genome should result in an increased separation of genes mapping on opposite sides of its site of integration; this has been confirmed. Further, the insertion changes the sequence of genes in the prophage (*cdab*) compared with its vegetative form (*abcd*); the linear order of genes in the prophage is a circular permutation of the gene order in the vegetative map. This too has been confirmed experimentally. Also, phages with deletions have been found which lack ability to insert into the host genome; this is interpreted as a loss of the regions responsible for homologous pairing.

Transduction

Recombination can occur between two mutant strains of *Salmonella typhimurium* when a culture of one mutant is treated with a cell-free

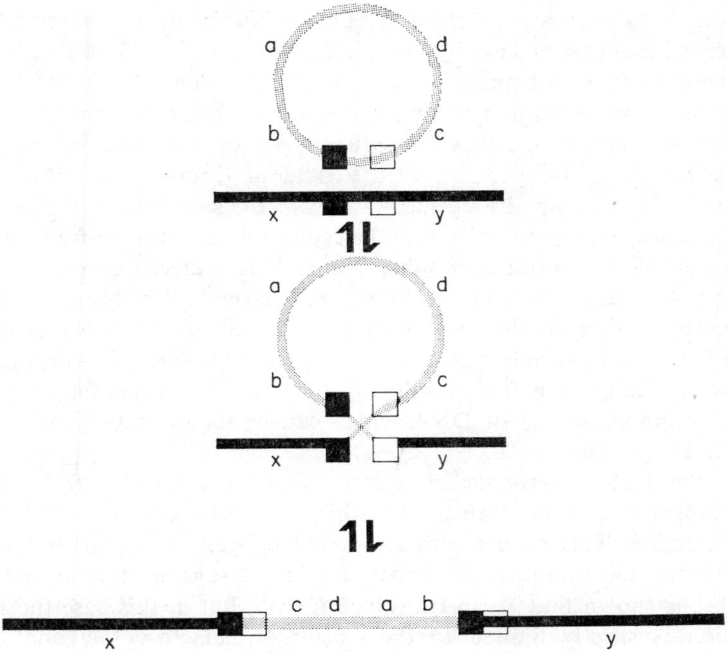

Figure 3.2. The Campbell model for lysogenic integration of bacteriophage into the bacterial chromosome to form prophage. A circular form of the phage pairs with the bacterial chromosome at two regions of homology (indicated by squares). A recombination event takes place between these two segments, and the breakage and reunion results in insertion of the phage as a linear segment of the bacterial chromosome. This increases the distance between the bacterial markers *x* and *y*, and also changes the order of phage markers from *abcd* to *cdab*. Lysogeny is reversible, and prophage may be released as phage by a reversal of this sequence of events

extract of the other. The *transducing agent* proved to be the temperate phage P22, which lysogenizes one parent strain and infects and lyses the other. However, its transducing activity is independent of its ability to lyse or lysogenize the recipient bacteria, since the transduction can occur when the recipient is already lysogenic for this phage (and thus immune to further infection), or when the phage is a mutant unable to lysogenize the recipient. Essentially, the phage acts as a *vector* in providing a means for carrying genetic information (DNA) from one bacterium to another, and virtually any genetic characteristic or biochemical requirement can be transduced (see reviews by Campbell, 1964; Thomas, 1968).

The frequency of transduction is low; only about 10^{-5} to 10^{-7} of the progeny phage emerging from lysis of donor bacteria convey markers to the recipient, which they subsequently infect. On average, the phage transports about 1–2 per cent. of the bacterial genome—this corresponds to some fifty to a hundred genes—and is of the same order of size as the phage genome. Thus only closely linked genes will undergo co-transduction. It is possible to transduce prophages as well as bacterial genes, and using doubly lysogenic bacteria as donors can give doubly lysogenic transductants, so that the phage particle must be able to act as vector of two complete phage genomes, in addition to a length of bacterial chromosome equivalent to its own. This would suggest that the limiting factor may be the continuous length of DNA which can be incorporated, rather than the total quantity.

In the P22–*S. typhimurium* system any small region of the donor chromosome can be transduced—this is *generalized*, or *unrestricted*, transduction. This occurs with temperate phage, either during the lytic cycle after infection or after induction of the lysogenic state. Subsequent work has shown that λ can transduce *E. coli*, but in this case the transduction is strictly limited to the cluster of galactose loci and biotin genes which map adjacent to the prophage site. This is *restricted*, or *localized*, transduction. However, unlike the P22 system, phage emerging from the lytic infection of sensitive bacteria are inert, and transduction is only shown after ultraviolet induction of lysogenic cultures. This implies that, in restricted tranduction, the phage genome in the state of prophage picks up a segment of bacterial chromosome lying near to its location.

When a *gal*$^-$ *E. coli* culture in infected with λ obtained by induction of a lysogenic *gal*$^+$ strain, the *gal*$^+$ transductants fall into two classes. About one-third have been stably converted from *gal*$^-$ to *gal*$^+$. Induction of these strains (which are nearly always lysogenic after infection) gives lysates whose transducing activity is similar to that of lysates derived from the original lysogenic *gal*$^+$ parent. This is the same type of result as is found in generalized transduction; the *gal*$^+$ genes transferred by the phage appear to have replaced their *gal*$^-$ homologues in the recipient genome.

The other two-thirds of the transductants are unstable for the *gal* character and segregate *gal*$^-$ progeny at about 10^{-3} per bacterium per division. This implies persistence of the recipient *gal*$^-$ gene, so the transductant must be diploid for the *gal* region. In this heterogenote, the added chromosomal fragment is known as the *exogenote* and its homologue in the recipient as the *endogenote*. When cultures of the heterogenotes (which are usually lysogenic) are induced by ultraviolet, the yield of phage is lower than normal. However, some half of this lysate can now accomplish transduction for *gal*$^+$, and the frequency of transduction is raised from the normal low 10^{-6} to as high as almost one per particle. These are known as h.f.t. (high frequency of transduction) to distinguish them from the original l.f.t. (low frequency of transduction).

The mechanism of restricted transduction can be explained on the basis of the Campbell model as being due to a recombination event between the bacterial and prophage genomes. Usually, the induction of vegetative λ from the prophage restores its normal genome by reversal of lysogeny (Figure 3.2). However, occasionally (10^{-6}), an illegitimate pairing between prophage and bacterial chromosome is followed by a single recombination event in the attachment region; this results in an exchange of part of the λ chromosome for some or all of the adjacent bacterial *gal* region, as illustrated in Figure 3.3.

The resultant phages lack part of their genome and have instead a part of the bacterial chromosome (in Figure 3.3 the phage region *b* remains on the bacterial chromosome and the phage acquires the bacterial region *x* in exchange). Phages which have collected bacterial genes then have a major region of homology with the next bacterium they infect; if the donor gave *gal*$^+$ genes, synapsis with a *gal*$^-$ recipient gives a *gal*$^-$/λ-*gal*$^+$ heterogenote. This is what has happened in the two-thirds of the transductants which subsequently segregate out *gal*$^-$. If a population of such heterogenotes is induced, every bacterium liberates a λ-*gal*$^+$ instead of only 10^{-6}, giving high frequency transduction.

This model has now been confirmed experimentally. The λ-*gal* phage particles have been shown to be defective in establishing lysogeny when there is only a low multiplicity of infection, as might result from the loss of phage gene functions. Such phages have been termed λ-d.g. (defective galactose). However, if bacteria are mixedly infected with λ-d.g. and normal λ, the latter 'help' the defective phage to establish lysogeny, and the bacteria become doubly lysogenic for λ-d.g. and λ. The transducing preparation then contains an excess of normal phages over the defective so that, generally, every bacterium infected by a λ-d.g. is also infected by at least one normal λ. Bacteria which *are* lysogenic for λ-d.g. alone are immune to λ, but upon induction lyse without releasing any phage particles; the λ-d.g. prophage is defective and unable to produce mature phage particles on its own. Bacteria lysogenic for both λ-d.g. and λ lyse after induction to release a mixture of equal numbers of λ-d.g. and λ phage particles.

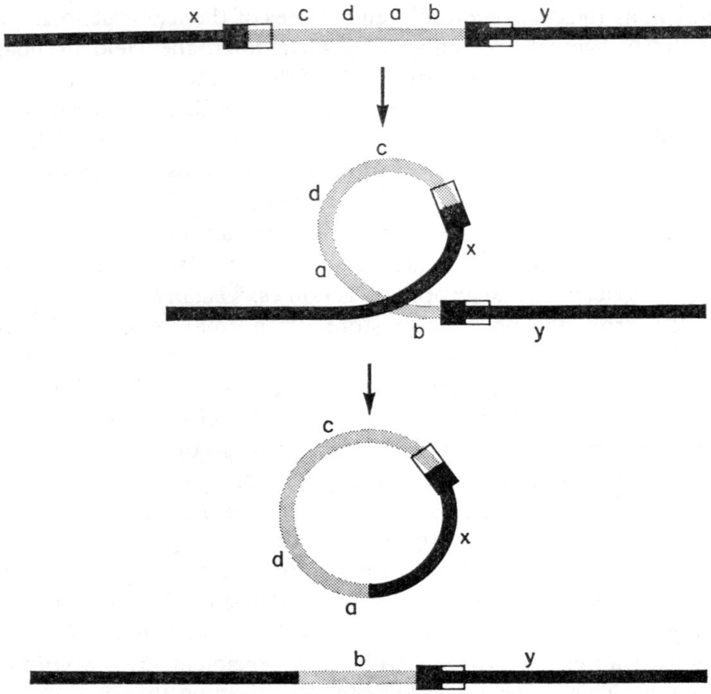

Figure 3.3. The Campbell model for transduction. An 'illegitimate' pairing between prophage and bacterial chromosome is succeeded by a recombination event between the regions of pairing. This generates a phage bearing part of the bacterial chromosome containing the marker x in exchange for part of its own genome (b), which remains on the bacterial chromosome

Marker rescue experiments show that defective phages indeed lack a variable length (between one-quarter and one-third) of their chromosome, and the missing segment is rather rigidly defined. In such experiments, a bacterium is mixedly infected with two phage types carrying genetic markers, one type being inactive. However, genes from the inactive class appear in the progeny phage—they have been 'rescued' by recombination. When heterogenotes carrying defective wild-type phage are induced and superinfected with phages carrying mutant markers, the progeny do not show wild-type alleles for certain markers; that is, these are lost in generating the λ-d.g. species. An examination of the densities of λ-d.g.

particles shows that they vary according to the amount of bacterial chromosome which has replaced the lost segment; some λ-d.g. species are lighter than normal, whilst some are heavier.

There are clear differences between generalized and restricted transduction. In the former, any bacterial gene can be transferred and can be picked up by lytic infection as well as after induction; thus vegetative phage is involved. The transferred piece of donor chromosome usually becomes integrated into the recipient and is not maintained as an exogenote, and most transductants show no trace of residual phage. Whereas in the generalized process fragments of host cell DNA are included in the phage particles, in the restricted process there is a single genetic element composed of both phage and bacterial genomes.

However, one theory suggests that the two mechanisms are related; cases are known in which generalized transduction results in production of defective phages, notably the PI–*E. coli* system in which a wide variety of PI–d.l. (defective lactose) types have been isolated. These have features in common with λ-d.g. phages. This proposal suggests that the generalized mechanism is similar to the restricted, except that there are many regions of the host chromosome with which recombination can occur during vegetative recombination. An alternative theory proposes that fragments of the bacterial chromosome are incorporated into phage heads in lieu of phage genomes during lytic infection by a process similar to phenotypic mixing. Physical analyses of transducing particles of phage PI support this; after radioactively labelling the chromosomes of the infected bacteria distinctly from the phage DNA, transducing particles were found with only fragments of the bacterial DNA and not phage DNA present.

The frequency with which bacterial chromosome fragments are transferred from donor to recipient by phages is considerably higher than that of the transductants recovered; there are many *abortive* transductants for every successful event. In these, only one daughter cell of the progeny of the recipient receives the transduced character. Hence the character, although expressed in the recipient, must lack the ability to replicate. This is shown visibly in the P22–*Salmonella* system, where one of the characters which can be transduced is the (dominant) ability to produce the flagella necessary for motility. If non-motile bacteria are seeded onto fairly soft agar, any motile transductants swim outward into the agar, and complete transduction of motility results in a swarm of bacteria coming out from the margin of the growth zone. By contrast, the 'trails' resulting from abortive transduction are some ten times more common. The trail marks the path of a single motile cell; each colony along the trail shows where a non-motile cell was produced at division, and subsequently underwent further divisions to yield a stationary colony.

Transduction can be used for the fine genetic mapping of closely linked loci, with the restriction that this applies only to those chromosome regions which happen to lie adjacent to the sites of insertion of a transducing phage. Recombination can occur between the homologous regions of the heterogenote to yield a phage with bacterial recipient markers, and to introduce the donor markers into the recipient genome. As with transformation, only co-transduced genes will show linkage. Further, λ-d.g. phages can be prepared from gal^- donor bacteria and can be used to infect other strains showing the gal^- phenotype. These will form heterogenotes of the form λ-gal_1^-/gal_2^-, and if the two mutants are in the same gene these will be gal^-, but if in different cistrons will show wild-type phenotype; transduction can therefore be used to provide a convenient *cis/trans* test for mutants in regions which are transduced by a particular phage.

Conjugation

Lederberg and Tatum (1946) found that when they mixed two multiple auxotrophic mutants of *E. coli* they could select for prototrophic recombinants. (*Auxotrophs* have some nutritional requirement; *prototrophs* for that character lack the requirement.) Although neither of two strains $++cde$ and $ab+++$ could grow in a mixture lacking all five of the growth factors *A* to *E*, a mixture of the two strains produced some $1/10^7$ recombinant cells capable of growth. Recombinant formation required direct contact between the two cell types.

Hayes (1952a, 1952b) showed that the parental cells play different roles in conjugation. Treatment of the cells of one parent with streptomycin drastically reduced viability, but did not affect the yield of recombinants; treatment of the other set of cells prevented recombinant formation. Whilst ultraviolet treatment of the first strain stimulated recombination, in the second it caused a drop in recombination formation parallel to the drop in viability. It was proposed that the first strain acts as donor of genetic information (so damage reducing its viability does not prevent successful recombination), whilst the second is the recipient which eventually gives rise to recombinant clones (so its viability is essential for survival of the progeny).

Cells can be divided into donor types and recipient types, and the recipients are converted to donors by the conjugation process. The ability to act as donor depends upon the presence of a factor (*F*) which is transmitted from donor F^+ to recipient F^- with a very high efficiency, in contrast to the low efficiency of transfer of genetic information. The *F* agent is a circular DNA of about $2{\cdot}5 \times 10^5$ base pairs (the same order of size as a T phage), and F^+ cells possess several copies of the agent. Studies on the physical properties of sex factors have been reported by Friefelder (1968). In infection, the circle could break at some characteristic point for a linear transfer to an F^- cell. Newly infected cells synthesize further copies of the factor prior to acting as donors to other cells.

There is a second type of donor cell, termed *Hfr*, which causes conjugation with a high frequency of genetic recombination; these are derived from F^+ cells and can revert to that state. Whereas an F^+ donor transfers *all* genetic markers with the same (low) frequency, recipients of transfer from an *Hfr* cell inherit genes from one region of the bacterial chromosome with high frequency, and genes from other regions with lower frequencies reaching down to that typical of transfer from F^+. Donor ability is transferred only very rarely, and the cells receiving it are invariably those which have received the genetic markers transferred at lowest frequency. If actively multiplying cells carrying F^+ are treated with acridine orange, they are converted to F^- cells due to loss of their *F* factors. However, acridine treatment does not reduce *Hfr* cells to the F^- phenotype, although they too contain the *F* factor.

The model accounting for this proposes that *F* is an extra-chromosomal element (*episome*) which is in an autonomous state in F^+ cells, but integrated into the bacterial chromosome in the *Hfr* species; the transition from F^+ to *Hfr* and vice versa corresponds to the chromosomal attachment and detachment of the *F* factor. *Hfr* cells have no free *F* factors because once the agent is integrated into the bacterial genome it no longer replicates autonomously. The transfer of genetic information is due to *Hfr* cells only; there are various *Hfr* strains, and each will transfer with high frequency markers from the chromosome region adjacent to one side of its site of integration. Genetic transfer from an F^+ strain is probably due to the presence of occasional *Hfr* cells in the population; the high frequencies of transfer corresponding to their different sites of attachment will average out, so that amongst the population as a whole, all markers are transferred at about the same low frequency.

Conjugation can be considered as occurring in the three stages of effective contact, genetic transfer, and formation of recombinants with subsequent segregation. There is considerable evidence that the first stage is random collision between cells of opposite mating type, after which contact is maintained through the presence of specific surface components on the male (F^+) cell. One suggestion (Brinton, Gemski and Carnahan, 1964) has been that the *F* factor causes synthesis of special pili on the surface, and these would not only maintain contact with the recipient but also provide a means for the transfer of DNA through an axial hole. The overall process is fairly complex since it depends not only on contact, but also on removal of the membrane and cell wall barriers to transfer.

Three methods have been used for analysing the process of genetic transfer. The simplest is to measure the overall numbers of recipients receiving various genetic markers from the *Hfr* donor; the different markers of any *Hfr* strain are transmitted to recombinants with different and characteristic frequencies. Jacob and Wollman (1961) showed that these markers can be arranged in a continuous gradient with regard to their frequency of transmission to progeny. High frequency recombinant formation is shown for markers close to the integration site of the *F* factor, and the frequency decreases as markers are located farther away. This gives a genetic map order based upon frequency of transfer.

This gene frequency gradient suggests that chromosome transfer is an orientated process in which an origin point at the *F* factor integration site first enters the recipient, and the rest of the chromosome follows in linear array. The process of transfer is subject to spontaneous interruption, so that genes near the origin, which are transferred sooner, are less likely to be prevented from entering the recipient by loss of contact between the two bacteria. Thus the zygotes formed by conjugation are generally only partially diploid.

The order of transfer of genes can be determined directly by *interrupted mating* experiments. Mating can be interrupted at various times after it has commenced—by killing the donors with male specific phage, or by mechanical interruption—and the times at which different genetic determinants enter the zygotes established. Transfer of the complete chromosome takes about ninety minutes, at an approximately constant rate, so a map in time units can be converted into more conventional terms. (One time unit is about twenty recombination units.) Markers which enter the recipient early attain a high final frequency, and markers which enter later reach only a lower final frequency. There is good correlation between the frequency and order of transfer analyses.

Recombination can be used to obtain map distances by establishing linkage relationships between unselected markers of the *Hfr* strain and those selected for. For example, if *Hfr*++ cells are crossed with F^- *ab*, the progeny can be selected for b^+ by growth on medium lacking *b*, and the number of these which have gained an a^+ allele as well as b^+

determined. The degree of joint transfer of selected and unselected markers is a measure of their linkage. The marker selected for must be the one entering the recipient later, since otherwise there is no certainty that both donor alleles will have been transferred.

The linkage group of any particular *Hfr* strain can be represented as a linear structure, and the *relative* arrangements between donor markers are the same for all strains. The maps obtained from different *Hfr* strains show that such two species may introduce a given set of markers in reverse order, and allowing for this, the maps are circular permutations of one another. Consideration of all the *Hfr* strains known suggests that the bacterial chromosome is circular, with the various *Hfr* types transferring segments of this circle, whether clockwise or anticlockwise, from their sites of insertion of the *F* factor. The *Hfr* character itself is transferred only rarely, always together with the most distal markers, and this can be accounted for if the chromosome breaks at the *F* factor before its linear transfer to the recipient, the factor itself being transferred only at the very end.

After chromosome transfer, the exogenote segment of DNA present in the recipient does not consist of a length of the pre-existing donor chromosome duplex, but comprises one donor strand and one newly synthesized strand (for review see Hayes, 1968b). Gross and Caro (1966) grew donor cells in a medium containing radioactive thymine so that their DNA was labelled with the isotope. They were then mated with unlabelled recipient cells either in a normal unlabelled medium, or in a medium containing the radioactive thymine. The latter acts as a control since under these conditions the DNA received by the recipient must be entirely labelled, whatever the mechanism of transfer. Autoradiography showed that the amount of label transferred to recipients under unlabelled conditions was half of that transferred when the medium was labelled. This suggests that one of the strands found in the recipient after conjugation is pre-existing, and the other synthesized during the mating.

By subjecting bacteria to a mutagenic treatment before mating, Bonhoeffer and Vielmetter (1968) were able to demonstrate that genetic information corresponding to only one strand of the donor duplex is transferred to the recipient. This treatment produces heterozygosity in the donor DNA since only one strand of the duplex is likely to be mutated in any given region. Thus if the donor is x^+/x^- and only one strand is transferred, each recipient must receive either x^+ or x^-. On the other hand, if both strands are transferred, the recipient must be heterozygous since it receives both x^+ and x^-. Experimentally, virtually no heterozygotes were found after conjugation. However, when the mutagenic treatment was performed after mating, a high proportion of heterozygotes resulted. This implies that the DNA introduced into the recipient must be in a duplex form *after* the conjugation.

Conflicting theories have been proposed to account for the mechanism of chromosome transfer. Jacob, Brenner and Cuzin (1963) proposed a model in which transfer requires concomitant replication of the donor genome, one of the two replica duplexes being transferred to the recipient as it is synthesized. The geometry of contact between the two bacteria could be defined so that the replication process itself drives one of the daughter chromosomes across the conjugation bridge between them. Certainly, it is clear that inhibition of DNA synthesis prevents successful conjugation. However, more recent work has suggested that it is DNA synthesis in the recipient, rather than in the donor, which is required.

Bonhoeffer and Vielmetter (1968) made use of temperature-sensitive mutants in replication to investigate this problem; such bacteria replicate normally at 37°C, but are unable to synthesize DNA when culture conditions are raised to 42°C. These experiments showed that inhibition of DNA synthesis in the donor does not prevent successful conjugation, but recombinants cannot be formed if the inhibition is in the recipient. This suggests that only a single strand of DNA is transferred from the donor, and its complement is synthesized in the recipient. Ihler and Rupp (1969) were able to demonstrate by the use of donors carrying prophage λ that only one of the λ strands can be recovered from recipients after conjugation. Opposite strands of the phage were recovered by using donors injecting in opposite directions. This suggests that one particular strand of the sex factor is broken, and transfer initiated at this site. The donor strand transferred always commences with a free 5′ end.

Finally, once genetic information has entered the F^- recipient, it must be incorporated into its genome if it is to be inherited by future descendents; a comparison of the frequency of entry with the frequency of appearance in recombinants shows that in *E. coli* the efficiency of incorporation is about 50 per cent.

Sexduction

Adelberg and Burns (1960) discovered an intermediate (*I*) donor strain which showed as a variant of an *Hfr* strain in three respects. Chromosome transfer retained the same orientated sequence with high efficiency of transfer, but at about 1/10 of the former *Hfr* efficiency. The sex factor was transferred to recipients with the same high efficiency as from an F^+, but the recipients were converted to *I* donors and not to the

F^+ state; this suggests that the sex factor has been modified so as to retain a memory of its location on the bacterial chromosome. When the original donor strain was treated with acridine orange, it was converted from *I* to F^- through loss of its sex factor, but on reinfection with a normal factor from an F^+ strain, it regained the *I* donor rather than the F^+ properties. This can be accounted for if the original site of location of the *F* factor on the bacterial chromosome has been modified so as to retain a high affinity for the sex factor.

These properties are analogous to those of the transducing phage particles in an h.f.t. preparation, and can be explained if the intermediate donors arise as the result of a process similar to the formation of transducing phage by lysogeny. A genetic exchange between sex factor and bacterial chromosome at regions adjacent to the site of insertion of *F* in the parent *Hfr* strain would produce a sex factor carrying some of the bacterial genome, and leave the latter with a segment of sex factor in place of lost gene(s). The altered sex factor (F') would thus retain a high affinity for this section of the genome when transferred to other strains of F^-, and its location would be homologous with a region of any further infecting *F* factor. The F' factor consists of part *F* episome and part bacterial genes, just as λ-d.g. phage has both phage genes and bacterial genes, although, as yet, no impairment of the functioning of F' factors compared to the normal *F* has been observed.

Sexduction is used in a very similar manner to transduction for fine structure mapping and investigating *cis/trans* relationships in a specific heterogenote. It has, however, an advantage over any single phage system in that F' species can be obtained for any of the several sites of attachment of the *F* factor in *Hfr* strains, and many of these have been used extensively.

Fine structure mapping

In organisms where recombination occurs between two complete chromosomes, only a single crossover is necessary to generate recombinants. In bacterial systems, however, when the genetic contribution of the donor parent is incomplete, one of the genetic factors involved may be only part of a genome. In this case, two crossovers (or more precisely, an even number) are required to maintain the integrity of the whole chromosome—one crossover results in its fragmentation. This restriction makes interpretation of genetic crosses more complex than in diploid organisms.

The distance between pairs of mutants can be mapped by two-factor crosses in which two mutant types are crossed and the progeny grown under selective conditions where only the wild type survive. The number of wild-type recombinants is a measure of linkage, and can either be used directly as such or related to some standard. The most satisfactory method is to measure also the number of donor-type recombinants emerging from the same cross. In the cross $+b \times a+$, plating on unsupplemented medium allows growth of only $++$ recombinants; there must have been a crossover in between the two markers and also one to the left of *a*. On medium supplemented with *B*, both the wild-type class and the $+b$ recombinants resulting from a crossover to the left of *a* and another to the right of *b* can grow. Thus the frequency of both donor-type ($+b$) and wild-type ($++$) recombinants can be obtained, and the latter expressed as a percentage of the former. Essentially, this approach measures recombination by using only one of the reciprocal recombinants and only one of the two parental types.

Two-factor crosses are subject to errors introduced by variation in growth conditions, and when very small distances are involved, the characteristics of bacterial recombination (such as negative interference which tends to enhance the occurrence of multiple crossovers) makes three-factor crosses necessary to determine the order of sites. For example, if three sites *a*, *b* and *c* are to be mapped where *a* lies to the left of *b* and *c*, but the order of the latter two is not known, the cross $+(+c) \times a(b+)$ is performed with each type in turn as donor. If the order is *abc*, both crosses will require two crossovers to generate wild-type recombinants: for $++c \times ab+$, one to the left of *a* and one between *b* and *c*, and for $ab+ \times ++c$, one between *b* and *c* and one to the right of *c*. However, if the order is *acb*, the cross $a+b \times +c+$ requires one crossover in between *a* and *c* and one between *c* and *b*, but the cross $+c+ \times a+b$ requires a further two, one to the left of *a* and one to the right of *b*, that is a total of four. So in this case, the frequencies of wild-type progeny emerging from the two crossovers will be distinctly different, whereas in the first arrangement they will be the same.

Mapping many mutants by three-factor crosses is a lengthy process, and *deletion mapping* is an alternative and shorter procedure. Deletion (or *multisite*) mutants have lost a segment of their chromosome, and its extent can be found through crosses with other mutants whose positions have already been determined. Wild-type recombinants can only be obtained if the mutant site lies outside the length of the deletion—when it lies within its extent the deletion strain has no corresponding wild-type site. Generally, a range of strains is established carrying a series of partially overlapping deletions of the order of size of a cistron or less. Any new mutant to be mapped is crossed with each of these, and if it gives wild-type progeny with one deletion but not with another partially overlapping one, it must lie in the region not covered by the overlap. If there are sufficiently many such partially overlapping sequences, its position can be accurately identified. The recombination values obtained from crosses yielding wild type give the distance of the mutant from the end of the deletion concerned.

II

Information Transfer from Gene to Protein

CHAPTER 4

The genetic code

Colinearity of gene and protein

Chromosomes consist of a linear array of genes, but this does not make any implications as to the structure of the gene itself. The simplest and most obvious way for a stretch of nucleic acid to specify some protein would be for them to be colinear; in other words, the sequence of nucleotides along the gene would specify the unique sequence of amino acids along the protein. This is equivalent to saying that the linear structure of the chromosome applies not only between but also within genes; together with the demonstration of recombination within a gene this leads to the picture of chromosomes as long lengths of DNA, successive segments coding for different proteins. There have been two successful approaches to the problem of demonstrating experimentally this equivalence between the map of a nucleic acid and the protein which it specifies.

Yanofsky and coworkers (1964) mapped the gene controlling the *A* chain of the tryptophan synthetase protein of *E. coli*. As well as the genetic mapping of mutants by recombination frequencies, deletion mapping was used for determining the positions of mutants close to each other. The various deletion strains used all had deletions extending into the *A* gene from one end, but differing in their end points within the gene. Recovery of a tryptophan-independent recombinant (that is one with a functional *A* gene) from a transduction cross between any *A* mutant and a deletion, placed the mutant site as outside the region deleted, and the recombination value gave the position of the mutant.

The altered protein resulting from each particular mutation was degraded into peptides by digestion with proteolytic enzymes, and the peptide differing from wild type isolated and its amino acid sequence determined. Primary structure studies established the sequence of peptides containing some seventy-five amino acid residues. The positions of the sites of amino acid replacement were established for seven mutations, and the map of the protein this gave was compared with the positions of the mutants on the genetic map of the gene. Yanofsky and coworkers

(1967) further reported determination of the complete sequence of the 267 residues in the *A* protein, and the finding of colinearity with the gene for virtually the entire length of the protein.

Sarabhai, Stretton and Brenner (1964) used another approach, involving mutants affecting the head protein of phage T4. *Amber* mutants cause a premature termination of protein synthesis; thus if gene and protein are colinear, such a mutation half-way along the gene should cause only the first half of the protein to be made. Ten different mutants were induced in a cistron controlling head protein synthesis by the use of chemical mutagens, and a genetic map was established by recombination analysis. The proteins synthesized in the phage-infected bacteria were labelled with radioactive amino acids, the cells lysed, nucleic acid removed, and the mixture degraded with trypsin or chymotrypsin without further purification. The peptides produced were characterized by high-voltage electrophoresis on paper and located by autoradiography. In controls, the peptides characteristic of the head protein of wild phage could easily be recognized in such digests, eliminating any need for their purification from the other proteins present.

It was found that different *amber* mutations caused termination of protein synthesis at different points along the head protein. Proteins are synthesized from the $-NH_2$ to the $-COOH$ terminus, and in each case the N-terminal peptide of the protein was present, but the number of peptides successive to it varied with the particular mutant. Different mutants had different peptides present, but when the peptides were ordered into a 'hierarchy' from the N- to C-terminal ends of the protein, these could be seen to represent the synthesis of different lengths of the protein, as illustrated in Figure 4.1. The length of protein present correlated with the genetic map position of the mutant, as would be expected from colinearity.

The triplet code

As twenty amino acids are commonly found in protein but there are only four bases in a nucleic acid, each amino acid must be specified by more than one base. The number of bases representing an amino acid has been referred to as the *coding ratio*, and this topic has been reviewed by Crick (1963), Lanni (1964) and Woese (1967). The minimum number of bases able to provide sufficient information would be three per amino acid:

Coding ratio	1	2	3	4
Amino acids specifiable	$4^1 = 4$	$4^2 = 16$	$4^3 = 64$	$4^4 = 256$

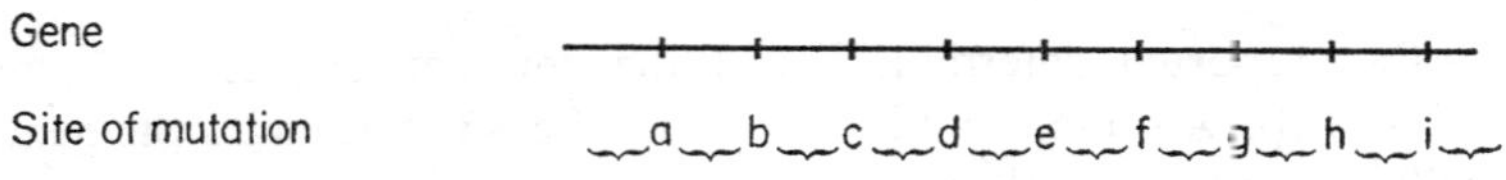

Peptide fragment	1	2	3	4	5	6	7	8	9	10
	N-terminus									C-terminus
Mutant a	+	−	−	−	−	−	−	−	−	−
b	+	+	−	−	−	−	−	−	−	−
c	+	+	+	−	−	−	−	−	−	−
d	+	+	+	+	−	−	−	−	−	−
e	+	+	+	+	+	−	−	−	−	−
f	+	+	+	+	+	+	−	−	−	−
g	+	+	+	+	+	+	+	−	−	−
h	+	+	+	+	+	+	+	+	−	−
i	+	+	+	+	+	+	+	+	+	−
Wild type	+	+	+	+	+	+	+	+	+	+

Figure 4.1. Correlation between the position of an *amber* mutant in the gene and the extent of protein synthesized (+ signifies peptide present, —indicates absence). Each *amber* mutant directs the synthesis of the extent of the protein from its N-terminal end shown by the peptides present. The length of protein synthesized corresponds to the genetic map position of the mutant. This is interpreted as demonstrating that a gene is colinear with the protein which it specifies

Early proposals for the form of the code envisaged some sort of stereochemical relationship between a nucleic acid and the amino acids specified. The nucleic acid would act as a physical *template* upon which the amino acids would first be assembled and subsequently covalently linked together. The distance between adjacent nucleotides in a polynucleotide chain is fairly close to the distance between amino acid residues in a fully extended protein chain, and this led Gamow (1954) to suggest an *overlapping code* in which the sequence

AGUCA

would specify the three amino acids

$$\alpha = \text{AGU}$$
$$\beta = \text{GUC}$$
$$\gamma = \text{UCA}$$

Overlapping and non-overlapping codes can be directly distinguished by the effects of a single base change in the nucleic acid. In an overlapping code, this would result in an alteration of more than one amino acid in the protein; in a non-overlapping code only one amino acid would be changed, as illustrated in Figure 4.2.

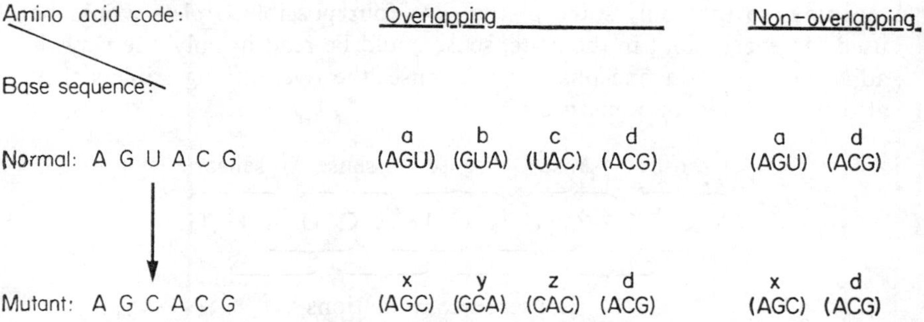

Figure 4.2. The effects of a single base mutation upon overlapping (left) and non-overlapping (right) codes. With an overlapping code, the alteration of one base alters three of the amino acids specified. With non-overlap, there is only one alteration in the protein

Ingram (1957) showed that the mutation in hemoglobin which causes sickle cell anaemia involved a change in only one amino acid: a glutamic acid was replaced by a valine residue. Wittman (1963) induced mutations in TMV-RNA (tobacco mosaic virus) by treating the nucleic acid with mutagens. The proteins of the mutant strain were separated from the nucleic acid, digested with trypsin, and the tryptic peptides isolated and analysed. In all cases only a single amino acid was substituted at any given site. Similarly, Yanofsky (1963) found that when a single amino acid of the *E. coli* tryptophan synthetase protein was altered by mutation, its neighbours on either side remained unaltered.

If the code is non-overlapping, there must be right and wrong ways of reading it, since otherwise the sequence

A G U A G U A G U A G U A G U

could be read in any one of three phases as

AGU AGU AGU AGU AGU

GUA GUA GUA GUA GUA

UAG UAG UAG UAG UAG

depending upon the starting point. This difficulty could be overcome either by reading the sequence from some fixed starting point, or by the nature of the code being such as to prevent this ambiguity. Crick and coworkers (1957) proposed a code 'without commas', so called because the phase would be inherent in the sequence. Certain sequences of three bases would represent an amino acid, but others would have no such meaning, so that only some of the sixty-four possible triplets would be used. At every point in the code, sense could be read in only one way; if adjacent triplets in one phase make sense, the overlapping triplets they also form would be nonsense:

sense sense sense sense sense

A G U A G U A G U A G U A G U

nonsense combinations

The four possible XXX triplets would not represent amino acids, since if these were to code for successive identical amino acids, the sequence XXXXXX would make sense in the wrong phase. The sixty remaining triplets were grouped into twenty sets of three, each set comprising cyclic permutations of the same three nucleotide bases; only one of these three could represent an amino acid. For example, if

$$\text{A G U A G U} = \alpha\text{–}\alpha$$

$$\left.\begin{array}{l}\text{G U A}\\ \quad\text{U A G}\end{array}\right\} = \text{nonsense}$$

Crick demonstrated that there are eight possible basic solutions.

The formulation of non-overlapping codes, relying on the similarity between internucleotide distance and amino acid spacing, suggested a 1 : 1 spacial relationship. Non-overlap would require each amino acid to be related to a different set of three nucleotides, and Crick suggested that this might be achieved through sequences coding for an amino acid (that is, the sense triplets) recognizing an appropriate 'adaptor' molecule. This would be a trinucleotide attached to the amino acid, and would hydrogen-bond with the correct sequence because its bases would be complementary to it. In its correct position, the adaptor would be held by hydrogen bonds between each of its three bases and those in the codon; in an incorrect position only two such attachments would hold, so that it would remain permanently attached to only its sense triplet.

Despite the elegance with which this model accounts for the excess number of triplets (sixty-four) over amino acids (twenty), Crick and coworkers (1961) showed that the code is read in triplets from a fixed point, with the phase of reading defined by the starting point. Genetic experiments were performed on the *r*II system of phage T4, using mutants induced by acridine. (The use of this system has been extensively reviewed by Barnett and coworkers, 1967). Wild-type phage grow on both the B and K12 strains of *E. coli*, but a phage which has lost the function of either of the two cistrons of the *r*II region (cistron *A* and *B*) will not grow on K12 and produces a different type of plaque on *E. coli* B (rapid lysis plaque):

	E. coli B	*E. coli* K12
Phage: wild	normal (small fuzzy plaque)	normal (small fuzzy plaque)
mutant	large plaque	no growth

Some types of mutant—termed *leaky*—show partial function in growing upon K12, but the plaque type on B is not true wild type.

Acridines (such as proflavin) act as mutagens by causing addition or deletion of a base in the nucleotide sequence of DNA (see Sesnowitz-Horn and Adelberg, 1968; Streisinger and coworkers, 1966). Whereas mutants produced by mutagens which cause the substitution of one base for another are often leaky (typically about half are), those produced by acridine mutation usually have a complete lack of the gene function, that is are non-leaky. If the code is read in non-overlapping triplets from a fixed point, misreading must occur after an addition or deletion because the phase of the message has been shifted by one nucleotide residue; this must cause a completely altered amino acid content of the protein from this point on (Figure 4.3), resulting in the complete lack of function found in acridine mutants. (This would not be the result if the phase was defined by the nucleotide sequence in a commaless code). In a base substitution mutant, where only a few bases in the nucleic acid have been altered, only a few amino acids will be changed in the protein. This may leave some residual degree of protein activity, giving the leaky phenotype.

Acridine mutants cannot be induced to revert by mutagens which cause base substitution, but can do so upon a further mutation with

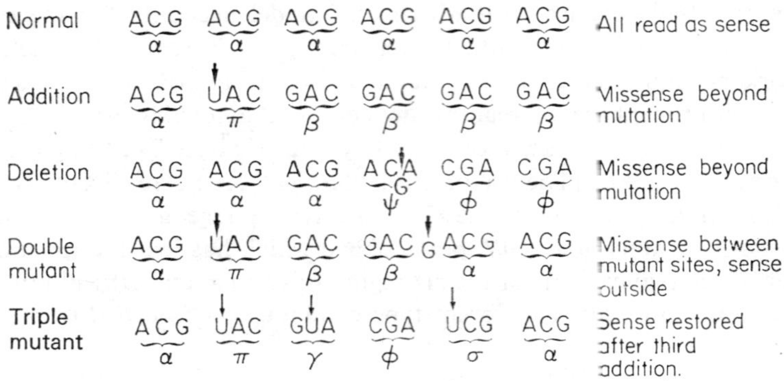

Figure 4.3. The effect of additions ($\overset{\downarrow}{X}$) and deletions ($\overset{\downarrow}{X}$) on the genetic code, showing that it is read sequentially from a fixed starting point (from left to right above). Either an addition or a deletion alters the reading frame so that the sequence to the right of the alteration is read out of phase as *missense*. When an addition is followed by a deletion, the correct phase is restored beyond the site of the second alteration and there is a missense region only between the two mutant sites. When there are three successive additions, the correct reading frame is restored beyond the last site.

acridine. If an acridine mutant is produced by, say, addition of a base, it should revert to wild type by its deletion. However, in FC_0, a mutant induced by proflavin, this reversion was shown to occur not by reversal of the original mutation, but by production of a second mutation at another site nearby in the same gene, this is termed a *suppressor*. If FC_0 was formed by a base addition, the original reading frame (phase) could be restored by deleting another base somewhere nearby. There would then be misreading only between the two mutant sites; the original phase would be restored to the parts of the gene outside them, as shown in Figure 4.3.

Experimentally, double mutants were formed by genetic recombination between FC_0 and a phage carrying the second mutation. Some eight suppressors of FC_0 were found, all located within a short region (1/10) of the *B* cistron. All these were themselves non-leaky mutants, so that their reversion to wild type could be studied by the same procedure used for FC_0 reversion. These too were found to revert by formation of a double mutant, their suppressors again comprising non-leaky *r*II mutants. A repeat on two of these suppressor–suppressors gave the same results.

Defining FC_0 as (+), that is base addition, and its suppressors as (—), that is base deletion, various double mutants consisting of + and —

combinations were made. No ++ or −− sequences showed any activity, but +− or −+ mutants were active. This is in accord with reading of the genetic code in groups of nucleotides from some fixed point, although it does not specify how many nucleotides represent each amino acid. However, when triple mutants of the type +++ or −−− were constructed, these proved to be wild type; this supports reading in triplets since three additions or deletions would then correspond overall to the insertion or removal of one amino acid, returning the reading frame to its normal phase after the third site of mutation (see Figure 4.3).

This theory predicts that the active double (or triple) mutants will differ from wild type in the sequence of amino acids corresponding to the stretch of the gene in between the two outside mutant sites. In fact, the multiple mutats which are active on *E. coli* K12 show a variety of phenotypes on strain B, raging from apparently wild type to those producing *r* mutant-type plaques. This can be explained by the presence of a sequence of altered amino acids in the protein (its length depending upon the distance apart of the sites of mutation), and gives the pseudo-wild phenotype.

The proteins synthetized by the *r*II cistrons have not been isolated, so direct confirmation of this theory is not possible in this system. However, work on the lysozyme synthesized by phage T4 (Streisinger and coworkers, 1966) has provided confirmation that the protein in a double mutant differs from wild type in only the amino acid sequence between the two sites of mutation. A deletion followed by an addition shortly after in the same gene resulted in a protein differing from wild type in only five amino acids. Further, using the codon assignments proposed on the basis of other studies, the wild-type and mutant sequences could be accounted for by a deletion and addition sixteen nucleotides apart.

The assignment of codons

Three types of system have been used to assign the actual nucleotide sequences within the triplet code-words (*codons*) to their respective amino acids (recent reviews have been by Woese, 1967, and Sadgopal, 1968). The first approach used cell-free extracts to achieve protein synthesis *in vitro* (the process of protein synthesis is discussed in the next chapter). Nirenberg and Matthaei (1961) used a system from *E. coli* comprising the protein synthetic components ribosomes and aminoacyl *t*RNA, with ATP as energy source. When a synthetic polynucleotide template was added to this, the incorporation of a specific amino acid into polypeptide was stimulated; the homopolymer poly-U promoted uptake of C^{14}-labelled phenylalanine, suggesting that the codon for phenylalanine is UUU.

Speyer (1963) used a similar system with ribopolynucleotide templates prepared by the enzyme polynucleotide phosphorylase; the base ratios of such copolymers closely resembles the composition of the nucleoside diphosphate incubation mixture in which they are prepared. The effect of a polynucleotide on the incorporation of a particular amino acid into protein was followed by using an incubation mixture in which this amino acid, labelled with C^{14}, was present together with the other (unlabelled) nineteen amino acids. Incorporation into protein was measured by precipitating the protein made and counting its radioactivity.

Experiments with homopolymers showed that poly-U directed incorporation of phenylalanine (and also a little leucine, presumably by mistake), poly-A encouraged lysine uptake, and poly-C stimulated proline incorporation. Poly-G could not be tested because of the difficulty of preparing a suitable physical form. Copolymers directed the incorporation of several amino acids; on the assumption that the base sequences of the copolymer are random—the nearest neighbour frequencies were shown to be so—the frequencies of the various triplets can be calculated from the nucleotide proportions of the template preparation mixture, and matched with the amino acid composition of the polypeptide synthesized. This approach has the limitation of giving only the nucleotide composition of a codon and not the sequence within it. For example, the calculation for poly-AC (5/1 ratio) assumes that on average 5/6 nucleotides are A and 1/6 are C. Thus the AAA triplet has a frequency of $(5/6)^3 = 125/216$; other frequencies are expressed as a percentage of this. Each of the triplets AAC, ACA and CAA has a frequency of $(5/6)^2 \times 1/6 = 25/216$, that is 20 per cent. relative to AAA. Similarly, each of ACC, CCA and CAC has a relative frequency of 4 per cent., and CCC of 0·8 per cent. Since poly-A directs the synthesis of polylysine, AAA must represent lysine, and the amino acid composition of the polypeptide synthesized as a proportion of its lysine showed that asparagine, glutamic acid and threonine were each about 20 per cent., and proline and histidine each about 4 per cent. Thus each of the first three must be represented by a 2A,1C codon, and each of the last two by a 1A,2C codon. Only six amino acids are incorporated, because two of the amino acids are each represented by two codons and, in fact, their frequency in the polypeptide corresponds to the sum of the frequencies of their two codons (threonine = ACA+ACC = 24 per cent.; proline = CCA+CCC = 4·8 per cent.).

Nishimura, Jones and Khorana (1965) and Jones, Nishimura and

Khorana (1966) reported the synthesis of polynucleotides containing a known sequence of nucleotides. A copolymer containing two bases in alternating sequence caused the interdependent incorporation of only two amino acids when used as template in a cell-free system. In such experiments, incorporation of the other eighteen unlabelled amino acids was tested by providing one labelled and nineteen unlabelled amino acids in the reaction mixture; no third amino acid was incorporated to any significant extent. The kinetics of incorporation of the two amino acids were similar, and the total amounts of each incorporated were approximately equimolar; the incorporation of either amino acid was stimulated in the presence of the other and very much reduced by its absence. There was a relatively low incorporation of one in the absence of the other, but this was probably due to an endogenous content in the extract system. One such experiment was the stimulation of valine and cysteine incorporation by poly-UG, as shown in Figure 4.4.

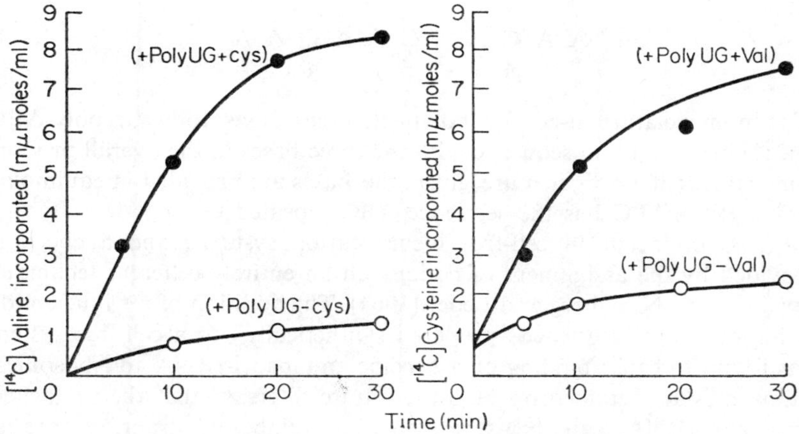

Figure 4.4. Stimulation of amino acid incorporation by poly-UG. Equimolar quantities of cysteine and valine are incorporated. Incorporation of each amino acid is interdependent, that is each can only be incorporated into polypeptide if the other is present. No third amino acid was incorporated to any significant extent. (After Jones, Nishimura and Khorana, 1966)

If the nucleotide sequence

U G U G U G U G U G U G U G U

directs synthesis of the polypeptide

cys – val – cys – val – cys

one of UGU and GUG must represent cysteine and the other valine. The use of other data, such as from random sequence polynucleotide templates, can distinguish the two possibilities. These results also provide a demonstration that the coding ratio must almost certainly be three, and in all events an odd number multiple of it.

Further work (Khorana and coworkers, 1966; Kossel, Morgan and Khorana, 1967) has made use of polynucleotide sequences consisting of repeating tri- and tetranucleotides, and analysis of the amino acids incorporated has defined the meaning of about half of the sixty-four codons. For example, Nishimura, Jones and Khorana (1965) reported that poly-*r*AAG directs the synthesis of the three homopolypeptides, polylysine, polyarginine and polyglutamate. This is explained if reading starts at a random triplet at the beginning of the message thus:

A A G A A G A A G
lys – lys – lys –

A A G A A G A A G A
arg – arg – arg –

A A G A A G A A G A A
glu – glu – glu –

(The nomenclature used for polynucleotides is as follows: poly-ABC (1:2:3) has a random sequence with the three bases in the overall proportions stated; if no figures are given, the bases are provided in equimolar ratios. Poly-*r*ABC has the sequence ABC repeated.)

Errors do occur in cell-free incorporation systems, and a check is provided by the assignment of codons on an entirely different technique developed by Nirenberg and Leder (1964). The addition of a trinucleotide codon of known sequence—prepared synthetically—to a cell-free system was found to cause binding of a specific aminoacyl-*t*RNA to ribosomes. Ribosomes are retained on cellulose nitrate filters, so that their attached aminoacyl-*t*RNA can be separated from unbound transfer species, which are removed by a wash of the filters with salt solution. Any triplet combination can be tested by synthesis of the appropriate trinucleotides and assaying against all the transfer species, one loaded with a radioactive amino acid and the remainder unlabelled. The radioactivity retained on the filter is determined for each of twenty reaction mixtures, each of which has a different amino acid radioactively labelled. This has been done for most of the triplets, and the results are in fairly good agreement with the data from polynucleotide directed amino acid incorporation experiments (for a summary see Brimacombe and coworkers, 1965, and Nirenberg and coworkers, 1966). The code now generally accepted from both sets of studies is shown in Figure 4.5.

First	Second				Third
	U	C	A	G	
U	phe	ser	tyr	cys	U
	phe	ser	tyr	cys	C
	leu	ser	ochre	CT	A
	leu	ser	amber	try	G
C	leu	pro	his	arg	U
	leu	pro	his	arg	C
	leu	pro	gln	arg	A
	leu	pro	gln	arg	G
A	ile	thr	asn	ser	U
	ile	thr	asn	ser	C
	ile	thr	lys	arg	A
	met,fmet	thr	lys	arg	G
G	val	ala	asp	gly	U
	val	ala	asp	gly	C
	val	ala	glu	gly	A
	val	ala	glu	gly	G

Figure 4.5. The genetic code. Codon meanings have been derived mostly from studies with *E. coli* systems, but appear to apply universally to all organisms. *Ochre*, *amber* and CT indicate codons which cause termination of protein synthesis. The fmet is a species used to initiate protein synthesis in *E. coli* (see next chapter), but probably does not exercise this function in higher organisms, although it may do so in some other bacteria

A third approach has concentrated on confirming that these codons are actually used *in vivo*. One basis for this comes from studies on the amino acid substitutions which result from the mutation of a single nucleotide, induced by a mutagen causing a specific type of base change. For example, if tryptophan is known to be coded by UGG and it mutates to arginine by induction of a U⟶C base change, CGG must be a codon used *in vivo* for arginine. A considerable amount of such work has been performed on the tryptophan synthetase *A* gene protein of *E. coli* (Berger, Brammar and Yanofsky, 1968).

Another basis for *in vivo* assignments comes from the analysis of amino acid sequences in double acridine mutants of the T4 lysozyme (Streisinger and coworkers, 1966). Using the codons proposed on the basis of *in vitro* studies, they found that for every double-mutant studied, a unique sequence of bases could be assigned which codes for the wild-type amino acid sequence and, with appropriate addition and deletion, for the double-mutant amino acid sequence. An example is given in Figure 4.6.

	thr	lys	ser	pro	ser	leu	asn	ala
Wild	AC-	AA$^{A}_{G}$	AGU	CCA	UCA	CUU	AAU	GC-
		↓					↓	
Mutant	AC-	AAA	GUC	CAU	CAC	UUA	AU$^{A}_{G}$	GC-
	thr	lys	val	his	his	leu	met	ala

Figure 4.6. Amino acid sequences in a wild-type and double-acridine mutant of T4 lysozyme. The two amino acid sequences detected experimentally can be related only by the nucleotide sequences shown, with the deletion and addition sixteen nucleotides apart. This confirms the use *in vivo* of these codon assignments. (At the sites of addition and deletion it is not known whether A or G is present; these are shown as $^{A}_{G}$)

Finally, direct evidence for the assignment and use of codons has been provided by the sequencing of a length of fifty-seven nucleotides of the coat protein gene of an RNA bacteriophage. Adams and coworkers (1969) have shown that these could code for the nineteen amino acids in positions 81–99 of the coat protein. Overall, therefore, a considerable proportion of the codon assignments of Figure 4.5 appear to be used *in vivo*.

Nonsense codons

Certain codons do not stimulate the binding of any aminoacyl *t*RNA in the ribosome binding assay, and when present in templates used to direct protein synthesis *in vitro*, cause termination of synthesis of the polypeptide chain. The action of one such class (*amber*) was confirmed *in vivo* by Sarabhai, Stretton and Brenner (1964) in their work on the colinearity of gene and protein, and two other classes have since been shown to behave similarly.

Benzer and Champe (1961, 1962) constructed a test to identify these *nonsense* mutants and distinguish them from *mis-sense* mutants. (The latter contain the wrong amino acid in a protein as the result of a base substitution, whereas the former terminate protein synthesis at the mutation.) They made use of a deletion in the *r*II region (*r*1589) which joins the *A* and *B* cistrons together; this region normally produces two proteins, one from each cistron, but only one joint protein is produced in the deletion mutant. However, this protein has *B* activity, so that its *B* part is presumably unaffected by the attachment of the *A* part. Various base-substitution mutants were tested in this system to see whether they had resulted in the production of nonsense codons. The procedure used was to insert a mutant in the *A* cistron in series with *r*1589 by genetic recombination and to test the double mutant for *B* cistron activity; a nonsense mutant gives no such activity because the polypeptide chain has been terminated in its *A* region, but a missense mutation in the *A* region would not affect *B* activity. This system is illustrated in Figure 4.7.

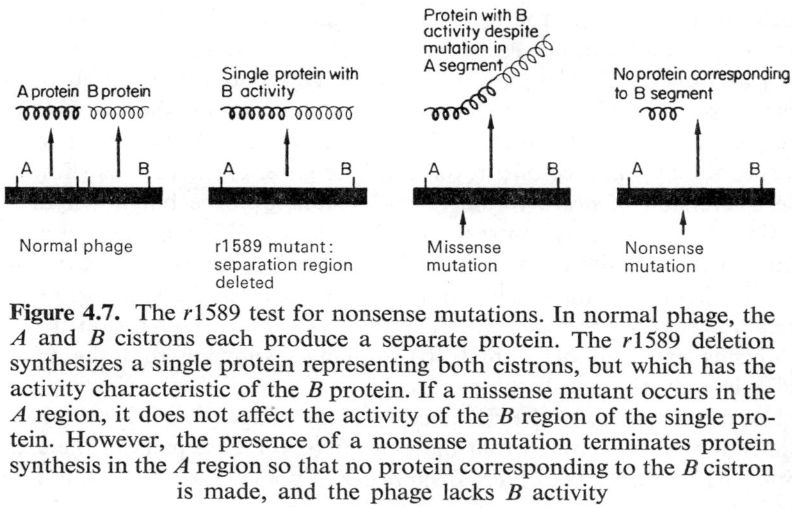

Figure 4.7. The *r*1589 test for nonsense mutations. In normal phage, the *A* and *B* cistrons each produce a separate protein. The *r*1589 deletion synthesizes a single protein representing both cistrons, but which has the activity characteristic of the *B* protein. If a missense mutant occurs in the *A* region, it does not affect the activity of the *B* region of the single protein. However, the presence of a nonsense mutation terminates protein synthesis in the *A* region so that no protein corresponding to the *B* cistron is made, and the phage lacks *B* activity

The nonsense mutants isolated were ambivalent according to their host strain of *E. coli*; in *non-permissive* strains the phage could not grow because of premature chain termination, but in *permissive* (*suppressor*) strains, growth occurred. Benzer and Champe suggested that this could be accounted for if the suppressor strain contains an aminoacyl-*t*RNA which responds to the nonsense codon by inserting an amino acid into the growing protein chain, whereas the non-permissive strain lacks this species and terminates protein synthesis at this codon. (This is discussed in detail in Chapter 9.) There are various suppressor strains, and sets of nonsense mutants can be classified by the suppressor locus to which they respond. They fall into three classes. The mutants found by Benzer and Champe all respond to the same suppressor and have been termed *amber*. Brenner and Beckwith (1965) found another class termed *ochre*; these are not suppressed by *amber* suppressors, but *ochre* suppressors can suppress *amber* as well as *ochre* mutants, so that, although distinct, the two types must be related. A third class was isolated by Sambrook, Fan and Brenner (1967) and by Zipser (1967); its suppressors vary in their ability to suppress the other two types.

Brenner, Stretton and Kaplan (1965) used two types of experiment to deduce the nucleotide sequences of *amber* and *ochre* codons. The first was to induce and revert *ambers* and *ochres* by the chemical mutagens hydroxylamine and 2-aminopurine. The former reagent causes the replacement of G/C base pairs by A/T base pairs, and the latter causes both this reaction and the reverse replacement of A/T by G/C. Hydroxylamine induced both classes of nonsense mutant; thus each must contain (at least) one A/T pair. However, neither type could be reverted by this reagent so that either the codons do not contain a G/C pair, or if one of them does, its conversion to A/T gives the other, and therefore does not lead to the appearance of a revertant. The latter is the case as shown by the conversion of *ochre* to *amber* with 2-aminopurine but not hydroxylamine; the *ochre* must possess an A/T pair which is matched by a G/C pair at the same site in the *amber*. (The predicted *amber* → *ochre* induction by hydroxylamine could not be demonstrated because *ochre* suppressors also suppress *ambers*.) The *ochre* → *amber* conversion had the frequency expected of a single base change, showing that the triplets differ in only the one base converted.

Use of a property of the *r*II function of the phage (that it is used before replication) demonstrated that there are two A/T base pairs in the *amber* triplet. Since the *amber* differs from *ochre* in possessing a G/C pair where the latter has an A/T, the *ochre* must thus have the same two A/T base pairs, and also a third. Wild-type phage was treated with hydroxylamine; the consequent conversion of C to U′ (which behaves as thymine in subsequent replication, thus causing the characteristic transition) could take place on either strand of the DNA. Only one strand—the coding—is transcribed into RNA, whilst the other—anticoding—is not, so the effect depends upon which strand carries the mutation. Figure 4.8 illustrates the effect of hydroxylamine mutation in these two instances.

When mutation is on the anticoding strand, the immediate messenger species is unaffected, whereas when the coding strand is mutated, an A

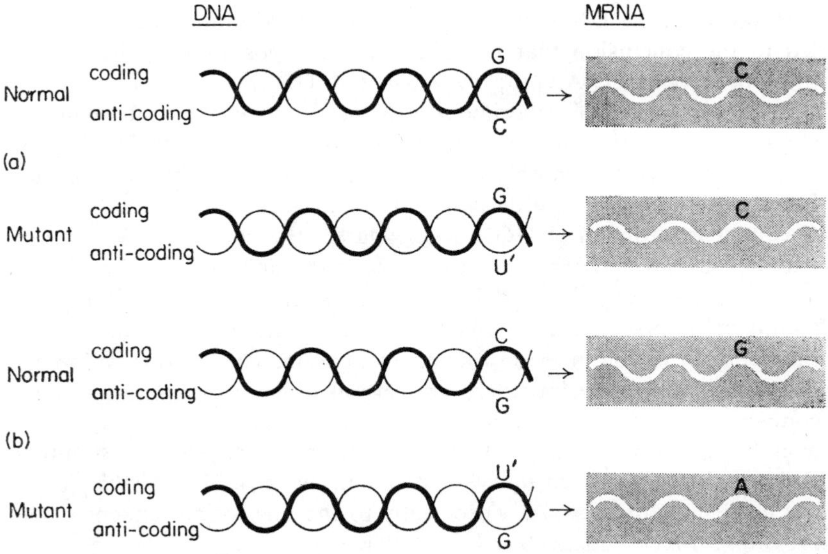

Figure 4.8. The effect of hydroxylamine on *m*RNA synthesis by the *r*II region of T4. (a) If the mutagenic interaction converts a cytosine on the anticoding strand to U′, the *m*RNA immediately synthesized is not affected since the coding strand from which it derives retains its normal nucleotide sequence. A mutation will be shown only after the next round of replication has made a mutated coding strand for the daughter duplex derived from the parental anticoding strand. (b) If a cytosine on the coding strand is converted to U′, the *m*RNA immediately synthesized will bear an adenine instead of a guanine nucleotide. (After Stretton, 1965)

replaces a G and can thus result in production of *ambers* and *ochres*. With K12 bacteria as hosts, wild-type expression of the *r*II region is essential for survival, and phages mutated on the coding strand will not grow (although they survive to give mutants on *E. coli* B where r^+ expression is not necessary). Because the *r*II region is used before replication, mutation on the anticoding strand gives normal phage particles on both B and K12, although this altered strand will affect future generations. When hydroxylamine-treated phages infect B and K12 strains, two classes of *amber* mutants are produced in about equal frequencies. In one set, *r*II *ambers* found when grown on B do not appear during growth on K12, and in the second, *ambers* are found with the same frequency in both bacterial strains. The former correspond to coding strand mutation, the latter to alteration of the anticoding strand. Thus *amber* triplets must contain two bases derived from A/T pairs in opposite orientation in the DNA, that is one A and one U on the messenger. This led to the conclusion that the nucleotide compositions of these codons must be: *amber*—U,A,G/C; *ochre*—U,A,A/U.

A second type of experiment was used to resolve the sequences; studies on mutation in which a purine is replaced only by the other purine, or a pyrimidine only by the other pyrimidine, showed that in T4 head protein, *ambers* are derived only from glutamine codons (CAA or CAG) or tryptophan (UGG). This suggests the codon UAG for *amber* and hence UAA for *ochre*. Studies on the suppression of *amber* and *ochre* triplets (see Chapter 9) have confirmed that this occurs through an aminoacyl-*t*RNA molecule which has been mutated so as to recognize the nonsense codon instead of its normal codon. The various amino acids inserted by these species (there are several different suppressors) are all connected by a single base change to the proposed nonsense triplets, and would be consistent with the UAG and UAA codons. One such suppressor aminoacyl-*t*RNA, which inserts tyrosine in response to the *amber* codon, has been shown to respond specifically to the nucleotide sequence UAG, providing direct evidence for this triplet.

It is now also known that the codon UGA is nonsense too; Brenner and coworkers (1967) showed that it is unacceptable in *r*II cistrons and Sambrook, Fan and Brenner (1967) demonstrated that it causes loss of activity in the *E. coli* β-galactosidase gene in a manner similar to *amber* or *ochre*. However, it seems propable that although all three of the codons UAG, UAA and UGA cause chain termination, only one, UAA (*ochre*), is actually used for this purpose *in vivo* (see pages 133–134).

Evolution of the code

The assignment of amino acids to codons has been worked out mostly in *E. coli*, but so far as is known at present, it applies to all living organisms. This conclusion is supported by the responses of iso-accepting *t*RNA molecules of different species to trinucleotides in the Nirenberg ribosome binding assay (see page 207), and strong evidence for the universality of the code has been provided by the synthesis of normal globin protein (as judged by its tryptic digest fingerprint) under direction from rabbit reticulocyte *m*RNA in an *in vitro* protein synthetic system derived from *E. coli* (Laycock and Hunt, 1969). (A single exception is that UGA may represent cysteine in vertebrates, rather than nonsense as in bacteria—Ycas, 1969). The code is highly *degenerate* in that sixty-one of the possible sixty-four triplets code for only twenty amino acids. Indeed, it is only because most amino acids are represented by more than one codon that it is possible for all organisms to use the same code, despite the drastic variations which exist in the base compositions of their DNA—the two extremes are *Tetrahymena pyriformis* which has only 25 per cent G+C, and *Micrococcus lysodeikticus* which has as much as 72 per cent. Nevertheless, the amino acid composition of proteins does not show such a range of variation, but tends to remain roughly similar for all species. Of course, this implies that although the code itself remains the same, different organisms will use to different extents the various codons available to represent any particular amino acid.

There is a pattern to the degeneracy of the code in that the arrangement of codons representing the same or similar amino acids is not random. The third nucleotide of the codon is the most degenerate; either the base in this position is irrelevant to the meaning of the codon, or a distinction is made only between purine and pyrimidine. This can be accounted for by the nature of the base pairing between the codon on the *m*RNA and the anticodon on the *t*RNA (and is discussed in detail on page 206). As a consequence, alteration of the final nucleotide in a triplet very often does not result in any change in the amino acid coded. As can be seen from Figure 4.5, the codons are thus grouped in sets representing a single amino acid.

Even allowing for the grouping of codon meanings, there is a further aspect to the degeneracy in that the codons for functionally related amino acids appear to be closely related themselves. For example, amino acids with U as the second base of their codon (first column of Figure 4.5) are the most hydrophobic, whilst those with A as the second base (third column of Figure 4.5) have polar side chains. The result of this relationship is to increase the probability that a single base mutation will result either in no change in the amino acid coded or in the substitution of a related amino acid.

Another facet of the code is that the various amino acids are represented by different numbers of codons, ranging from one to six:

1 codon—met, try
2 codon—glu, gln, asp, asn, cys, lys, his, tyr, phe
3 codon—ile, (CT)
4 codon—val, pro, thr, ala, gly
6 codon—leu, ser, arg

Mackay (1967) and King and Jukes (1969) examined the amino acid compositions of various unrelated proteins and observed that the frequency of occurrence of an amino acid is in reasonably good agreement with its number of codons (see Figure 4.9). In fact, given the triplet base mechanism, the natural genetic code appears to come close to the optimal coding proportions on a frequency basis. King and Jukes suggested that, rather than regarding codon assignments as attaining their present meanings in order to best match the amino acid composition of living material, we should view the relationship as the converse—the composition of proteins as reflecting the relative availability of amino acids, given their codon assignments.

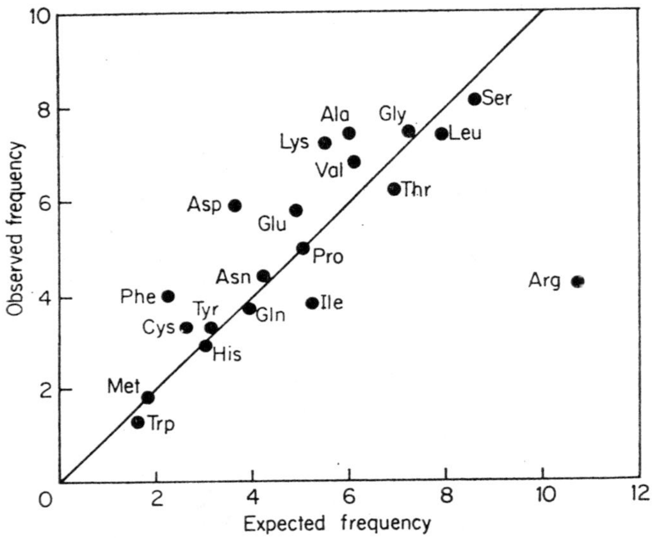

Figure 4.9. Amino acid frequencies in proteins compared with the frequencies expected from their number of codons. There is good agreement with the sole exception of arginine. (See King and Jukes for discussion). (From King and Jukes, 1969. Copyright 1969 by the American Association for the Advancement of Science)

Two types of theory have been proposed to account for the evolution of the code (for reviews see Woese, 1969, and Ycas, 1969). *Mechanistic* models suggest that the original code was based upon some physico-chemical relationship between an amino acid and its codon. This implies that the code could be constructed in only one, or at most a very few, ways and its universality results from this defined relationship. *Stochastic* models do not depend upon a fixed relationship between a codon and an amino acid, but argue that the initial set of codon assignments was free to vary. Selective forces would act to modify this starting set during the course of evolution, and the final set of codons must thus be optimal with respect to the criteria used by selection. As such, stochastic theories do not account for the universal nature of the code; to explain this they require the assumption that the code developed at an early stage during evolution, and all organisms descended from a single cell line, or more precisely, from a single in-breeding population. Of course, pure mechanistic and pure stochastic models represent extremes, and there are possible intermediate models with features of both. To date, one mechanistic and three stochastic models have been proposed; the stochastic models differ in the nature of the selective forces which they postulate as responsible for the evolution of the code.

Woese (1967, 1968) has proposed that the original code might have been determined by a stereochemical fit between an amino acid and its codon, the nucleic acid acting as a physical template for assembly of the protein. The complex system now used for protein synthesis would evolve at a later stage to improve the efficiency of the process. Initially, the code might commence in an autocatalytic cycle in which a polynucleotide and polypeptide assisted each other to replicate, so that at first their relationship would be reciprocal, only subsequently becoming unidirectional. Indeed, simulations of the conditions thought to have comprised the prebiotic environment have suggested that polymers of amino acids are likely to have preceded the formation of polynucleotides. A difficulty of conducting any experimental analysis is that the nucleotides and amino acids involved under primitive conditions may not have been the same as those employed today. However, there is some evidence that the sort of complexes which could have been involved can be formed between the present day components of the code. Lacey and Pruitt (1969) have observed the formation of a complex between poly-L-lysine and mononucleotides; the data is consistent with a model in which the nucleotides form an array of two chains which resembles duplex DNA, except that the structure of each nucleotide chain is stabilized by an interaction with a polylysine chain instead of by covalent linkage between the nucleotides. Woese (1969) has suggested that the physicochemical properties of trinucleotides may be correlated with the physicochemical relationships between the amino acids for which they code. At present, however, it is controversial as to whether an amino acid could interact with its codon with sufficient stereochemical specificity of fit to account for the development of the code. Nevertheless, it cannot be denied that it is an attractive hypothesis to visualize some system of physical interaction as the basis for evolution from the prebiotic environment.

Most simple mutations (that is, not gross chromosomal rearrangements) are due to the substitution of a single nucleotide by another. Mutations tend to exert a deleterious effect upon the cell, and Sonneborn (1965) has suggested that cell lines which can reduce their burden of harmful mutations are likely to be at a selective advantage. The development of such a code 'buffered' against mutational change would require many of the features actually exhibited by the genetic code. It must be highly degenerate since unassigned codons would very probably cause termination of protein synthesis. Further, there should be 'maximal connection' between the various codons assigned to any particular amino acid, that is, they should differ by as little as possible (one base of the triplet). This achieves the maximum probability that a mutation will be 'silent'—that it will not result in an alteration of the amino acid specified, because the original and mutated triplets code for the same amino acid. The burden of mutation can be further reduced by increasing the likelihood that if there is to be a change in the amino acid specified, the replacement should have similar properties to the original species; this is achieved through the representation of functionally related amino acids by related codons. Epstein (1966) has observed that it is particularly important that the code should minimise replacements which involve changes between hydrophobic amino acids (which tend to lie in the interior of protein molecules) and hydrophilic (which tend to be positioned on the surface). In fact, the actual dictionary of codon meanings comes close to the maximum connection theoretically possible. The mutational-buffer theory is purely stochastic since it is in no way dependent upon the absolute relationship between a codon and an amino acid, but demands only that the codons for similar amino acids should relate to each other.

However, it is precisely this facet of the model which comprises its main drawback; it does not account for the universal nature of the code because it appears unlikely that only one set of codon assignments should be optimal. As Woese (1967) has observed, even under similar conditions one might expect independently evolving species to evolve different optimal sets, and considering the wide variety of conditions actually encountered, it seems even less likely that evolution would result in only one code. A further disadvantage is that the model requires codons to change their meanings during the course of evolution. This is likely to have considerable effects, since whilst a given amino acid substitution may benefit some proteins, it will create a series of probably deleterious mutations in the other proteins coded by the genome (but see also page 92).

Woese (1965, 1967, 1968) has proposed an alternative model which envisages the selective forces of evolution as acting to minimize errors made in the translation of codons into amino acids. This model is not entirely stochastic since it depends to some extent upon the mechanistic features involved in translation. Woese argues that translation is at present accomplished by a very specific and thus a very highly evolved system, so that it follows that the processes involved in translation in primitive cells must have been very different. He suggests that the translation apparatus of a primitive cell is likely to have been relatively inaccurate, with a high degree of ambiguity in the assignment of amino acids to codons. Errors would be so common that the protein molecules produced by any one gene would have no unique structure, but rather would comprise a group of 'statistical proteins' related to some theoretical primary structure, and hence to each other, by varying degrees of closeness. Because of this, it would be comparatively easy to alter codon assignments—this would have little harmful effect upon such an error-prone system. At this early stage, the system would probably involve groups of related amino acids, for example hydrophobic and hydrophilic, rather than specific species, which would be introduced later.

It appears that the likelihood of an error occurring in the translation of a codon varies somewhat with the codon, and Woese suggests that evolution would favour a dictionary in which the more error-prone codons are represented by amino acids which are functionally less important. Thus amino acids which tend to be involved in the mechanisms of enzyme catalysis (those with polar side chains) would tend to be assigned to the least error-prone codons, while the less functionally important (the non-polar) would be assigned to the more error-prone codons. Also, codons which are likely to be mistaken for each other in the translation process would tend to be assigned to the same, or at least to functionally related, amino acids. Woese observes that the probabilities of misreading occurring at each of the three nucleotides in a codon are 100:10:1 from position III:I:II. In the genetic code, position three, the most error-prone, is the most degenerate, so that codons differing in this base position tend to be assigned to related amino acids. The second base is the least error-prone, and probably the least involved in such restraints.

As evolution progresses, the cell would make more precise distinctions amongst amino acids and would recognize individual species rather than just functionally similar groups. Selection would favour the development of a less error-prone and more complex translation apparatus; and only when the codon catalogue has become (virtually) completely ordered as to its amino acids would it be possible to evolve a really efficient translation apparatus. (The ways in which this might be achieved have been discussed by Woese, 1967, and Orgel, 1968). Any cell succeeding in this development would have an enormous selective advantage, and as a result all living cells would be descendents of this single cell line. As translation evolves into a more sophisticated form, the effect of changing any particular codon assignment increases greatly, so that finally such change becomes lethal and no longer possible. Evolution must thus have been confined to the earlier and more fluid stage of codon assignments.

Crick (1968) has proposed a model in which he suggests that the code might start in a primitive form in which a small number of triplets coded for comparatively few amino acids. The assignment of codons to amino acids would be purely random—so this is a purely stochastic model. Crick argues that it is most likely that the code started as triplet since if it commenced with some other coding ratio all previous messages would become nonsense, and presumably lethal to the cell, with the switch to triplet reading. It is not clear exactly which amino acids would be present in this initial code, although some of the more complex species present today are rather unlikely to have existed under prebiotic conditions. In an intermediate phase, these primitive amino acids would take over most of the triplets of the code in order to reduce to a minimum the number without meaning; the extension of codons for any particular amino acid would very probably be to those related to the currently assigned codon. The final code would form as new amino acids replaced some of the primitive ones. This would happen most easily if the new amino acid was related to that previously coded by this triplet and if the organism coded only a few rather crude proteins. As the number and sophistication of proteins increased, a situation would arise when the substitution of new amino acids would disrupt too many preexisting proteins and would be lethal. At this point, the code would be 'frozen'; this theory has been termed the 'frozen accident model'. In order to explain universality, it is again necessary to postulate that this must have happened at an early stage of evolution, with all cells descended from the one population involved.

CHAPTER 5

Protein synthesis

The sequential synthesis of proteins

Since the genetic code is read as non-overlapping triplets from a fixed point, the most likely mode of assembly of amino acids into protein must be by sequential synthesis of the peptide bonds between successive amino acids, starting from one end of the polypeptide chain. The alternative, that all the amino acids are positioned on the template, after which peptide bonds are synthesized to connect them, would reopen the question of how only triplets in the correct phase are recognized. Dintzis (1961) and Naughton and Dintzis (1962), working with reticulocytes synthesizing hemoglobin, distinguished experimentally between these possibilities. If some amino acid is suddenly replaced by an isotopic radioactive form, prior positioning would result in an equal incorporation of the label into all parts of the protein. For sequential synthesis, the amino acids comprising the chain lengths completed before the switch to radioactive conditions would be normal, and the part of the chain synthesized after the switch would contain the isotope. This point will be different for each chain under synthesis. However, if the radioactive content of completed hemoglobin molecules is analysed, after a short period of incubation only the far end will be labelled. As incubation proceeds and more polypeptide chains are completed, the radioactive label will move nearer to the starting end, as illustrated in Figure 5.1.

The procedure used was to change the conditions of incubation by replacing C^{14}-leucine with H^{3}-leucine, after which incubation was allowed to proceed for various lengths of time. The hemoglobin proteins were then isolated and degraded to peptides by trypsin, followed by finger printing on two-dimensional chromatography. The peptides were then characterized according to their position in the polypeptide chain. The levels of radioactivity in each peptide due to C^{14} and H^{3} were counted, and the H^{3} expressed relative to the C^{14}, which acts as a control of the absolute level of leucine radioactivity (some peptides might have one or two leucines, others three or four). As expected, it was found that the

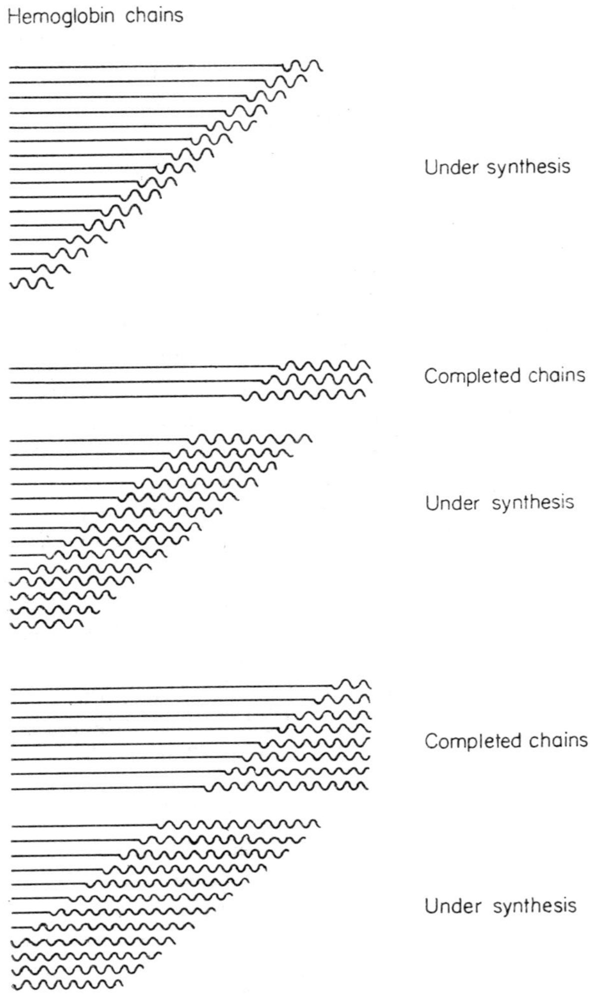

Figure 5.1. Synthesis of hemoglobin chains after increasing periods of incubation with labelled amino acids. Radioactively labelled portions of the chain are indicated by a jagged line. As the period of incubation increases, a greater extent of the chain from the far terminus is labelled and the radioactivity approaches the starting end

peptides toward one end of the protein—the C-terminal—proved to have a higher (relative) level of tritium; this fell steadily toward the other end, indicating that protein synthesis proceeds sequentially from the N- to C-terminus. As the period of incubation was increased, the level of radioactivity rose in all peptides, but the gradient from the *N*- to *C*-terminus was always maintained. The experimental results are shown in Figure 5.2.

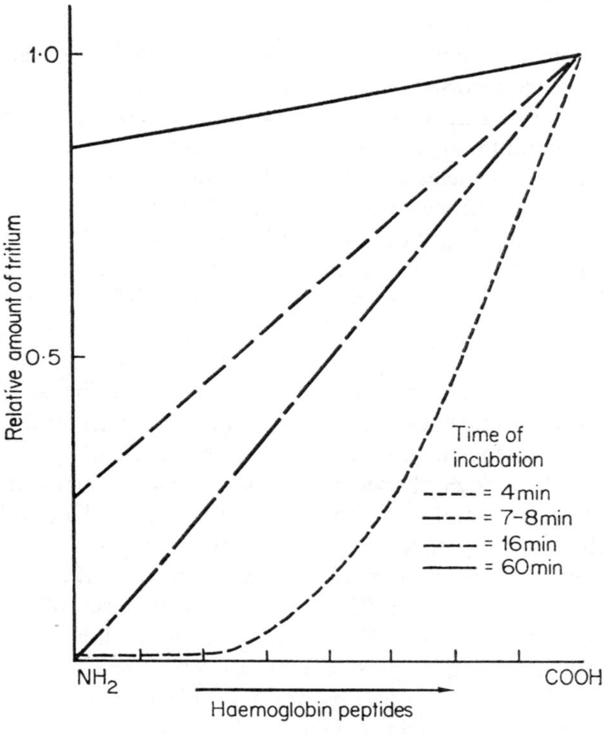

Figure 5.2. The incorporation of radioactive leucine into hemoglobin after various periods of incubation. There is a gradient from the far (COOH) terminus to the starting end (NH_2 terminus), in agreement with sequential synthesis from the NH_2— to —COOH terminus. As the time of incubation is increased, the total amount of radioactivity incorporated increases, but the gradient of label remains. (After Dintzis, 1961)

The ribosome

As proposed by the one cistron–one polypeptide chain hypothesis, protein biosynthesis is under the control of the DNA comprising the genome. In higher organisms, although the chromosomes are restricted to the nucleus, protein synthesis proceeds in the cytoplasm through the mediation of ribonucleic acid. A connection between protein synthesis and the presence of RNA has been noted for some time—cells which are very active in synthesizing proteins have high levels of RNA, whilst in less active cells there is a corresponding reduction in RNA content. Cytoplasmic RNA provides three major components of the protein biosynthetic apparatus.

Ribosomes are compact ribonucleoprotein particles which account for a considerable proportion of the RNA in the cell (80–90 per cent. in exponentially growing *E. coli*). In eucaryotic cells they are attached to the endoplasmic reticulum of the cytoplasm and appear in electron micrographs as spherical particles attached to the membrane surface. A fraction containing both membranes and ribosomes (the *microsome* fraction) can be obtained by centrifugation of homogenized cells. Chemically, ribosomes consist purely of RNA and protein; in bacteria, the RNA/protein ratio is about 6/4, but the ribosomes of higher organisms have about equal proportions of each. They are characterized as to size by their sedimentation properties: bacterial ribosomes sediment at 70S; the mammalian ribosomes are slightly larger, sedimenting at 80S. Their properties are dependent upon the magnesium ion concentration; if this is lowered they dissociate into two constituent subunits (this starts at 4–5 mM Mg^{2+} for bacterial species). The subunits sediment at 50S and 30S for the bacterial, and 60S and 40S for the mammalian species. The larger, which is roughly spherical, is about twice the size of the smaller, which is rather more asymmetrical (Tissières and coworkers, 1959).

Littlefield and coworkers (1955) injected pulse doses of C^{14}-leucine intraveneously into rats, and followed the fate of the label as the amino acid was incorporated into protein in the liver. When a short dose of radioactivity is followed (*chased*) by normal amino acids, its passage can be followed through the various stages of its incorporation into protein. Each of the three lobes of the liver was removed at various times after the injection and homogenized to yield cell sap (soluble proteins of the cytoplasm) and microsomes. The microsomal fraction was extracted with deoxycholate to separate it into the deoxycholate insoluble (ribosomes) and deoxycholate soluble (lipoprotein membranes of the endoplasmic reticulum). The radioactivity was rapidly (2–3 minutes) incorporated into the ribosome fraction, after which it fell in this and began to appear associated with membranes. Over a longer period, the label was found in the soluble proteins of the cell sap. The interpretation of these results was that proteins are synthesized at the ribosomes, after which they are passed by the endoplasmic reticulum to the cell sap.

A very early suggestion was that amino acids could be assembled into proteins on a nucleic acid template which would carry the sequence of nucleotide triplets specifying them. The ribosomes, as particles composed of roughly equal amounts of RNA and protein, were at first thought to be ideally suited for this role. However, since proteins vary so greatly in size and amino acid composition, a similar degree of variety would be expected in the nucleic acid template specifying them, but ribosomal RNA (*r*RNA) fulfills neither demand—it is fairly constant in both size and composition. Further, Davern and Meselson (1960) demonstrated that the template responsible for protein synthesis in bacteria is fairly short lived, but ribosomal RNA is stable over many generations. Brenner and coworkers (1961) found that after infection of *E. coli* by phage, the new proteins specified by the phage (through the production of new nucleic acid templates) were synthesized on the preexisting bacterial ribosomes, excluding the possibility that these act as the template. Ribosomes (in the 70S form) are now seen as part of the machinery of the cell responsible for assembling amino acids into protein on the template carrying the codon sequences.

Messenger RNA

This template is known as messenger RNA (*m*RNA); it is a species of (single-stranded) RNA which bears an exact copy of the nucleotide sequences of the genes on the DNA specifying the particular proteins for which it codes (reviewed by Lipmann, 1963). In higher organisms, it is synthesized (*transcribed*) from the DNA in the nucleus and subsequently moves through the nuclear membrane into the cytoplasm (see Georgiev, 1967) where it acts as the template for the assembly of amino acids into protein (this is termed *translation*). In such organisms, the different cell types synthesizing different proteins have qualitatively different messenger species, as shown by the technique of competitive hybridization with DNA (Church and McCarthy, 1968). Since all cells have the same DNA content, the transcription of these different messengers provides a level of control over which particular proteins are synthesized (although which messengers are translated may also be controlled). Generally, such messengers are believed to be relatively stable in higher organisms, whereas they appear to be unstable in bacteria.

In bacterial cells infected with phages, protein synthesis is not accompanied by net synthesis of RNA, but Volkin and Astrachan (1957) observed that a minor fraction (about 3 per cent.) of the RNA was synthesized and rapidly degraded—its rate of incorporation of P^{32} was extremely rapid (half-saturation was achieved in less than 30 sec) and its decay rate was also high. Using *E. coli* infected with either T2 or T7, extraction of the RNA from infected bacteria incubated with P^{32} showed that the base composition of the radioactive (that is newly synthesized) RNA was what would be expected if it had been synthesized from the phage DNA. Thus the infecting phage genome directs the production of proteins through the synthesis of unstable short-lived *m*RNA. Confirmation that the intermediate in information transfer from gene to protein must be unstable came from Davern and Meselson (1960), who incorporated 5-fluoruracil into the RNA of bacteria; this causes mistakes to be made when the RNA subsequently directs the assembly of amino acids, so that altered proteins are produced (β-galactosidase activity was greatly reduced). The effect was almost immediate and the molecular population of proteins was homogeneously abnormal; a stable intermediate would result in a heterogeneous population with a progressive increase in the proportion of abnormal protein.

Direct evidence for the production of unstable messengers in micro-organisms came from the work of Brenner, Jacob and Meselson (1961) and Gros and coworkers (1961). The former workers found that in T2 phage-infected *E. coli* cells a new RNA with a relatively rapid turnover was synthesized after phage infection. This species had a nucleotide composition corresponding to that of the phage DNA (in the DNA of most organisms, the two strands have the same proportions of the four bases; thus although RNA is transcribed from only one strand it should possess these characteristic proportions). It had a sedimentation constant in the range 14–16S, and although interpretation of sedimentation data is somewhat controversial, this should be the right order of magnitude to code for a long polypeptide chain. It became associated with pre-existing ribosomes, and was continually being replaced. The latter workers detected a comparable species in uninfected *E. coli* which became pulse-labelled and associated with ribosomes engaged in protein synthesis.

Experiments involving the annealing of messenger RNA to denatured DNA have shown that, as expected, it possesses a copy of the nucleotide sequence of the DNA. Hall and Spiegelman (1961) extracted the *m*RNA synthesized by bacterial cells shortly after infection with T2 (it was identified by the use of a P^{32} radioactive label) and added it to heat-denatured phage DNA which had been labelled with H^3. Equilibrium centrifugation on a CsCl density gradient was used to separate the resulting DNA–RNA double-stranded hybrid, which was identified unambiguously by its double radioactive label. Controls, when T2-specific *m*RNA was slowly cooled with heterologous DNA, gave no such complex, even if this DNA had the same overall base composition as the T2 DNA.

Hayashi and coworkers (1963) demonstrated the formation of specific hybrids between newly synthesized RNA and chromosomal DNA in uninfected bacteria. In a fast growing bacterial culture it is difficult to distinguish the small amounts of unstable *m*RNA from the other, stable species of RNA which are synthesized in much larger quantities (*m*RNA represents only some 2–3 per cent. of the cellular RNA). This was overcome by growing the bacteria in a medium devoid of amino acids; under such conditions few new ribosomes are synthesized, whilst messenger synthesis continues. CsCl density gradient centrifugation showed that *m*RNA from *E. coli* would form hybrids with denatured DNA from the same organism, but not with DNA from a different bacterial species.

The transfer of information encoded in a gene into protein through an RNA intermediate appears to be unidirectional. This idea has been termed the 'central dogma' (Crick, 1958), and suggests that genetic information can be perpetuated through the replication of DNA, but its transfer to RNA and thence to protein is irreversible. When transcription copies a nucleotide sequence into the form of *m*RNA, or translation transfers this into protein, the information in either form cannot be returned to a higher level. In other words, DNA cannot be synthesized from an *m*RNA base sequence, and the amino acid sequence of a protein cannot be converted into information at the RNA level. This has been represented as:

$$\circlearrowleft \text{DNA} \longrightarrow \text{RNA} \longrightarrow \text{protein}$$

Transfer RNA

This is the smallest species of RNA in the cell and sediments at about 4S—this corresponds to a molecule some 75–80 nucleotides in length. [Its earlier name of soluble RNA (*s*RNA) has now given way to the nomenclature transfer RNA (*t*RNA).] This is the 'adaptor' species, first proposed by Crick (1957), which is responsible for fitting an amino acid to its correct nucleotide triplet on the messenger. Each transfer RNA has the two properties of specifically recognizing both its particular amino acid and the codon representing it. This latter feature is achieved by possession of an appropriate trinucleotide sequence of bases, the *anticodon*, complementary to the bases of the codon for the amino acid. Hydrogen bonding between codon and anticodon provides the mechanism by which a triplet of nucleotides directs the incorporation of a particular amino acid into protein.

All transfer RNA molecules have the same nucleotide sequence at their 3′ hydroxyl end, and this common (three base) sequence is required for their interaction with amino acids to form *aminoacyl-t*RNA. Hecht, Stephenson and Zamecnik (1958, 1959) and Hecht and coworkers (1958) found that in cell-free systems the labelled nucleic acid precursors C^{14}-CTP and C^{14}-ATP were added sequentially to transfer RNA. Such extract systems possess phosphodiesterases which remove the last few nucleotides of a *t*RNA molecule, and this degraded species is inactive in protein synthesis because it can no longer accept amino acids. The incorporation of the radioactive precursors represents replacement of the cleaved species, restoring biological activity to the *t*RNA. Two CTP molecules are incorporated, followed by one ATP, and other nucleotides cannot fulfill the specific requirement for these bases. Thus the 3′ hydroxyl terminal sequence of a biologically active *t*RNA is . . . . *p*C*p*C*p*A.

Preiss and coworkers (1959) used treatment with periodate to show that loss of the hydroxyl groups on the ribose of the 3′ terminal adenosine abolishes the ability to accept amino acids. If this is because the amino acid is covalently bound here, *t*RNA charged with amino acid should be resistant to such treatment, and it was found that aminoacyl-*t*RNA was indeed protected against such inactivation. Zachau, Acs and Lipmann (1958) showed directly that the amino acid is accepted into an ester link by charging *t*RNA with radioactively labelled leucine and then degrading it with ribonuclease. They were able to identify a small fragment as either 2′- or 3′-leucyladenosine (it was not known which at the time, but is now agreed to be the latter). The 3′ terminus of aminoacyl-*t*RNA is thus as shown in Figure 5.3.

The initial step in the assembly of amino acids into protein is their activation, followed by attachment to their particular transfer RNA (reviewed by Novelli, 1967). Activation by reaction with ATP is a common step in biosynthetic pathways, and Hoagland, Keller and Zamecnik (1956) showed that a soluble extract from rat liver catalysed an exchange of P^{32}-labelled pyrophosphate with ATP, subject to a several-fold enhancement by L-amino acids. AMP failed to inhibit the reaction and did not exchange with the ATP, so it cannot be released as a result of the reaction. This would suggest that the amino acids are activated through formation of an enzyme-bound aminoacyl-AMP complex, with release of pyrophosphate. Hoagland and coworkers (1957) confirmed this by

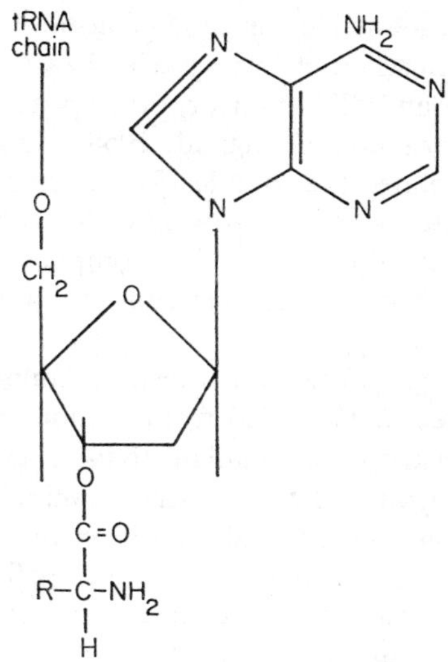

Figure 5.3. The 3′ terminus of aminoacyl-*t*RNA. An amino acid is linked through an ester bond to the 3′-hydroxyl position of the ribose of the terminal adenylate residue of the polynucleotide chain

cleaving the activation complex with hydroxylamine; when this was performed after using tryptophan labelled with O^{18} in the carboxyl group, the label was found in the AMP isolated after the hydrolysis. Direct proof was provided when Kingdon and coworkers (1958) isolated a complex of tryptophan–AMP. The reaction is quite specific for ATP and no other nucleoside triphosphate can be substituted.

Hoagland and coworkers (1958) found that cytoplasmic *t*RNA became labelled with C^{14}-amino acids upon incubation in the presence of ATP and the amino acid activating system. The aminoacyl-*t*RNA subsequently formed transferred the amino acid into protein in the presence of GTP and an energy-generating system. They proposed that aminoacyl *t*RNA formation could be represented by the two reactions:

$$\text{(i) ATP} + \text{amino acid} + \text{enzyme} \xrightarrow{Mg^{2+}} \text{enzyme–amino acid–AMP} + PP_i$$

$$\text{(ii) enzyme–amino acid–AMP} + t\text{RNA} \longrightarrow \text{amino acyl-}t\text{RNA} + \text{enzyme} + \text{AMP}$$

The same enzyme is responsible for catalysing both reactions; when the protein fraction with this activity was purified over a hundred-fold, the amino acid-activation and *t*RNA-incorporation activities were purified to the same extent (Berg and Ofengand, 1958). Lagerkvist, Rymo and Waldenstrom (1966) isolated the complex formed in reaction (i) for valine, and showed that addition of the appropriate *t*RNA resulted in the latter becoming associated with this in a further complex for step (ii) to occur. The charging of *t*RNA with amino acid can be represented as illustrated in Figure 5.4.

Amino acid-activating enzymes (also termed aminoacyl-*t*RNA synthetases) have been isolated and purified from a number of sources, ranging from *E. coli* and yeast to mammalian tissues. Generally, they have molecular weights of around 100,000. Each amino acid has (at least) one specific enzyme responsible for placing it on the appropriate transfer molecule, and these enzymes usually have a high affinity for their amino acid substrate, with K_m values in the range 10^{-3} to 10^{-5}. They have a high degree of specificity, and each synthetase forms an aminoacyl *t*RNA for only a single amino acid. (For discussion see pages 202–203.) However, the two reactions which it catalyses have different specificities; Bergmann, Berg and Dieckmann (1961) showed that the *E. coli* isoleucyl-*t*RNA synthetase converts either isoleucine or the related amino acid valine to the aminoacyl-adenylate, but transfers only the former onto *t*RNA. It is this high specificity for placing amino acids onto only their correct *t*RNA which appears to be the predominant factor in the accuracy of protein synthesis. Loftfield (1963) calculated that the frequency of errors of isoleucine replacement by valine in this system were 1/50 for reaction (i), but only 1/5000 for reaction (ii). This corresponds fairly well with the error rate found for the *in vivo* synthesis of chick ovalbumin, which is 1/3000.

According to the adaptor hypothesis, it is the anticodon nucleotide sequence of aminoacyl-*t*RNA which is responsible for recognition of the codons on the messenger. Chapeville and coworkers (1962) confirmed this by demonstrating that the only stage in protein synthesis at which the identity of the amino acid plays a role is the charging of *t*RNA; after this it is the transfer molecule which is recognized. It is possible to perform a reductive desulphuration of cysteine with Raney Nickel to give alanine whilst the amino acid is still attached to its transfer molecule (*t*RNA$_{cys}$). This produces the hybrid species alanyl-*t*RNA$_{cys}$. Using poly-UG (which directs the incorporation of cysteine and valine) this hybrid results in the incorporation of alanine in place of cysteine; thus it is the *t*RNA, and not the amino acid, which is recognized by the messenger.

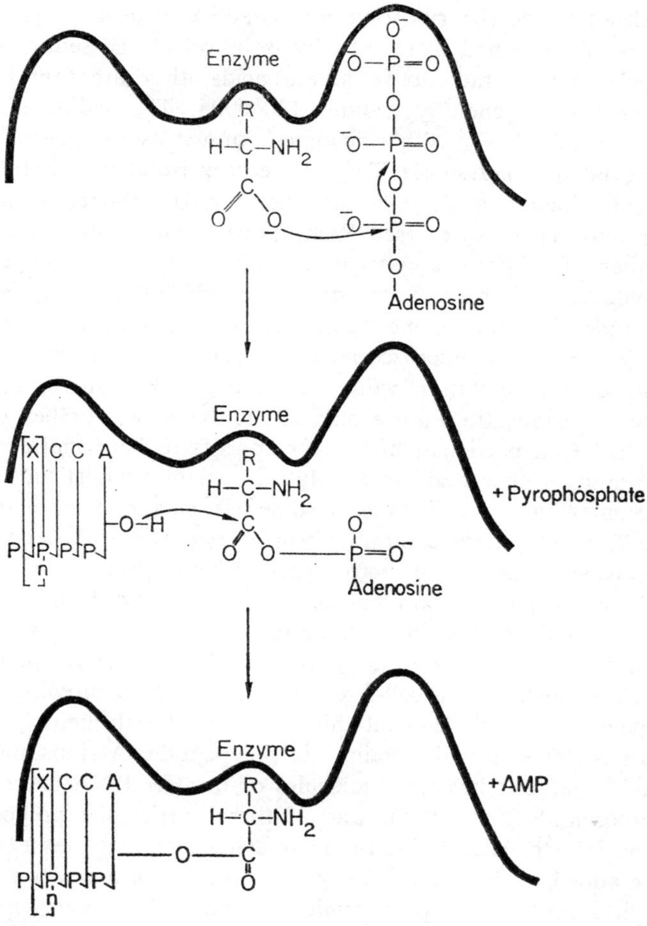

Figure 5.4. The formation of aminoacyl-*t*RNA by aminoacyl-*t*RNA synthetase. The first reaction takes place between the amino acid and ATP to yield an enzyme-bound aminoacyl adenylate moiety. The activated amino acid is then transferred to the *t*RNA with the release of AMP

The polysome

The process of peptide bond formation between the amino acids assembled through their transfer RNAs at the appropriate codons on *m*RNA is catalysed by the ribosome. Ribosomes have an affinity for single-stranded nucleic acid, that is *m*RNA, and also for *t*RNA.

Although the 70S ribosome unit is active in protein synthesis, if ribosomes are obtained from cells by isolation of the sedimenting fraction associated with radioactive amino acids, they are found to sediment rather faster, generally around 140–200S. The sedimenting unit is a complex of *m*RNA with ribosomes, known as the *polyribosome* (often abbreviated to *polysome*). These have been isolated and studied in many systems, including *E. coli* and the poly-U cell-free system, and rat liver and reticulocytes (see Rich, Warner and Goodman, 1963). The number of ribosomes associated with a single messenger varies; in reticulocytes (Warner, Knopf and Rich, 1963) the major species has five ribosomes, although some messengers might have only four, or as many as six. Electron microscopy confirmed this interpretation of the range of sedimentation values obtained, and a micrograph of pentasomes is shown in Figure 5.5a (Plate II). The number of ribosomes attached to a particular messenger appears to be a matter of statistical probability rather than an absolute invariable; whilst most reticulocyte polysomes have five ribosomes, some will have one more and some one less. The characteristic number of ribosomes depends on the particular messenger, according to factors such as its length.

If the magnesium ion concentration is lowered, the ribosomes on polysomes dissociate into subunits, just as the free ribosome does. Gilbert (1963) showed that when this dissociation is performed after protein synthesis (in a cell-free system for poly-U directed polyphenylalanine synthesis), the 30S subunit is associated with the poly-U messenger whilst the 50S subunit retains the polypeptide. A ribosome engaged in protein synthesis has two molecules of transfer RNA associated with it (Warner and Rich, 1964); one of these carries the polypeptide chain synthesized to date, whilst the second has attached the next amino acid to be added to the chain. This gives a picture of a ribosome engaged in protein synthesis as a 30S subunit attached to the messenger and covering at least two codons, with the two transfer species representing the last and next amino acids incorporated, attached at their codons through the 50S subunit.

Polypeptide synthesis starts by binding of the aminoacyl-*t*RNAs bearing the *N*-terminal and adjacent amino acids to their appropriate codons. Peptide bond synthesis between these involves reaction of the carboxyl group of the first amino acid with the amino group of the second, so that the dipeptide formed is attached to the second *t*RNA. The first *t*RNA, no longer bearing an amino acid, then falls away from the ribosome, which moves three nucleotides further along the message. The *t*RNA for the next amino acid approaches and the sequence is then repeated, the dipeptide being transferred onto this *t*RNA to give a tripeptide. The ribosome site bearing the *peptidyl-tRNA* is referred to as the P site, and the site entered by the incoming aminoacyl-*t*RNA as the A site. (See Figure 5.7 which illustrates the process of peptide bond synthesis.)

As the ribosome moves along toward the end of the message the polypeptide chain progressively lengthens with each amino acid addition. When the ribosome has moved far enough away from the start, another may attach and start synthesis of the next polypeptide chain. Thus a messenger will bear a series of ribosomes, each successively carrying a greater length of the polypeptide chain under synthesis, as shown in Figure 5.5b.

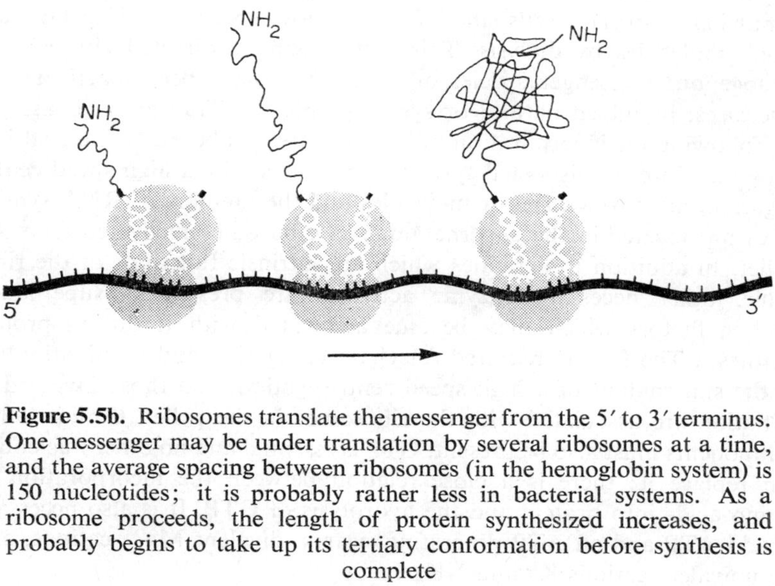

Figure 5.5b. Ribosomes translate the messenger from the 5′ to 3′ terminus. One messenger may be under translation by several ribosomes at a time, and the average spacing between ribosomes (in the hemoglobin system) is 150 nucleotides; it is probably rather less in bacterial systems. As a ribosome proceeds, the length of protein synthesized increases, and probably begins to take up its tertiary conformation before synthesis is complete

The messenger is always read in triplets from its 5′ end to its 3′ end (the same direction as it is itself synthesized). Salas and coworkers (1965) used polynucleotides with from twenty-one to twenty-three nucleotide residues of the form

$$pApApAp..............pApApCp(Ap)_nC$$

to direct protein synthesis. The $-NH_2$ terminal amino acid of the product was lysine and the *C*-terminal asparagine; thus if the polypeptide is assembled sequentially from the *N*- to *C*-terminus, the polynucleotide must be read as from 5′ to 3′. The work on frameshift mutants (page 82) has also shown that the amino acid sequences in double mutants and wild type can only be related if the triplets are orientated so that their 5′ to 3′ polarity parallels the *N*- to *C*-terminal protein polarity.

The ribosome cycle in chain elongation

Most experiments on the processes involved in protein synthesis as the ribosome translates a messenger have been performed with *in vitro* systems (for review see Lengyel and Soll, 1969). Crude cell-free extracts capable of protein synthesis can be prepared by breaking cells, and removing unbroken cells and debris by low speed centrifugation and small molecules by dialysis. This system can be directed either by its endogenous messenger species or, after these have been inactivated by nucleases, by added natural or synthetic *m*RNA. Translation is assayed by following the incorporation of a radioactively labelled amino acid into protein. More sophisticated systems are achieved by a high speed centrifugation, after which *t*RNA molecules and the aminoacyl-*t*RNA synthetases are located in the supernatant, and ribosomes recovered from the pellet. In addition to activities which are intrinsic functions of the ribosome, other necessary enzyme activities are present as supernatant protein factors which may become associated with it during protein synthesis. The factors required for chain elongation and termination are in the supernatant of a high speed centrifugation, and those involved in initiation are associated with the ribosomes in the pellet. Certain other components must also be present. GTP is essential and no other nucleotide can replace it; there is a close relation between the incorporation of amino acids into protein and the hydrolysis of GTP. It is also necessary to add ATP and an ATP generating system, divalent Mg^{2+} cations, and the univalent cations K^+ and NH_4^+.

Puromycin is an antibiotic inhibitor of protein synthesis, and undergoes a reaction with polysomes synthesizing protein which has been used as a model system to investigate the reaction mechanisms of protein synthesis. There is a chemical resemblance between puromycin and an amino acid attached to the terminal adenosine of *t*RNA, as shown in Figure 5.6. Puromycin reacts with the peptidyl-*t*RNA on the ribosome as would the incoming aminoacyl-*t*RNA, so that the peptide so far synthetized becomes attached to the analogue. Its release as polypeptidyl-puromycin

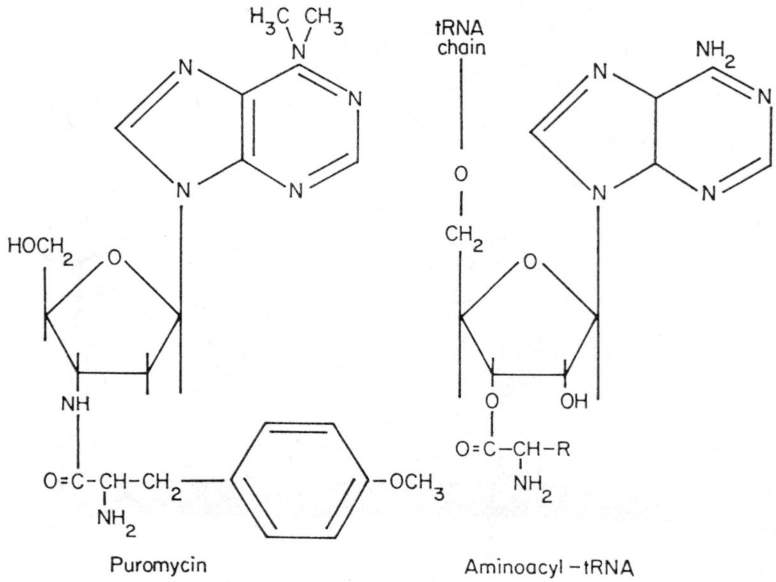

Figure 5.6. Aminoacyl-*t*RNA and its analogue puromycin

then terminates protein synthesis prematurely, and this accounts for the inhibitory effect of the antibiotic (Allen and Zamecnik, 1962; Nathans, 1964). The puromycin reaction has similar requirements and is inhibited in the same manner as protein synthesis, justifying its use as a model system.

Traut and Monro (1964) found that with aminoacyl-*t*RNA as substrate, the puromycin reaction was partly dependent upon the presence of GTP and supernatant factors, but the extent of this dependence varied amongst different ribosomal preparations. They proposed a scheme in which the ribosome can exist in either of two states according to the stage of the synthetic process, conversion between these demanding GTP and supernatant factors. Nishizuka and Lipmann (1966) have also suggested a cycle for ribosome behaviour in peptide bond formation on a similar basis, as shown in Figure 5.7. When a ribosome has a peptidyl-*t*RNA in its P site it is ready to accept an aminoacyl-*t*RNA (or the puromycin analogue) into its A site. In this state, condensation takes place between the peptide chain and the next amino acid to yield a ribosome with the peptide chain attached to the *t*RNA in the A site. Such a ribosome cannot accept another aminoacyl-*t*RNA until after a *translocation*, when it

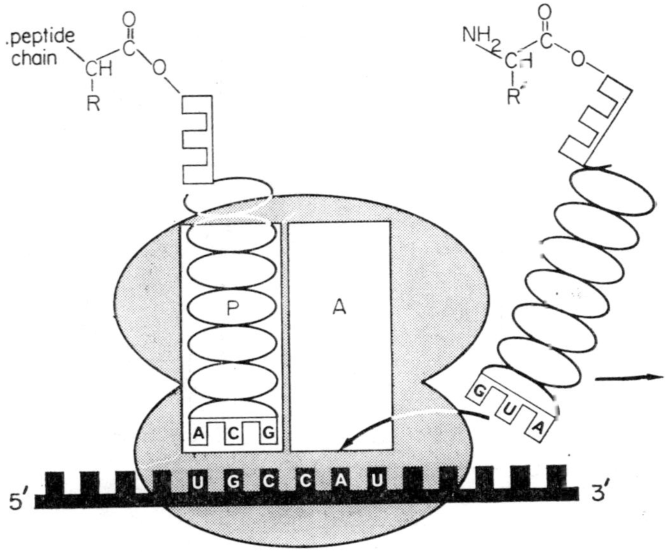

peptide chain
CH
R
C
O
O
NH2
CH
R′
C
O
O
P
A
A C G
G U A
5′
U G C C A U
3′

(a)

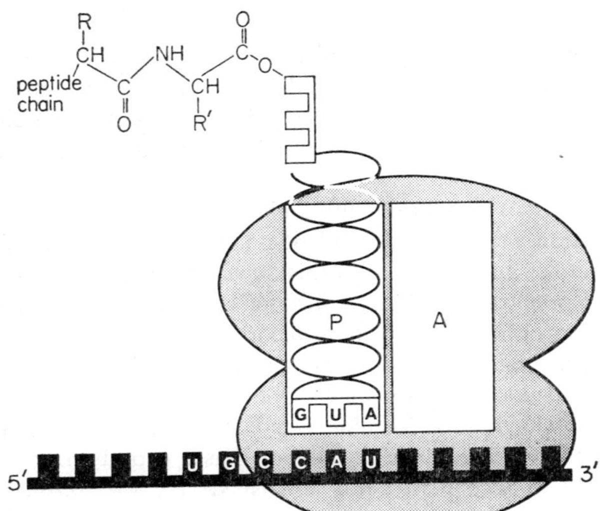

R
CH
peptide chain
C
O
NH
CH
R′
C
O
O
P
A
G U A
5′
U G C C A U
3′

(d)

Figure 5.7

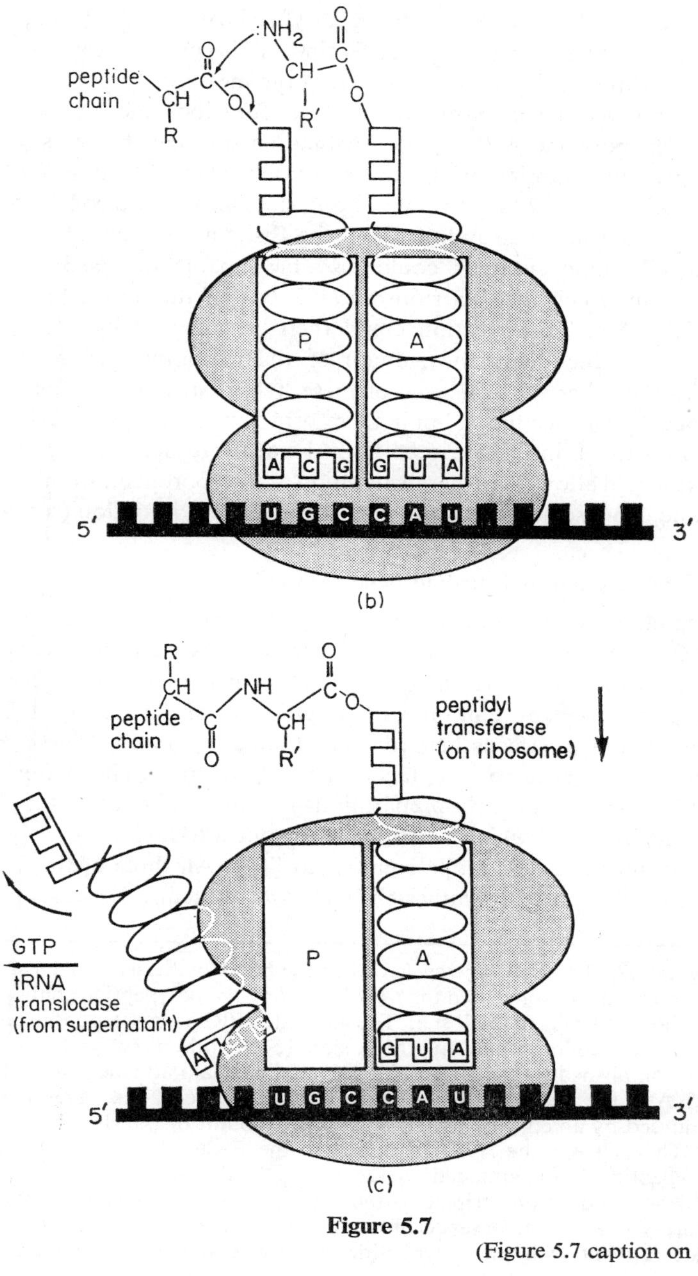

Figure 5.7

(Figure 5.7 caption on page 110)

advances one triplet to move the peptidyl-*t*RNA into its P site. This restores the same state as at the beginning of the cycle, except that the ribosome is three nucleotides further along the message.

Bretscher (1968a) has proposed a model for ribosome action which suggests that both the A and P *t*RNA-binding sites can be envisaged as consisting of two subsites which are separated when the ribosome subunits dissociate. In a 70S ribosome, the A site can thus be considered to comprise the subsites 30a and 50a, and the P site as 30p and 50p. In this case, the ribosome subunits could also be capable of existing in an association in which the 70S ribosome has one of the hybrid sites 30a; 50p or 30p; 50a. Translocation could then be achieved by a two-step movement in which first one subunit of the ribosome 'swings round' relative to the other to create one of these hybrid structures, after which it is joined by the second subunit to reform the normal A and P sites. This is illustrated in Figure 5.8. Path 1 seems rather more probable because it would allow an earlier recognition of the approaching aminoacyl-*t*RNA, and can also account more plausibly for chain initiation (see later).

The binding reaction and peptide bond synthesis

Supernatant protein factors required for protein synthesis have been isolated from a variety of systems, ranging from bacteria and yeasts to reticulocytes and rat liver. In *E. coli* the factors have been separated into three fractions by elution with a gradient of KCl from DEAE-Sephadex columns (Lucas-Lenard and Lipmann, 1967). These have been termed as G-factor, T_s-factor and T_u-factor. (The latter two are so called because in early preparations they were obtained as a single fraction, the T-factor, which has since been split into a stable component [T_s] and an unstable heat-labile component [T_u]). All three of the factors are required for protein synthesis by *in vitro* systems. Because of the

---

Figure 5.7. The ribosome cycle of peptide bond synthesis. (a) This represents a ribosome with the *t*RNA carrying the polypeptide chain synthesized so far in the P site. The aminoacyl *t*RNA corresponding to the next codon is entering the A site. (b) Peptide bond synthesis is accomplished by transfer of the polypeptide chain attached to the *t*RNA in the P site to the aminoacyl-*t*RNA in the A site. This is catalysed by an enzyme bound to the 50S subunit of the ribosome. (c) This releases the *t*RNA formerly in the P site, and leaves the polypeptide chain attached to the *t*RNA in the A site. (d) The ribosome moves one triplet to the right in a step requiring GTP hydrolysis and the presence of supernatant factors. This restores state (a), but with the polypeptide chain one amino acid longer.

heat-lability of factor T_u, attempts have been made to prepare supernatant protein factors from the bacterium *Bacillus stearothermophilus*, which lives at higher temperatures than *E. coli.* Ono and coworkers (1969a) reported that three factors are required in this system, and the use of heterologous reaction mixtures in which these factors were used to replace the *E. coli* factors has shown that S_1, S_2 and S_3 are equivalent to T_s, G and T_u respectively. Indeed, the functions of the supernatant factors appear to be fairly similar in most systems; a rat liver extract has two components, TF_1 and TF_2, which probably have comparable roles to factors T and G respectively (Ibuki and Moldave, 1968; Schneider, Raeburn and Maxwell, 1968) However, despite the similarities between the properties of the supernatant factors derived from different species, only those from closely related organisms can function in a heterologous system.

The two T factors of the *E. coli* system act to form a complex with aminoacyl-*t*RNA and GTP, and it is this complex which is responsible for transferring the aminoacyl-*t*RNA to the ribosome. The G factor is concerned with promoting the translocation of the ribosome which must succeed peptide bond formation (for review see Lipmann, 1969). An early step in the action of the supernatant proteins is the formation of a complex with GTP, and Allende and Weissbach (1967) showed that it is T_u which is responsible for binding the nucleotide; the GTP binding activity of column fractions followed the T_u elution curve, proved to be heat-labile, and was retained on sucrose-washed ribosomes but extracted by an ammonium chloride wash, as is T_u. The binding was specific for GTP, and other nucleotides bound only at much lower levels.

The conditions required for formation of the complex with GTP were investigated by employing adsorption on millipore membrane filters to detect the complex. This showed that GTP and T_u can interact with ribosomes in the absence of aminoacyl-*t*RNA. By contrast, Gordon (1967) found that when gel filtration was used to test for complex formation, aminoacyl-*t*RNA was required as a component. These contradictory results were reconciled when Gordon (1968) found that the presence of aminoacyl-*t*RNA reverses the ability of the GTP acceptor protein to bind to millipores, so that upon its addition the T factor (T_s+T_u) activity appeared in the filtrate. This suggests a sequence of reactions in which a first complex is formed by the binding of GTP to T_u; this is then converted into a second complex by the addition of aminoacyl-*t*RNA. Allende and Wissbach were detecting the first complex, Gordon the second.

Ertel and coworkers (1968) found that the amount of GTP binding by T_u was considerably enhanced by the addition of T_s. Studies on the

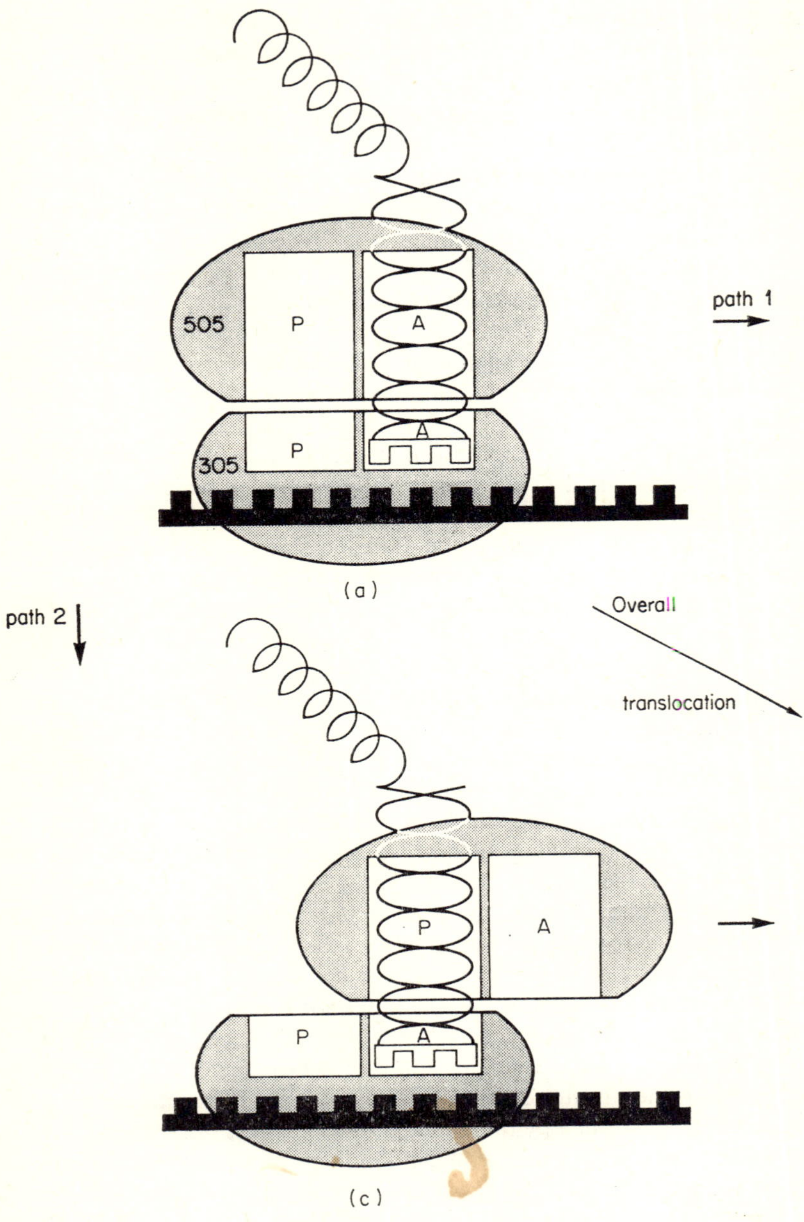

Figure 5.8

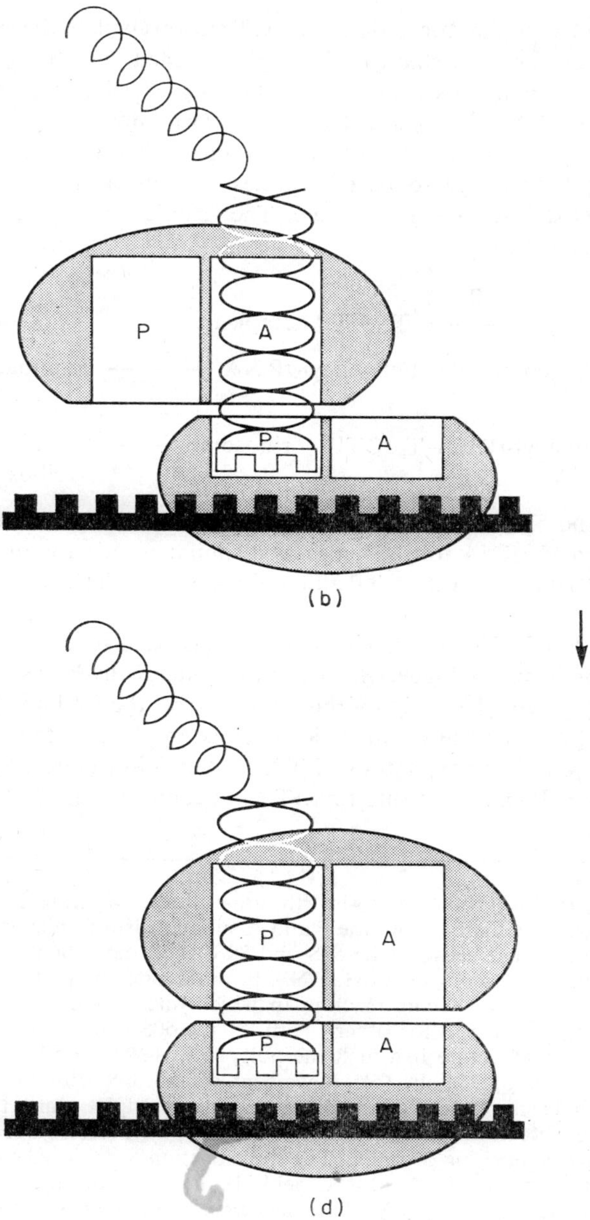

Figure 5.8

(Figure 5.8 caption on page 114)

kinetics of complex formation with GTP showed that the concentration of T_u determines the *amount* of complex formed, and the quantity of T_s present determines its *rate* of formation by exerting a catalytic action. Aminoacyl-*t*RNA is not involved in the formation of the first complex, but reacts with it to form a further complex, which Lucas-Lenard and Haenni (1968) found to be the species responsible for transferring the aminoacyl-*t*RNA to the ribosome. The reaction sequence may thus be summarized as:

(i) $GTP + T_u \xrightarrow{T_s} T_u$–GTP complex

(ii) T_u–GTP complex + aminoacyl-*t*RNA ⟶ aminoacyl-*t*RNA–T_u–GTP complex

(iii) aminoacyl-*t*RNA–T_u–GTP + ribosome ⟶ aminoacyl-*t*RNA–ribosome complex

The S_1 and S_3 factors of *Bacillus stearothermophilus* promote the binding of aminoacyl-*t*RNA to ribosomes by a similar reaction sequence in which S_3 forms a complex with GTP and aminoacyl-*t*RNA (Ono and coworkers, 1969a).

The role of GTP in protein synthesis has been investigated by substituting for it an analogue which cannot undergo hydrolysis; GMP–PCP (5′-guanylylmethylene diphosphonate) was synthesized by Hershey and Monro (1966), and has a methylene bridge in place of the oxygen which links the β and γ phosphates in GTP, as shown in Figure 5.9.

GMP–PCP can substitute for GTP in promoting the factor-dependent

---

Figure 5.8. The hybrid site ribosome model for translocation in protein synthesis. Both the P and A sites consist of units which can be separated between the 50S and 30S ribosome subunits. Translocation of a polypeptidyl-*t*RNA in the A site to the P site takes place by a two-stage reaction in which one of the subunits first moves relative to the other. Path 2: the 50S subunit moves first, giving a structure in which the polypeptidyl-*t*RNA is located in a hybrid 50p; 30a site (c); movement of the 30S subunit follows. Path 1: movement of the 30S subunit one triplet along the messenger creates hybrid (b) where the polypeptidyl-*t*RNA is located in a site comprising 50a; 30p. The translocation is achieved by movement of the larger subunit to yield (d). This path has the advantage that the next incoming aminoacyl-*t*RNA could be recognized by hybrid (b) before completion of the translocation. (After Bretscher, 1968a)

PLATE I

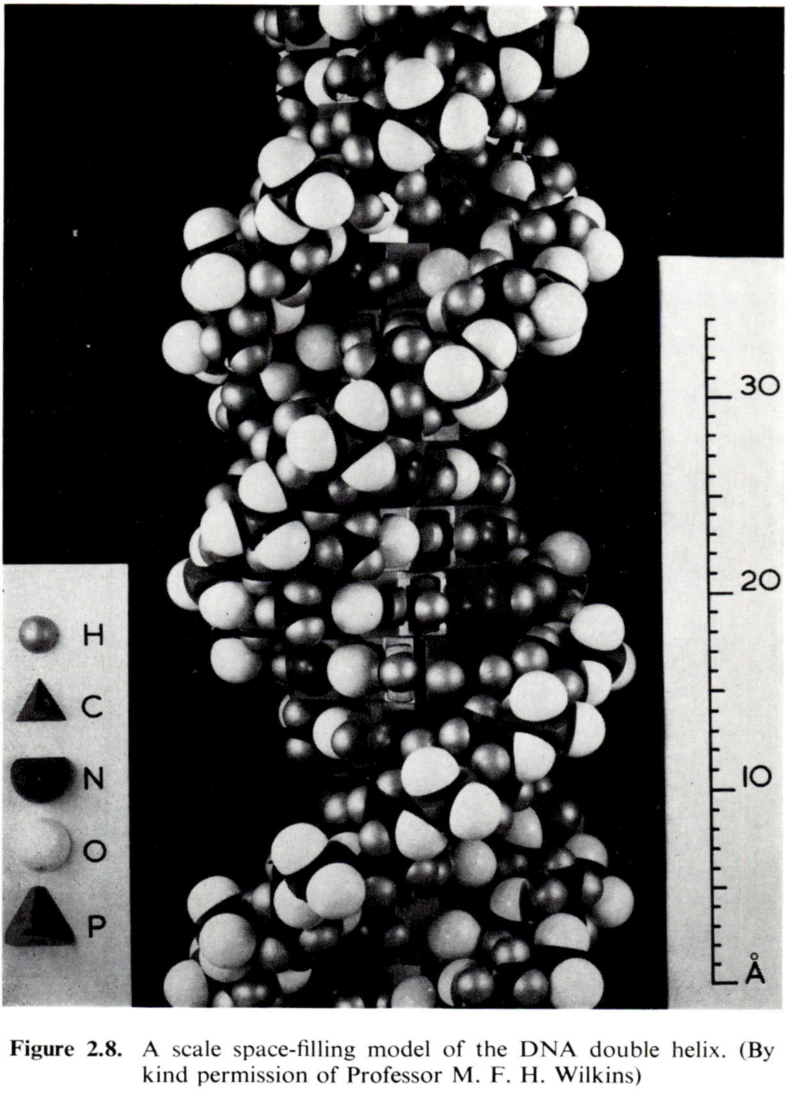

Figure 2.8. A scale space-filling model of the DNA double helix. (By kind permission of Professor M. F. H. Wilkins)

PLATE II

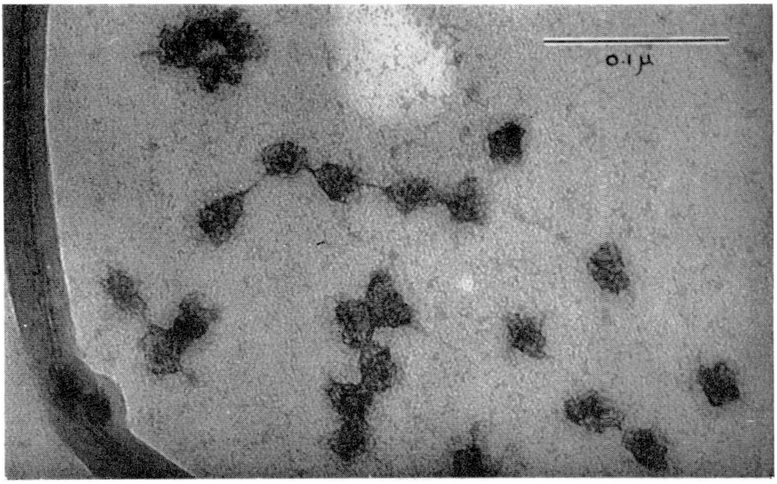

Figure 5.5a. Electron micrograph of pentasomes synthesising hemoglobin in reticulocytes. (From Slayter and coworkers (1963), courtesy Dr. H. S. Slayter)

PLATE III

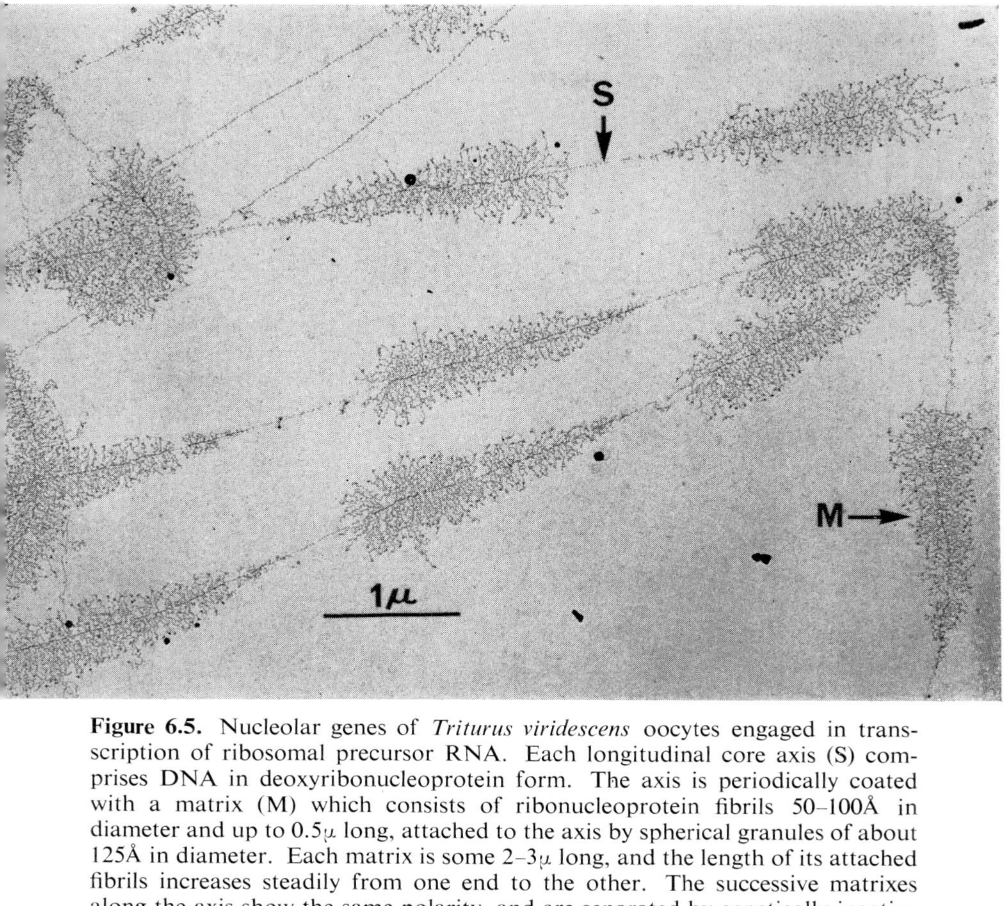

Figure 6.5. Nucleolar genes of *Triturus viridescens* oocytes engaged in transcription of ribosomal precursor RNA. Each longitudinal core axis (S) comprises DNA in deoxyribonucleoprotein form. The axis is periodically coated with a matrix (M) which consists of ribonucleoprotein fibrils 50–100Å in diameter and up to 0.5μ long, attached to the axis by spherical granules of about 125Å in diameter. Each matrix is some 2–3μ long, and the length of its attached fibrils increases steadily from one end to the other. The successive matrixes along the axis show the same polarity, and are separated by genetically inactive segments of DNA whose function is unknown. Each matrix represents a gene engaged in the simultaneous transcription of about a hundred RNA molecules; the shortest of its fibrils is attached to a polymerase molecule that has just initiated transcription, and the longest to one that has just completed synthesis of its RNA. Since the maximum length of the fibril is much less than that expected of an RNA corresponding to a gene 3μ long, the RNA must be coiled within the ribonucleoprotein structure to give a ratio of about 12 RNA : RNP fibril length. (From Miller and Beatty, 1969a. The photograph was taken by O. L. Miller Jr. and Barbara R. Beatty at the Biological Division, Oak Ridge National Laboratory, USA. Copyright by the American Association for the Advancement of Science, 1969)

PLATE IV

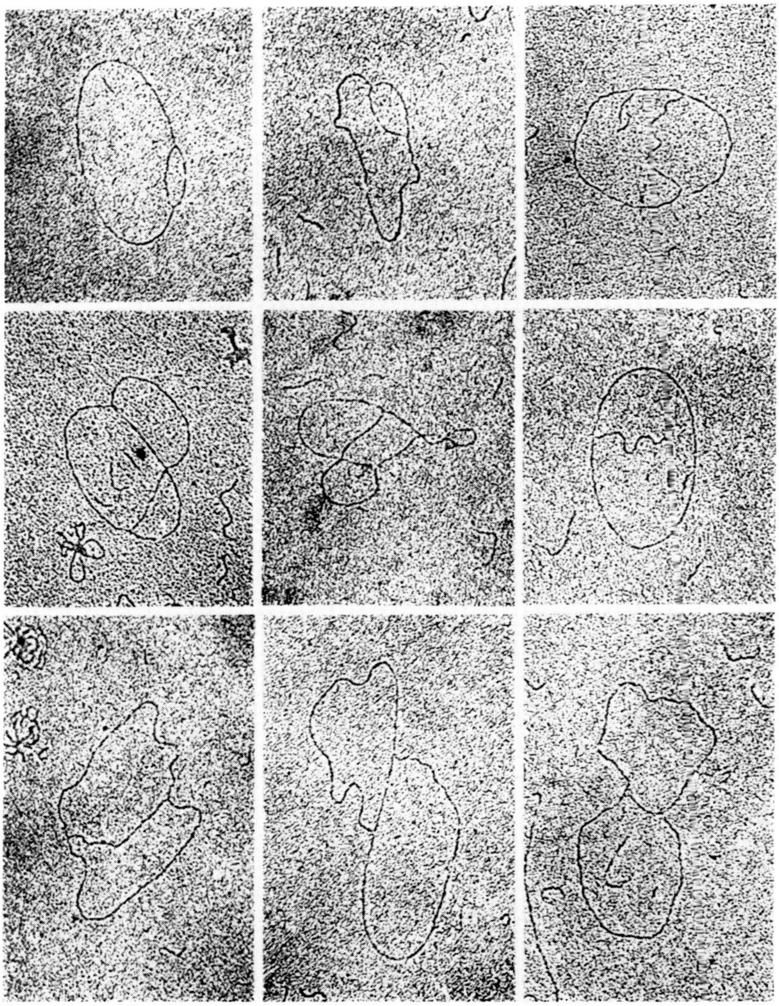

Figure 13.3. Electron microscopic observations of replicating polyoma virus DNA. The growing point moves sequentially round the circle until two chromosomes have been synthesized. Each visible strand represents a duplex of DNA. (From Hirt, 1969)

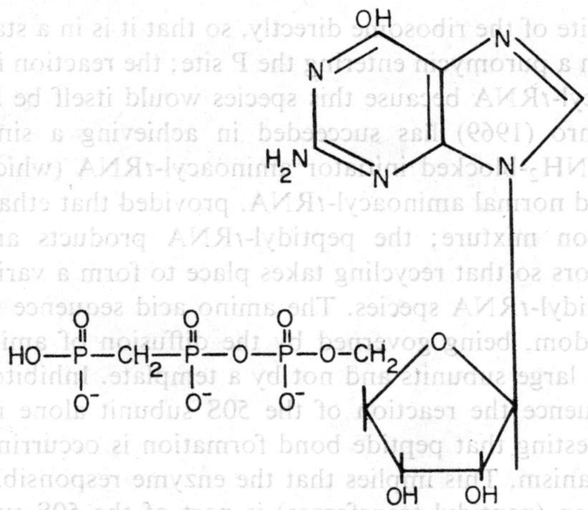

Figure 5.9. The GTP analogue 5′-guanylylmethylene diphosphonate (GMP–PCP)

binding of aminoacyl-*t*RNA to ribosomes, so that although the presence of GTP is necessary if the binding is to take place, its hydrolysis is not required. However, after such substitution, peptide bond synthesis is no longer possible. S_1 and S_3 exert a ribosome-dependent GTPase activity, and this is stimulated by the presence of *m*RNA and aminoacyl-*t*RNA. Neither the antibiotic sparsomycin, which blocks peptide bond formation, nor the translocation inhibitor fusidic acid interferes with the hydrolysis. This suggests that the cleavage takes place on the ribosome after attachment of the S_3–GTP–aminoacyl-*t*RNA complex, but before or concomitant with peptide bond formation (Ono and coworkers, 1969b). Quantitative experiments with poly-U directed polyphenylalanine synthesis have shown that aminoacyl-*t*RNA and GTP are present in equimolar amounts in the complex with S_3, and the GTP molecule cleaved by the factor is that present in the complex. Thus one GTP molecule is cleaved for each phe-*t*RNA bound to the ribosome poly-U complex. When GTP is present in the reaction mixture, S_3 is recovered from the supernatant after protein synthesis, but when GMP–PCP is substituted the factor remains bound to the ribosome; this suggests that its release may occur at the step which hydrolyses the triphosphate.

Traut and Monro (1964) reported that when polypeptidyl-*t*RNA is employed as substrate rather than aminoacyl-*t*RNA, the puromycin reaction can occur on the 50S subunit alone without need for the 30S subunit, *m*RNA, GTP or supernatant factors. (Polypeptidyl-*t*RNA enters the P site of the ribosome directly, so that it is in a state in which it can react with a puromycin entering the P site; the reaction is not possible with aminoacyl-*t*RNA because this species would itself be located in the A site). Monro (1969) has succeeded in achieving a similar reaction between the NH_2-blocked initiator aminoacyl-*t*RNA (which also enters the P site) and normal aminoacyl-*t*RNA, provided that ethanol is present in the reaction mixture; the peptidyl-*t*RNA products are themselves active as donors so that recycling takes place to form a variety of di-, tri- and tetrapeptidyl-*t*RNA species. The amino acid sequence synthesized is probably random, being governed by the diffusion of aminoacyl-*t*RNA species to the large subunits and not by a template. Inhibitors of protein synthesis influence the reaction of the 50S subunit alone in their usual manner, suggesting that peptide bond formation is occurring through its normal mechanism. This implies that the enzyme responsible for peptide bond formation (peptidyl transferase) is part of the 50S subunit, and is stimulated by the presence of the alcohol. Normally, of course, 50S subunits are inactive in the absence of the other components of the synthetic apparatus, and, indeed, this must be a necessary control since independent activity would presumably result in a random polymerisation of free aminoacyl-*t*RNA molecules in the cell.

Translocation

The processes involved in the next step of protein synthesis—translocation—have been investigated by performing peptide bond synthesis both under normal conditions, when translocation is necessary, and with a system in which one aminoacyl-*t*RNA is introduced directly into the P site so that the need for translocation is obviated. The more stringent requirements of the former situation can only be attributed to the need for translocation. Pestka (1968) compared the conditions required for formation of diphenylalanine (phe_2) and triphenylalanine (phe_3) under direction from a poly-U template. The former does not depend upon either GTP or factor G, whereas both are required for the synthesis of phe_3. This can be accounted for if synthesis of the first peptide bond does not require translocation (one phe-*t*RNA could enter the P site directly, and the other the A site normally), but formation of ph_3 requires translocation of the dipeptide product in the A site to the P site. Haenni and Lucas-Lenard (1968) confirmed this finding by use of the peptidyl-*t*RNA analogue *N*-acetyl-phe-*t*RNA, which is known to enter the P site on the ribosome. Although the hydrolysis of GTP was required for the formation of *N*-acetyl-phe_2-*t*RNA, the presence of factor G was not necessary.

However, factor G has to be added for any further peptide bond formation, presumably since the need then arises for translocation of the *N*-acetyl-phe_2-*t*RNA product. Erbe, Nau and Leder (1969) obtained similar results by using the amino-blocked initiator aminoacyl-*t*RNA which enters the P site directly. They found also that factor G appears to stabilize the peptidyl-*t*RNA–ribosome-*m*RNA complex—this would be compatible with its assisting the translocation of peptidyl-*t*RNA from a site of lesser to a site of higher affinity, perhaps from the A to the P site.

Leder, Skogerson and Nau (1969) have purified factor G from *E. coli.* The pure translocase has a molecular weight of 72,000 and appears to consist of only a single polypeptide chain. It comprises more than 2 per cent. of the soluble protein of exponential growth bacteria—a considerable proportion. The purified protein has a ribosome-dependent GTPase activity, and protein synthesis in a cell-free poly-U directed system depends upon the quantity of translocase added, up to a saturation level corresponding to one molecule of protein per ribosome. The enzyme then acts in a catalytic manner. Purification of the protein has permitted the preparation of a specific antibody (Leder, Skogerson and Roufa, 1969), and this anti-translocase has been used to achieve a specific inhibition of the action mediated by factor G in protein synthesis. As expected, the antibody has no effect upon formation of the first peptide bond between an aminoacyl-*t*RNA species in the P site and one in the A site, but inhibits any further peptide bond formation as judged by the puromycin reaction (the peptidyl-*t*RNA cannot react with puromycin to release polypeptidyl puromycin until it has been translocated to the A site, clearing the A site for entry of the puromycin). Its inhibitory effect can be overcome by the addition of increasing concentrations of translocase enzyme.

Removal of free *t*RNA from the P site of the ribosome after peptide bond formation between its polypeptide and the aminoacyl-*t*RNA in the A site appears to be tightly coupled to the translocation step (Lucas-Lenard and Haenni, 1969). When ribosomes carrying *N*-acetyl-C^{14}phe-H^3 *t*RNA were incubated with C^{12}phe-*t*RNA in the presence of factor T and GTP, reaction took place to yield ribosomes carrying both H^3-*t*RNA and *N*-acetyl-C^{14}phe-C^{12}phe-*t*RNA. The latter species did not react with puromycin, suggesting that although peptide bond formation has occurred through the transfer of *N*-acetyl-C^{14}phe to the second (C^{12}) phe-*t*RNA, translocation has not taken place. After incubation of these ribosomes with factor G and GTP to allow translocation, there was a simultaneous release of H^3-*t*RNA and C^{14}phe-C^{12}phe-puromycin in stoichiometric proportions.

Although the presence of GTP is necessary to bind factor G to ribosomes, its hydrolysis is not required in this action since it can be replaced by GMP–PCP. However, translocation in the puromycin analogue system cannot be achieved after the substitution of GMP–PCP, suggesting that the hydrolytic action may be necessary for movement of the ribosome. Overall, therefore, two triphosphates are cleaved for each amino acid added to the polypeptide chain, one after the binding of aminoacyl-*t*RNA and a further one at translocation. The energy required for the peptide bond formation itself appears to be provided by cleavage of the energy-rich ester link between the peptidyl carboxyl group and the 3′ terminal adenosine of *t*RNA, so that the energy derived from the hydrolysis of GTP is probably applied elsewhere (Lipmann, 1969). Although it is not clear just how the energy is fed into protein synthesis, Nishizuka and Lipmann (1966) have suggested that there may be an analogy between the ribosome-linked GTPase activity and the ATPase-linked contraction of muscle; a pulsating ribosomal contraction could occur after the peptide bond formation reaction to achieve the translocation, and this might also help to expell the uncharged *t*RNA species remaining in the P site.

Thus the aminoacyl-*t*RNA passes through three distinct stages in protein synthesis (for review see Lipmann, 1969). First, in the binding reaction it becomes bound to its codon at the ribosome through mediation of a complex with factor T_u (S_3) and GTP; formation of this is catalysed by factor T_s (S_1). Then, in the transfer reaction, the polypeptide chain synthesized to date is transferred from the *t*RNA in the P site to the aminoacyl-*t*RNA in the A site; formation of the covalent bond is catalysed by the peptidyl transferase of the 50S subunit. This may require the hydrolysis of GTP at some stage. Finally, a translocation occurs in which the ribosome moves a codon further along the message, the uncharged *t*RNA in the P site is expelled, and the polypeptidyl-*t*RNA previously in the A site is transferred to the P site. This stage requires factor G (S_2) and the hydrolysis of GTP.

Initiation of the polypeptide chain

At first, it was thought that there was no special mechanism for initiating synthesis of polypeptide chains; a ribosome would just attach to the 5′ end of the messenger and commence reading the triplets. However, this leaves the possibility that the message might be read in the wrong phase, and it appears that, in *E. coli* at least, there is a specific mechanism for initiation.

Waller (1963) observed that the *N*-terminal amino acid composition of *E. coli* proteins is not random but comprises only a few species, largely:

met	45%
ala	30%
ser	15%

The finding of *N*-formylmethionyl-*t*RNA (fmet-*t*RNA) in *E. coli* extracts by Marcker and Sanger (1964) suggested that this species might be involved in some special mechanism for initiation. This was confirmed by Adams and Capecchi (1966) and by Webster, Engelhardt and Zinder (1966). The former demonstrated that labelled formyl groups from fmet-*t*RNA were incorporated into at least two, if not all three, of the proteins coded by phage R17 RNA when used as messenger template in an *E. coli* cell-free extract system. The amino acid adjacent to the *N*-formylmethionine was alanine, which is the physiological N-terminal group. The latter workers found that when phage f_2 RNA was used in an *in vitro* system to direct synthesis of coat protein, the normal *N*-terminal group (alanine) was preceded by *N*-formylmethionine.

The *N*-terminal sequence of phage R17 coat protein synthesized *in vitro* is fmet–ala–ser, and comparison with the *N*-terminal frequency in *E. coli* might suggest that all protein chains commence with this sequence, after which a variable length of the peptide chain is removed by hydrolysis. However, this was excluded by the finding (Gussin and coworkers, 1966) that in the proteins commencing with methionine, the amino acid in the second position is not uniquely confined to alanine. But it does appear that all *N*-terminal amino acids can be accounted for by removal of either the formyl group (to leave methionine) or formyl-methionine (for other amino acids).

Gussin and coworkers (1966) reported the presence of an enzyme in *E. coli* extracts which liberated amino groups from the dipeptide *N*-formylmethionylalanine; its action was confined to attack on the first peptide bond. Adams (1968) used the peptides fmet–ala and fmet–ala–ser as model systems to test for deformylating activity in such extracts. He was able to characterize an enzyme which would remove the formyl group within two minutes of incubation, without attacking any other bond. Although the system was specific in that it would not attack formylmethionine alone—there had to be a second amino acid attached—the nature of the amino acid adjacent to methionine did not affect the hydrolysis. An assay for formate showed that the formyl group is released in this free form rather than transferred to an acceptor molecule. The enzyme activity was extremely labile under all experimental conditions (pH, ionic strength, etc.), and this probably explains why the formyl group remains attached to proteins synthesized under *in vitro* conditions. A similar enzyme from *B. stearothermophilus* is more stable than that from *E. coli*, and attempts are currently proceeding at its purification. Takeda and Webster (1968) found that an analogous deformylase in *B. subtilis* is specific for methionine as the *N*-terminal amino acid, and it seems probable that it removes the formyl group from all nascent fmet-polypeptides.

A second activity must then be required to remove methionine from some of these nascent proteins, and Takeda and Webster detected an aminopeptidase activity which was inactive on the *N*-terminal analogue formylmethionyl puromycin (fmet-puro), but would hydrolyse the met-puro product after the deformylase had removed the formyl group. An *N*-terminal coat protein fragment synthesized *in vitro* by phage f_2 RNA was tested with these enzymes. The aminopeptidase by itself was inactive on this substrate, but on joint incubation with deformylase a large amount of free methionine was produced. A small amount of alanine was also released in this reaction, so the aminopeptidase may have a lower degree of specificity than the deformylase. It is not known what is its mode of discrimination on deformylated nascent chains—possibly this might depend on the subsequent residues.

The fmet-*t*RNA is not formed by loading of formylmethionine onto the transfer molcule, but by enzymic action to formylate the methionyl-*t*RNA already formed. The reaction is catalysed by a specific enzyme, and Marcker, Clark and Anderson (1966) and Dickermann and coworkers (1966) reported the formylation reaction as:

$$\text{met-}t\text{RNA} \xrightarrow[\text{formyl-THF-met-}t\text{RNA transformylase}]{\text{10-formyltetrahydrofolate}} N\text{-formyl-met-}t\text{RNA}$$

The enzyme is comparatively small, with a molecular weight of about 25,000. Dickermann and Smith (1967) studied the effect of substrate protection on its heat or trypsin inactivation; met-*t*RNA conferred protection against loss of activity, uncharged $t\text{RNA}_{\text{met}}$ conferred less protection, and other *t*RNA species were not active. This protective effect is consistent with specific direction of the activity of the enzyme toward met *t*RNA. Further, interaction between enzyme and met-*t*RNA in the absence of formyl donor requires the aminoacyl ester bond.

Clark and Marcker (1966) demonstrated that the methionine-accepting activity of *E. coli* transfer RNA could be fractionated into two species, one of which could be formylated (met-*t*RNa$_f$), and one which could not (met-*t*RNA$_m$). The binding of each charged species to ribosomes was measured under direction of a number of trinucleotides, and it was found that whilst the methionine codon AUG stimulated binding of both species, the fmet-*t*RNA$_f$ initiator also responded to the related codon GUG (which codes for valine). Although at first UUG was thought to behave similarly, this does not appear to be so. Use of the synthetic polymer poly-AUG (Clark and Marcker, 1966; Khorana and coworkers, 1966) has shown that both of the two methionine transfer species respond. Analysis of the incorporation of sulphur from S^{35}-met *t*RNA$_f$ into polypeptide showed that at least 70 per cent. was incorporated into the *N*-terminal position, whereas by contrast, incorporation from the non-formylatable species was almost exclusively into internal positions. The use of poly-UG showed that the codon GUG recognizes fmet-*t*RNA$_f$ externally, but inserts valine internally. Thus the interpretation of the code for the triplets AUG and GUG depends upon their position in the message. Externally, both code for the initiator fmet-*t*RNA$_f$, but internally for their respective normal aminoacyl transfer species, met-*t*RNA$_m$ and val-*t*RNA. The discovery of only one anticodon in met-*t*RNA$_f$ species implies that there must be a 5′ nucleotide degeneracy in its recognition of either AUG or GUG as opposed to the usual 3′ wobble degeneracy (although this is shown only in the ribosomal P site, and not the A site, which is not used by this species).

The presence of an initiation codon in a message phases its translation of triplets into amino acids. Ghosh, Soll and Khorana (1966) found that when formylated met-*t*RNA$_f$ was used for initiation with poly-UG messenger, the message

U G U G U G U G U G U G U

(is read as) fmet – cys – val – cys

since reading starts in phase with GUG. Sundarajan and Thach (1966) found that if AUG is present at the beginning of long polynucleotide chains, it suppresses reading of codons which partially overlap it, and promotes reading of the subsequent 3′ codon.

There have been conflicting results on the necessity of formylation for the activity of met-*t*RNA$_f$. Early results suggested that formylation did not affect its ability to bind to ribosomes under direction from AUG (Clark and Marcker, 1966) or reaction of the bound species with puromycin (Bretscher, 1966). However, more recent results have shown that formylation is necessary for this activity. Anderson and coworkers (1967a, 1967b) demonstrated that formylation is necessary for AUG-directed ribosomal binding, and Economou and Nakamoto (1967) found that met-tRNA$_f$ would only function as initiator at low Mg^{2+} ionic conditions if formylated, although the unformylated species was active at a reduced rate if the magnesium ion concentration was raised.

The efficiency of amino acid incorporation in polypeptide synthesis directed by artificial messengers led to the idea that there was no need for a special initiation system. However, Ghosh, Soll and Khorana (1967) reported that these positive results were due to the use of a high (above 20mM) concentration of magnesium ions; this results in an artificial stabilization of the complex between the various components involved in translation, abolishing the need for proper initiation. The use of lower ionic concentrations demonstrated such a requirement. Leder and Nau (1967) and Anderson and coworkers (1967a, 1967b) showed that whereas various extra-ribosomal protein factors and GTP are required for initiation at low ionic concentrations, they are not necessary if the magnesium ion concentration is increased. In the presence of all the factors required, the optimum Mg^{2+} concentration proved to be about 10 mM, but in the absence of some or all of these, this had to be raised to a level nearer 14–16 mM for optimum ribosomal binding of fmet-tRNA$_f$. The Mg^{2+} dependence of the binding reaction is shown in Figure 5.10.

A variety of work has shown that at low magnesium ion concentrations the presence of fmet-tRNA$_f$ is essential for proper initiation. Khorana and coworkers (1966) found that whilst polypeptide synthesis proceeds well at 11mM Mg^{2+} under direction from poly-UC, -AG, -AC and -UG, at lower concentrations it is strikingly reduced in the first three, but continues even at 4mM Mg^{2+} with poly-UG. At this low level, protein synthesis proved possible only in the presence of fmet-tRNA$_f$. Thach and coworkers (1966) showed that at low concentrations (optimum 9 mM Mg^{2+}), AUG acts as a phase selector at the 5′ end of oligonucleotides, but as the ionic level increases the phase reading of the message becomes increasingly random. At low concentrations the AUG codon was able to exert its phasing function even when not strictly the 5′ terminal but merely located near this end.

Virtually all the work on chain initiation has been performed in bacterial systems, and almost all using *E. coli*. Studies on *B. subtilis* (Horikoshi and Doi, 1967, 1968; Takeda and Webster, 1968) have shown that it appears to possess an initiation system comparable to that of *E. coli*. Extracts contain a transformylase activity which converts one of its two methionine-accepting transfer species to fmet-tRNA and this shows characteristics similar to those of the *E. coli* species in its reaction

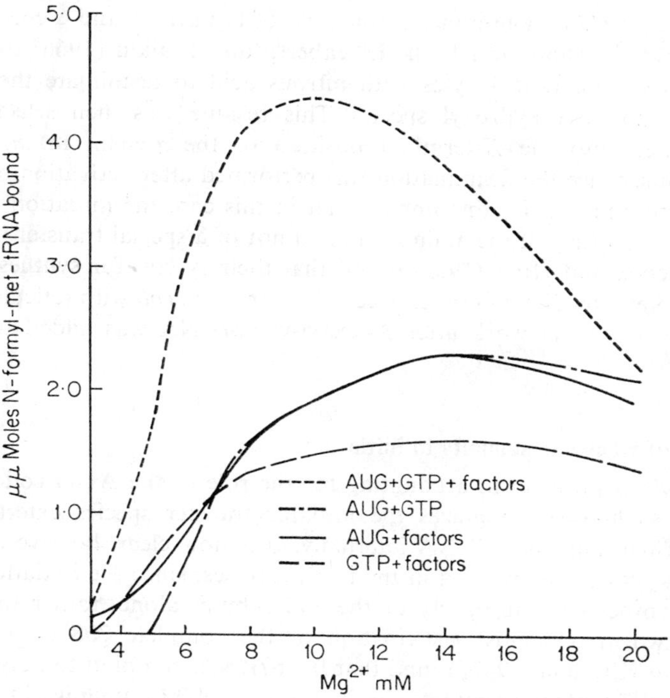

Figure 5.10. Mg^{2+} dependence of fmet-*t*RNA_f binding to ribosomes. (After Leder and Nau, 1967)

with puromycin when bound to ribosomes under direction from AUG. Alanine comprises about 90 per cent. of the *N*-terminal amino acids, and deformylase and aminopeptidase activities which could account for this have been found. Although this initiator does not appear to function in the cytoplasmic protein synthesis of higher organisms, the presence of fmet-*t*RNA has been reported in the mitochondria of Hela (human tumor) cells (Galper and Darnell, 1969), further emphasizing the similarities between the behaviour of this cell organelle and bacteria.

There is no evidence about the mode of initiation in higher organisms *in vivo*, but comparable mechanisms employing NH_2-blocked aminoacyl-*t*RNA initiators have been postulated for some *in vitro* systems. Lucas-Lenard and Lipmann (1967) reported that *N*-acetyl-phe-*t*RNA has an initiating effect in the poly-U directed phenylalanine incorporating system at low (4mM) Mg^{2+} concentration. (This system is usually operated at

20 mM Mg^{2+}.) The *N*-terminal amino acid of both the α and β chains of hemoglobin is valine, and Rich, Eikenberry and Malkin (1966) treated valyl-*t*RNA from reticulocytes with nitrous acid to deaminate the free amino group to a hydroxyl species. This residue was then selectively incorporated into the *N*-terminal position of the α-chain in *in vitro* experiments. Since the deamination was performed after activation of the *t*RNA with amino acid, this implies that in this case the initiation must be solely a property of the amino acid and not of a special transfer molecule. Laycock and Hunt (1969) found that their system for synthesizing globin protein, by directing a cell-free system from *E. coli* with reticulocyte *m*RNA, would only work after *N*-acetyl-valyl-*t*RNA was added to act as initiator.

The role of ribosome subunits in initiation

An obvious problem in accounting for the role of the AUG codon in initiation is how it recognizes the initiator transfer species externally, but a normal aminoacyl-*t*RNA internally. It is now clear, however, that different systems are involved in the two processes, since the initiation of protein synthesis is a property of the 30S subunit alone, rather than a function which demands participation of the complete 70S ribosome. Ghosh and Khorana (1967) found that fmet-*t*RNA_f can bind to poly-UG or poly-AUG synthetic templates in the presence of 30S subunits, but this initiation complex must be joined by a 50S subunit before other aminoacyl-*t*RNA species can bind (see Figure 5.11). Indeed, Nomura, Lowry and Guthrie (1967) observed that, whilst 30S subunits can bind fmet-*t*RNA_f to phage f_2 RNA messenger, association with 50S subunits inhibits this binding (presumably by removing free 30S subunits), but stimulates the incorporation of other aminoacyl-*t*RNA species. The inhibition can be overcome by the addition of initiation factors (discussed below); these may assist the dissociation of ribosomes into subunits, either directly or indirectly, and such an activity can also account for the apparent ability of 70S ribosomes to bind to messengers in the presence of the factors—binding could occur through a dissociation and subsequent reassociation of the ribosomes.

A critical test of subunit initiation was reported by Guthrie and Nomura (1968). Fmet-*t*RNA_f was allowed to bind to 70S ribosomes which were labelled with heavy isotopes; the reaction mixture contained initiation factors and also a large excess of light 50S subunits. If initiation demands 70S ribosomes, the initiator *t*RNA should bind directly to the heavy 70S ribosomes. By contrast, if the initiation complex is formed with only the small subunit, the fmet-*t*RNA$_f$ must bind to a dissociated heavy 30S subunit. In order to yield a 70S ribosome, this must then undergo reassociation with a 50S subunit; since there is an excess of light 50S subunits, the initiator *t*RNA must end up attached to a 'hybrid' ribosome. A possible flaw in this procedure is that the latter result could also be achieved if there were to be a rapid and spontaneous equilibrium rearrangement between the heavy ribosomes and the added 50S subunits before formation of the initiation complex. In order to exclude this, the binding of ordinary aminoacyl-*t*RNA was followed by the same procedure; this should become attached directly to heavy 70S ribosomes. As expected, whilst the aminoacyl-*t*RNA indeed associated with the heavy 70S particles, initiator fmet-*t*RNA$_f$ was found attached to hybrid ribosomes. There is now also some evidence that subunit initiation may apply to protein synthesis in yeast (Kaempfer, 1969) and to the reticulocyte hemoglobin synthesizing system (Colombo, Vesco and Baglioni, 1968).

Subunit initiation provides an explanation as to how AUG initiator codons are distinguished as such from those coding for methionine internally. Only fmet-*t*RNA$_f$ must be able to bind to 30S subunits in response to AUG, whilst with a 70S ribosome reading is only permitted as met-*t*RNA$_m$. This is achieved by the interactions with aminoacyl-*t*RNA of the extraribosomal protein factors involved in protein synthesis. Rudland and coworkers (1969) reported that normal aminoacyl-*t*RNA molecules, including met-*t*RNA$_m$, do not bind to initiation factors, which specifically promote the binding of fmet-*t*RNA$_f$ only to the 30S-*m*RNA initiation complex; thus only fmet-*t*RNA$_f$ is recognized as an initiator and other aminoacyl-*t*RNA species are excluded from external recognition. On the other hand, Ono and coworkers (1968) observed that fmet-*t*RNA$_f$ is the only aminoacyl-*t*RNA which is unable to form an intermediate complex with the elongation transfer factors ($T_u + T_s$), and this may serve as the mechanism for excluding it from internal recognition. It seems likely that both the presence of the blocked amino group, and features of the *t*RNA structure, play a role in its interactions with the extraribosomal factors.

Thus when the first AUG codon at the beginning of a messenger is encountered by a 30S subunit, it directs formation of the initiation complex. After attachment of the 50S subunit, subsequent AUG codons are only read by the 70S ribosome, and hence code unambiguously for met-*t*RNA$_m$. When a polycistronic messenger is translated, the ribosome (presumably) dissociates into its subunits upon chain termination at the end of a cistron. If the 30S particle remains attached to the messenger,

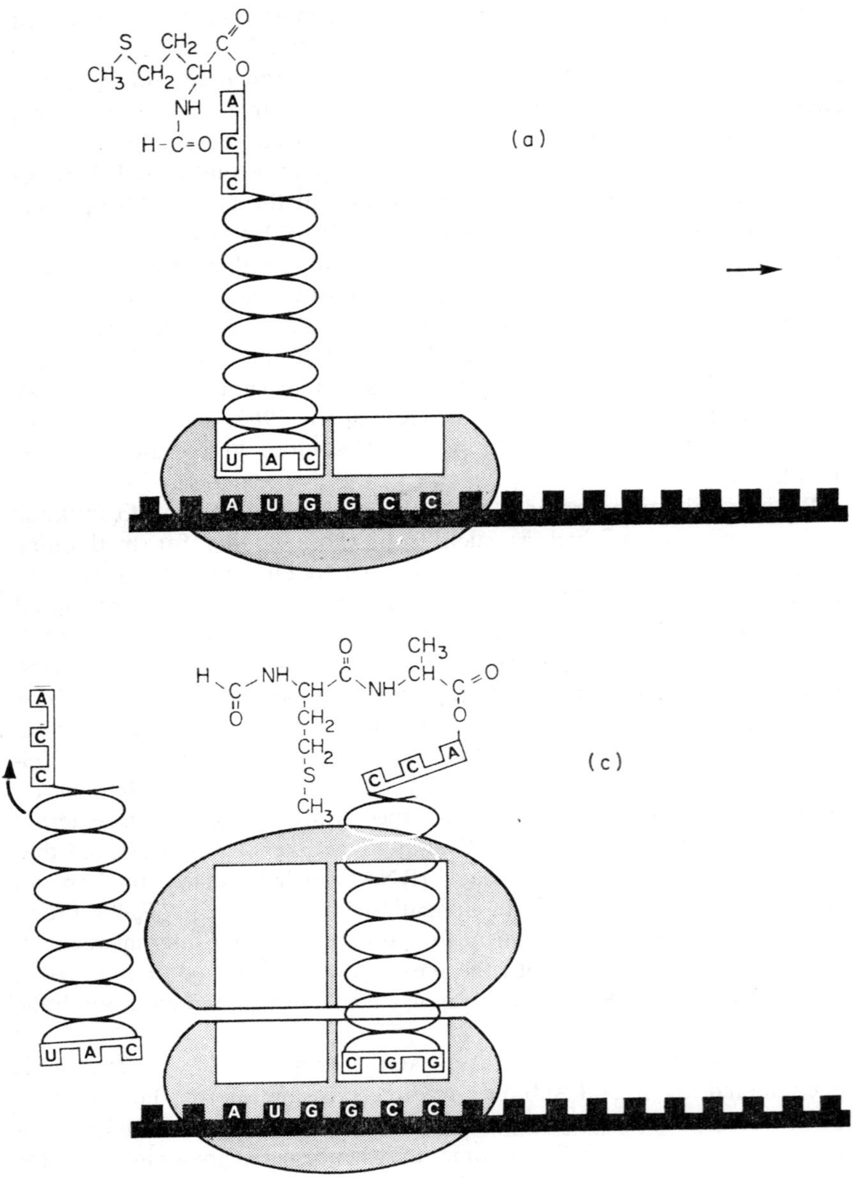

Figure 5.11

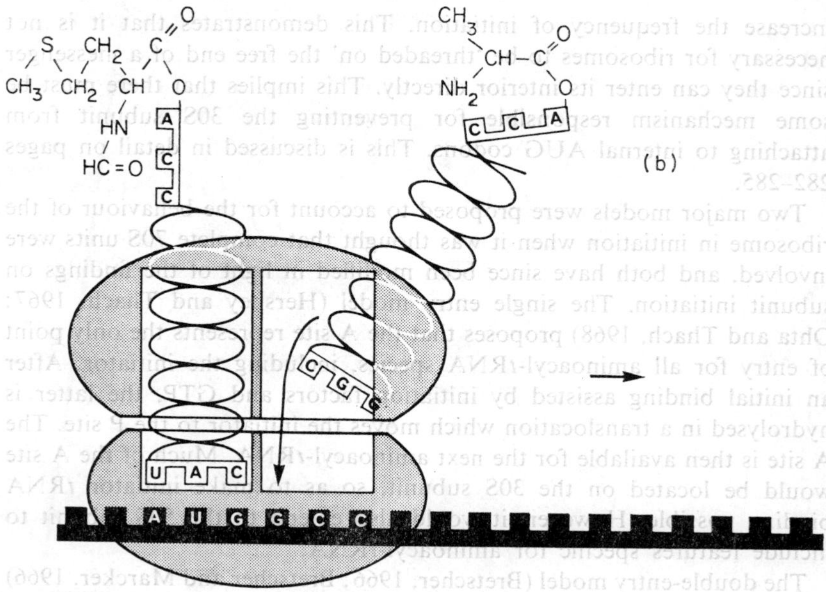

Figure 5.11. Initiation of protein synthesis. (a) An initiation complex is formed between *m*RNA bearing an AUG codon near its 5′ terminus and fmet *t*RNA; initiation factors and GTP are also necessary although not shown. (b) A 50S subunit attaches to the 30S and the second aminoacyl-*t*RNA (alanine in this case) can bind. (c) After peptide bond formation the free *t*RNA is released, and protein synthesis continues by chain elongation. The formyl group is subsequently removed from the protein.

it will then recognize the AUG codon at the beginning of the next cistron as an initiator, and commence synthesis of the next protein specified.

The most obvious way for a ribosome to undertake its initial attachment to a messenger would be for a 30S particle to attach at the 5′ end and move along to the first AUG codon, where it would form an initiation complex and commence translation. However, it appears rather that ribosomes have an ability to bind directly at the initiation site itself. This was tested experimentally by using a circular messenger species to direct protein synthesis; since there is no such *m*RNA, Bretscher (1968c) made use of the finding that single-stranded DNA from phage *fd* can be translated directly *in vitro* provided that neomycin B is also present. (Recent work has produced a system in which addition of the antibiotic is not necessary—see Bretscher, 1969). The DNA could be translated in its circular form; breakage of the circles to generate linear ends did not increase the frequency of initiation. This demonstrates that it is not necessary for ribosomes to be 'threaded on' the free end of a messenger since they can enter its interior directly. This implies that there must be some mechanism responsible for preventing the 30S subunit from attaching to internal AUG codons. This is discussed in detail on pages 282–285.

Two major models were proposed to account for the behaviour of the ribosome in initiation when it was thought that complete 70S units were involved, and both have since been modified in light of the findings on subunit initiation. The single entry model (Hershey and Thach, 1967; Ohta and Thach, 1968) proposes that the A site represents the only point of entry for all aminoacyl-*t*RNA species, including the initiator. After an initial binding assisted by initiation factors and GTP, the latter is hydrolysed in a translocation which moves the initiator to the P site. The A site is then available for the next aminoacyl-*t*RNA. Much of the A site would be located on the 30S subunit, so as to make initiator *t*RNA binding possible. However, it would also extend to the 50S subunit to include features specific for aminoacyl-*t*RNA.

The double-entry model (Bretscher, 1966; Bretscher and Marcker, 1966) ascribes different roles to the two *t*RNA binding sites on the ribosome. Normally, peptidyl-*t*RNA is bound at the P site and the next aminoacyl-*t*RNA can only enter the A site. However, the initiator transfer species is presumed to possess the ability to enter the P site directly; the penultimate *N*-terminal aminoacyl-*t*RNA then enters the A site, and the first peptide bond is synthesized. There is no ambiguity since only fmet-*t*RNA$_f$ can respond when the initiator codon is in the P site, and only the normal transfer species when it is in the A site. Experiments with puromycin support this idea, since ribosomal bound fmet-*t*RNA$_f$ is sensitive and reacts to give fmet-puro, whereas met-*t*RNA$_m$ is comparatively insensitive under identical conditions. This was taken to imply that fmet-*t*RNA$_f$ binds at the P site, and met-*t*RNA$_m$ at the A site. The hybrid ribosome model for protein synthesis (see page 110) accounts for initiation on the supposition that the initiator binds to the 30p site. Addition of the 50S subunit then forms a hybrid structure, probably the 30p; 50a (structure (b) in Figure 5.7). GTP is required for this, and its hydrolysis then converts (b) to (d), that is converts the fmet-*t*RNA$_f$ into the reactive state shown by the puromycin reaction.

Formation of the initiation complex

Formation of the initiation complex between a 30S subunit, *m*RNA, and initiator fmet-*t*RNA$_f$ requires the presence of GTP and initiation factors. The extent of the need for GTP varies greatly, and its effect on the binding of fmet-*t*RNA$_f$ to ribosomes varies according to the experimental conditions, for example which initiation factors are present. The analogue GMP–PCP can substitute for GTP in stimulating this binding, but whereas the fmet-*t*RNA$_f$ bound to 30S subunits in the presence of GTP can react with puromycin to yield fmet-puro after the addition of 50S subunits, it cannot do so when GMP–PCP is used (Anderson and coworkers, 1967a; Hershey and Thach, 1967). Similarly, investigation of the conditions required for binding aminoacyl-*t*RNA to the second codon of the messenger showed that GMP–PCP is quite inactive as a substitute for GTP (Ohta, Sarkar and Thach, 1967). This suggests that the sequence of events in initiation is first the formation of an initiation complex in which the presence of GTP is necessary to bind fmet-*t*RNA$_f$ to the 30S subunit; since it can be replaced by GMP–PCP its hydrolysis is not required. Association with a 50S subunit then yields a 70S ribosome with a high affinity for aminoacyl-*t*RNA; the inability of GMP–PCP to substitute shows that at this stage the hydrolysis of GTP is necessary to activate the fmet-*t*RNA$_f$ and render it sensitive to the second aminoacyl-*t*RNA (or to puromycin).

The protein factors required for chain initiation are distinct from those concerned with chain elongation, and comprise fractions derived from 30S subunits by washing with ammonium chloride. That they are concerned only with the initiation of protein synthesis was demonstrated by the finding that they are essential for the *in vitro* translation of natural messengers (which always require initiation); however, whilst they stimulate protein synthesis directed by artificial messengers under conditions requiring initiation, they can be dispensed with under conditions (such as high Mg^{2+} concentration) which obviate the need for initiation (Salas and coworkers, 1967; Stanley and coworkers, 1966). Two of the factors, f_1 and f_2, are required to bind fmet-*t*RNA$_f$ to ribosomes in response to AUG (Anderson and coworkers, 1967a), and the third fraction, f_3, stimulates the binding of 30S subunits to *m*RNA (Brown and Doty, 1968).

Revel and coworkers (1968a, 1968b) have used an alternative system in which natural *m*RNA is produced from phage T4 by infection of *E. coli* cells. A crude fraction of soluble proteins associated with the ribosomes was found to be necessary for protein synthesis with this messenger, although it was not required for the translation of a synthetic poly-U template. Fractionation on DEAE-cellulose yielded three peaks. Using the formation of fmet-puro as a model system, factor A proved necessary for the ribosomal binding of fmet-*t*RNA$_f$, factor B stimulated formation of the first peptide bond with the second aminoacyl-*t*RNA, and factor C was required for binding 30S subunits to the messenger. Since factor C also assisted the attachment of initiator *t*RNA to 30S subunits, it may perhaps consist of more than one protein. Wahba and coworkers (1968) have reported that f_3 may consist of two proteins, one of which is also present in the C fraction. Factor A is probably equivalent to f_1.

Both f_1 and f_2 have been subjected to extensive purification procedures and the purified fractions have been used to study their role in initiation. The homogenous f_1 protein is comparatively small, with a molecular weight of only 9000 (Hershey, Dewey and Thach, 1969). At the time of writing, it is not clear whether the f_2 fraction comprises only a single homogenous protein or two closely similar proteins with virtually identical biochemical and catalytic properties (Chae, Mazumder and Ochoa, 1969; Kolakofsky, Dewey and Thach, 1969), although it is clearly much larger than the F_2 protein, with reported molecular weights from 65–80,000.

The f_2 fraction has GTPase activity, and the first step in binding fmet-*t*RNA$_f$ appears to be the formation of a preliminary complex between f_2 and GTP (although this has not been isolated directly). The activity of f_2 is inhibited by reagents which interact with free sulphhydryl groups, but the addition of GTP or 30S subunits confers protection against the inactivation. This suggests that the factor has reactive -SH groups which become masked upon the formation of a complex with GTP or when it binds to 30S subunits. The idea that its first action is to complex with GTP is also supported by the finding that preincubation of f_2 and GTP before assay of its activity results in a faster initial binding of fmet-*t*RNA$_f$ to ribosomes.

The f_2 and GTP mediated binding of fmet-*t*RNA$_f$ to the *m*RNA–30S complex takes place in two stages (Wahba and coworkers, 1968). First, reaction of f_2 and GTP with the initiator species forms a complex, which although stable at 0°c, is rather less so at 25°c. The formation of this complex does not require f_1, but is succeeded by a second reaction in which it interacts with this factor to form the more stable preinitiation complex responsible for transferring fmet-*t*RNA$_f$ to the 30S subunit. In accord with this, Hershey, Dewey and Thach (1969) found that f_1 only binds to 30S subunits as an integral part of the initiation complex. The factor was prepared from bacteria grown in a medium with a radioactive label, and appreciable radioactivity was found to associate with the ribosomes only when the reaction mixture contained all the components required for formation of the GTP-factor–fmet-*t*RNA$_f$–30S-AUG complex. The addition of 50S subunits to the initiation complex formed a

70S ribosome, and the association resulted in release of almost all the bound radioactivity. This still happened when GMP–PCP was substituted for the GTP. The f_2 factor probably follows a similar cycle since it is not found in association with 70S ribosomes, but as might be expected from its GTPase activity, is probably released only after cleavage of the triphosphate. As is consistent with the results obtained on the respective abilities of GTP and GMP–PCP to assist in initiation, it must thus act at a later stage in initiation, since its GTPase activity is exerted only in association with 70S ribosomes and not with 30S subunits alone.

We should note that Mangiarotti (1969) has suggested a rather different role for initiation factors from that postulated above. Contrary to these results, he reported that initiation factors *are* present on 70S ribosomes, but appear absent because they are lost during the sucrose density gradient centrifugation used in the extraction procedure; when ribosomes were obtained by differential centrifugation they proved to be active, but upon purification on a sucrose gradient or by washing with NH_4Cl, became dependent upon added factors. It appeared that factors A and B were removed from the ribosomes, but C remained. Ribosomal subunits showed a similar behaviour in that after centrifugation on a sucrose gradient they could no longer yield factors when washed with NH_4Cl, but despite this, and unlike 70S ribosomes, remained active nevertheless. This would suggest that only factor C is required for initiation.

In further experiments, 70S ribosomes were obtained bearing C^{14}-labelled nascent polypeptide chains and H^3-labelled *m*RNA. When factors A and B were added, the labelled *m*RNA and peptidyl-*t*RNA were released from the ribosomes. Mangiarotti has therefore suggested that these factors are concerned with termination rather than initiation; their apparent role in initiation arises because they must release their resident polypeptidyl-*t*RNA before they are capable of initiation. In support of this, treatment of puromycin, which achieves this end, was found to obviate the need for factors A and B in initiation.

Chain termination

During protein synthesis, polypeptide chains are attached to *t*RNA molecules by the ester bond between the most recently added amino acid and the terminal adenosine of its *t*RNA. When the last amino acid has been added to the chain, this link must be broken in order to release the completed protein and allow re-use of the *t*RNA. However, the spontaneous rate of hydrolysis of polypeptidyl-*t*RNA is rather low under physiological conditions. Using *in vitro* systems for protein synthesis, most synthetic polynucleotide templates yield only polypeptidyl-*t*RNA species. In the poly-U directed system, the product is poly-phe-*t*RNA, and the 1 per cent. free polyphenylalanine can be accounted for by spontaneous hydrolysis (Bretscher, 1965). Thus chain termination is not automatic at the end of a segment of messenger, and some specific mechanism must operate *in vivo* to release the newly synthesized polypeptide chain from its transfer molecule.

The use of poly-U,X (X represents xanthanilic acid, which is quite meaningless) also yields polypeptidyl-*t*RNA, so that the presence of a sequence without meaning is not sufficient to achieve proper chain termination (Bretscher and coworkers, 1965). When phage f_2 RNA is used in an *in vitro* system lacking the aminoacyl-*t*RNA corresponding to its seventh codon, hexapeptidyl-*t*RNA is formed, and not free hexapeptide (Bretscher, 1968b). This implies that there must be an active mechanism for release of the chain, and not merely termination in default of amino acid recognition. However, when synthetic polymers with a high A and U content are used as messengers, a high proportion (as great as one-half) of the polypeptides synthesized are free from *t*RNA. The use of defined messengers with an AUG initiator followed by various A,U containing codons to a UAA nonsense terminator has shown that *ochre* causes an unequivocal termination, with release of the polypeptide chain from *t*RNA (Last and coworkers, 1967). Thus the specific mechanism which must be responsible for chain termination acts in response to nonsense codons to separate polypeptide and *t*RNA.

One mechanism which has been suggested is that a special *t*RNA species might recognize nonsense triplets; this could carry no amino acid, or possibly some molecule which would cause termination by interacting with the polypeptidyl-*t*RNA. However, no such *t*RNA has been isolated or even identified. Bretscher (1968b) has shown conclusively that termination does not depend upon a transfer molecule, by use of a phage f_2 mutant which has an *amber* triplet in place of the usual seventh codon of its messenger. This was allowed to direct protein synthesis in an *in vitro* system in which the only *t*RNA species present were those corresponding to the first six codons. Nevertheless, the hexapeptide synthesized was released from its *t*RNA.

Capecchi (1967a) used a similar mutant of phage R17 RNA; this also directs synthesis of a hexapeptide which is released as such. The translation of this messenger can be halted before it reaches the nonsense codon by starvation for one of the preceding aminoacyl-*t*RNA species required. It is then possible to isolate the *m*RNA—ribosome—*t*RNA complex carrying the chain so far synthesized, and study the requirements for its release. After starvation for the amino acid corresponding to the triplet immediately preceding the nonsense codon, addition of its aminoacyl-*t*RNA together with GTP and supernatant factors effected chain release. This suggests that chain termination, like elongation and initiation, may depend upon specific extraribosomal protein factors.

The factors responsible have been separated by DEAE-Sephadex column chromatography into two peaks of about the same size. Each behaved as a homogenous species through an extensive purification procedure, and appears to comprise a protein of molecular weight around 40–50,000. When tested for terminator ability, R_1 proved to be active under direction from either UAA or UAG, and R_2 with UAA or UGA (Scolnick and coworkers, 1968). Some 60 per cent. of the UAA-dependent activity was present in R_1. Milman and coworkers (1969) used a messenger comprising an initiation codon followed by a termination codon to bind f-H^3 met-*t*RNA_f, and then measured the release of the radioactive label. Both the actions of R_1 and R_2 were stimulated by the addition of a further factor, S, which proved to be a heat-labile protein. So far, the possibility that S may comprise one of the factors involved in some other stage of protein synthesis has not been completely excluded. It is worth noting that, whilst the degeneracy pattern of R_1 codon recognition resembles that of some aminoacyl-*t*RNAs (see pages 206–209), there is no known counterpart to that exhibited by R_2.

The characteristic patterns of suppression of the various nonsense codons suggest that it is only the *ochre* which is used *in vivo* for chain termination. Whilst *amber* and UGA suppressors work with an efficiency of up to 50 or 60 per cent., even the most efficient *ochre* suppressors are much weaker, at about 5 per cent. only. If a codon is used in the wild type for chain termination, its suppression would affect not only the mutant codon, but also the normal chain terminators. This would probably be lethal to the organism, so that only weak suppressors could be tolerated; thus the strong suppression of a nonsense codon suggests that it is not normally used as a chain terminator. That *ochre* is used *in vivo* is supported by the finding of Person and Osborn (1968) that, when an *E. coli* strain possessing an *amber* suppressor suffers its conversion to an *ochre* suppressor, the bacterial cells grow more slowly themselves and in addition can only support growth of T4 phage at a reduced level.

Of course, this raises the question of why the cell recognizes UAG and UGA as terminators, even though it does not normally use them as such; one would expect that it would be more advantageous for these codons to represent amino acids (see pages 86–92). So far as the *amber* codon is concerned, this may be a consequence of the mode of codon–anticodon recognition (see pages 206–207); it seems likely that any *t*RNA which responds to UGA would have to respond also to UAA. Since this would prevent the use of *ochre* as a chain terminator, it may be necessary for UAG to represent nonsense if UAA is to be able to do so. The nonsense role of the UGA codon, however, cannot be explained in these terms, but there are indications that, in any case, it may represent cysteine in mammals, so that its use as a nonsense codon may be somewhat limited.

The ribosome subunit cycle

If initiation requires 30S subunits alone, a ribosome which undertakes synthesis of more than one polypeptide must dissociate into its subunits at some point after the termination of one chain, but before the initiation of the next. This demands that ribosome subunits cannot remain associated permanently *in vivo*, but must exchange their partners at each round of protein synthesis. Experimentally, this was investigated by transferring bacteria labelled with heavy isotopes into a medium containing only light species (Kaempfer, Meselson and Raskas, 1968). The density distribution of ribosomes showed a progressive replacement of heavy ribosomes by the hybrid species 50S (L); 30S (H) and 50S (H); 30S(L); this indicates that dissociation and reassociation must have occurred. All 70S particles suffered subunit exchange. An unequivocal demonstration that this comprises an integral part of the mechanism of protein synthesis was provided by the development of a cell-free synthetic system requiring subunit exchange for activity (Kaempfer, 1968). Further, as would be predicted, subunit exchange ceased when protein synthesis was inhibited.

The precise role of the 70S ribosome with respect to cellular metabolism is currently controversial. Mangiarotti and Schlessinger (1966), using a rapid preparation of polysomes and ribosomes from *E. coli* cells, observed that virtually all the free ribosomes were in the form of individual subunits, with less than 2 per cent. in the form of free 70S monomers. Newly labelled ribosomal RNA appeared in subunits and polysomes at the same time, but not as 70S associated particles, implying that the ribosomes in polysomal form are in rapid exchange with the free subunits (Mangiarotti and Schlessinger, 1967; Schlessinger, Mangiarotti and Apirion, 1967). Pulse-labelling experiments showed that the radioactivity becomes trapped in a pool, indicating that used subunits are returned to this pool for reuse. They suggested that ribosome subunits undergo a cycle in which they are withdrawn from pools of free 30S and free 50S particles to associate at initiation, function as 70S monomers during chain elongation, and dissociate at termination to return to their pools. Thus the 70S form would exist only to undertake translation.

One possible cause for the presence of 70S ribosomes in extracts is degradation of polysomes to yield monomers. However, Kohler, Ron and Davis (1968) reported that, although the rapid lysis conditions used by Mangiarotti and Schlessinger produce no free 70S ribosomes, other methods which do not appear to cause degradation of polysomes do yield such free particles. When cells are deprived of amino acids, or treated with actinomycin or puromycin, ribosomes complete synthesis of their current polypeptide chains and run-off polysomes, but do not start synthesis of new chains. Under these conditions, 70S ribosomes remained in their associated state, and, in fact, the run-off yielded only 70S particles (the level of subunits remained essentially unaltered). There was a difference in the behaviour of polysomal and run-off ribosomes in that the latter were dissociated completely by 1 mM Mg^{2+}, but the former were not until the concentration was lowered to 0·5 mM (Ron, Kohler and Davis, 1968). The small peak of free 70S ribosomes found in cells proved to behave in a similar manner to the run-off ribosomes, suggesting that the latter indeed represent the behaviour of normal 70S ribosomes when they complete a round of protein synthesis.

By contrast, a substantial increase in the concentration of ribosomal subunits have been observed when *E. coli* are incubated at 8°C or below (Friedman, Lu and Rich, 1969). When the temperature was raised above 8°C, there was a dramatic decrease in their concentration. This can be accounted for if formation of the initiation complex is blocked below 8°C, although ribosomes which are currently synthesizing a polypeptide chain can complete it. This suggests that when ribosomes run-off messengers, they remain as subunits because they are unable to undertake initiation. It is, at least, now clear that different preparative methods yield different profiles for the relative distribution of subunits, ribosomes and polysomes. One explanation which has been suggested (Phillips, Hotham-Iglewski and Franklin, 1969) is that the 70S ribosomes may arise when initiation complexes are allowed to form after cell lysis.

The dissociation of ribosomes into subunits—whether it occurs at chain termination or only subsequently—appears not to be spontaneous, but, rather, to be mediated through the action of a protein factor. Schlessinger, Mangiarotti and Apirion (1967) observed a difference in behaviour between 'native' subunits (those obtained from cells as such) and 'derived' subunits (those obtained by dialysis of 70S ribosomal particles from polysomes). The latter could readily reassociate to 70S monomers in 10 mM Mg^{2+}, whilst the former were unable to do so.

Eisenstadt and Brawerman (1967) found that, with f_2 messenger, native subunits were some five to nine times more active than derived subunits or undissociated 70S ribosomes. However, when under direction from poly-U, there was no difference between them—this implies that the difference is concerned with initiation. In the presence of an initiation factor obtained from the supernatant of a high speed centrifugation of *E. coli* cell extracts, the difference between native and derived subunits under f_2 direction was abolished. They proposed that this factor might be released from the 30S subunit when the ribosome commences translation, and since it is not present in soluble extracts, it must presumably be transferred immediately to a new 30S particle about to undertake initiation.

Kohler, Ron and Davis (1968) have proposed a more detailed model for the role of such a factor in ribosomal dissociation, as illustrated in Figure 5.12. They suggest that dissociation requires stoichiometric complexing of 30S subunits with a dissociation factor (DF), and this reaction frees the 50S subunit. When the 30S particle forms an initiation complex, changes in its properties would cause release of the factor, and permit association with a 50S subunit. The cell would contain only a small supply of DF, and its level would regulate the supply of free subunits. Since ribosomal dissociation is a prerequisite for initiation, DF can be regarded as an initiation factor. Subsequently, Subramanian, Ron and Davis (1968) examined the properties of the total set of initiation factors extracted by an ammonium chloride wash of ribosomes. An extract from the 30S subunits proved to contain a protein with the expected properties, and this protein was not present in associated 70S ribosomes. When a crude preparation of DF was added to 70S monomers, it caused dissociation into subunits; the extent of dissociation was proportional to the quantity of DF added, and the reaction appeared to be stoichiometric rather than catalytic. Final elucidation of the metabolic role of 70S subunits must await a more precise definition of the behaviour of this factor in the subunit cycle.

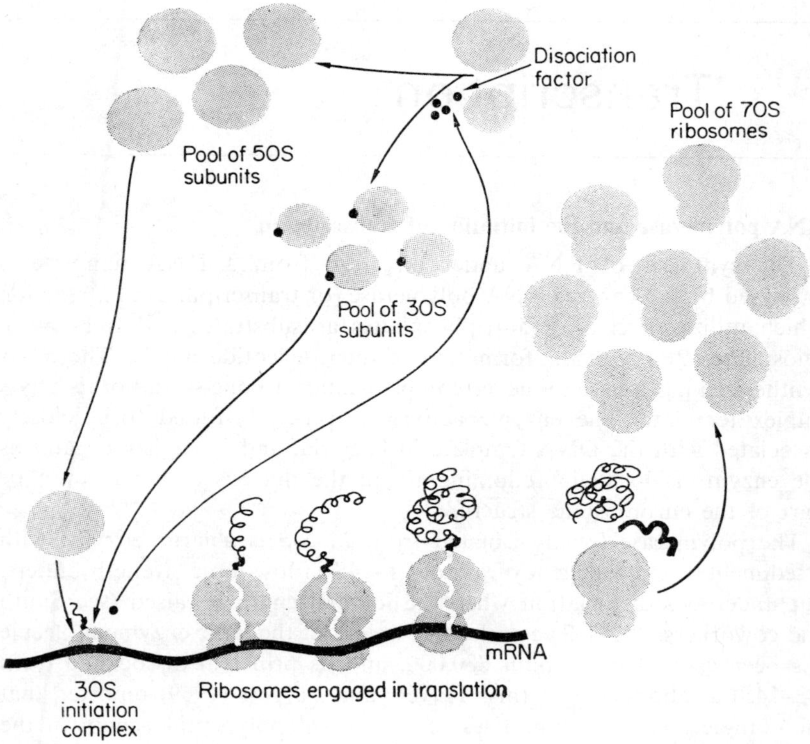

Figure 5.12. The ribosome subunit cycle. Upon the completion of translation of a messenger, the 70S ribosome, *t*RNA, and completed polypeptide chain are released. The ribosome joins a pool of 70S monomers. Ribosomes are withdrawn from the pool when they complex with a dissociation factor and this releases them into pools of free 30S and 50S subunits. When a 30S subunit complexes with a messenger to form an initiation complex, it releases its dissociation factor, and consequently can associate with a 50S subunit to form a 70S ribosome which undertakes translation. (Note: the presence of free 70S ribosomes is currently controversial)

CHAPTER 6

Transcription

RNA polymerase and the initiation of transcription

The synthesis of RNA under direction from a DNA template is catalysed by the enzyme RNA polymerase (or transcriptase) in a reaction which utilizes nucleoside-5′-triphosphates as substrates and frees pyrophosphate (PP_i) with the formation of internucleotide bonds. The RNA synthesized has a base sequence complementary to one strand of its DNA duplex template. The enzyme activity is generally found to be closely associated with the DNA template in bacteria, and in higher organisms the enzyme is located predominantly in the nucleus and may even be part of the chromosome structure.

The polymerase freshly isolated from *E. coli* is a large enzyme with a sedimentation coefficient of about 21–24S in low ionic strength buffers, but undergoes dissociation when the ionic strength is raised (see Smith and coworkers, 1967). The molecular weight of the large enzyme molecule has been estimated at about 880,000, and its principal dissociated form (13–14S) at about half of this. Travers and Burgess (1969) reported that the complete enzyme comprises four types of polypeptide chain in the ratios $2\alpha : 1\beta : 1\beta' : 1\sigma$. The α chain has a molecular weight of around 40,000, β of about 155,000 and β' of about 165,000, whilst the sigma molecule has a weight of around 95,000; this totals about 495,000 and corresponds to the 13–14S form of the enzyme.

There have been conflicting reports as to the size of the active species, and enzyme activity has been found in assay conditions that give rise to forms varying from 13S to 26S (for review see Richardson, 1969). The properties and activity of the polymerase are affected by the presence of polynucleotides and individual nucleotide species. Smith and coworkers (1967) found that polynucleotides cause a lowering of its sedimentation coefficient from the large aggregate to the 13S form. Single nucleotide species enhance the stability of the enzyme and protect against its loss of ability to bind DNA after comparatively slight increases in temperature (Stead and Jones, 1967). Purines are more effective than pyrimidines, and this is probably a reflection of their role in initiation (see below).

Under conditions of low ionic strength, the initiation of RNA chains ceases very rapidly after addition of the enzyme, which remains attached to its template in an enzyme–DNA–nascent RNA complex. At high salt concentrations the RNA chains are released, but at low concentrations they remain attached to the DNA—the reason for the rapid cessation of enzyme activity may be that the nascent RNA chains exert an inhibitory effect upon the polymerase. Travers and Burgess (1969) made use of the fact that this results in each enzyme molecule being able to initiate only one chain of RNA at low ionic levels; thus measurement of the number of RNA chains initiated by a known weight of enzyme gives the molecular weight of the enzyme species at initiation. This proved to be 480,000—in good agreement with the estimated value for the molecular weight of the 13–14S form of the enzyme.

At first, it was thought that the binding reaction was irreversible, but Richardson (1966) and Stead and Jones (1967) showed that it could be reversed in the absence of nucleoside triphosphates. Maitra and Hurwitz (1967) found that reversibility depends upon whether the enzyme is actively engaged in transcription; after RNA synthesis has started on the DNA the enzyme proves to be irreversibly bound, but in the absence of transcription shows a small degree of dissociation. The increased stability of the complex during transcription can be seen as a mechanism to help prevent the premature release of unfinished chains.

The growth of the RNA chain proceeds from the 5′ phosphate to the 3′ hydroxyl terminus; thus the DNA strand which is transcribed is copied in the direction 3′ to 5′. The initial nucleoside triphosphate incorporated in both DNA-dependent and RNA phage-induced production of messenger retains both its β and γ phosphate groups during the subsequent chain elongation (Maitra, Cohen and Hurwitz, 1966; Maitra, Nakata and Hurwitz, 1967). By contrast, only the α phosphate is retained by nucleotides incorporated internally. Thus the incorporation of a γ P^{32} radioactive label into DNA can be used to follow the chain initiation reaction, whilst C^{14}- or αP^{32}-labelling follows the overall synthesis of RNA.

RNA chains are preferentially initiated with purine nucleoside triphosphates in an *in vitro* system from *E. coli*, and the general characteristics of the process seem to be the same when the enzyme from *Azobacter vinelandii* is employed instead. Phage systems have similar characteristics to the DNA-dependent systems; the 5′ terminal sequences of several RNA phages have been determined, and all start with *ppp*G*p* . . . . (see Dahlberg, 1968). Roblin (1968) has sequenced the 5′ terminus of R17 RNA, which proved to be *ppp*G*p*X*p*Y*p* . . . . , where

X is a purine and Y a pyrimidine; this would exclude the possibility that a translation initiation codon occurs directly at the end of the chain.

The size difference alone between the DNA of the genome and the much smaller RNA species implies that there must be many points where transcription can commence. However, synthesis of RNA does not start randomly at any point along the genome, but there appear to be specific sites of attachment where RNA polymerase initiates transcription. Work on the control of protein synthesis in bacteria has defined certain regions (promoters) which probably correspond to these sites (see pages 232–234)—these are of the order of length of around a hundred nucleotide base pairs. The 13–14S form of the polymerase would cover roughly thirty base pairs when bound to DNA. Measurements of the amount of polymerase which can bind to a DNA template have shown that, under suitable conditions, it is considerably less than would be expected from a random binding. Pettijohn and Kamiya (1967) found that large amounts of polymerase would bind to polyoma virus DNA (about eight cistrons in size) in low ionic strength solutions, the stoichiometry suggesting that binding was limited only by space on the DNA. At higher ionic strengths, the maximum amount of enzyme bound was reduced to a level corresponding to from four to seven polymerase molecules for each polyoma genome, that is of the order of the number of cistrons. It seems likely that only this more stable binding represents attachment to true initiation sites.

Pyrimidine-rich clusters have been found to be asymmetrically distributed between the two strands of DNA isolated from a variety of organisms, including bacteriophages, bacteria and mammals (for review see Szybalski, Kubinski and Sheldrick, 1966). Clusters of deoxycytosine on one strand were detected by their rapid reaction in complexing with poly-G, although the lack of simple stoichiometry in the reaction does not permit exact measurement of the length of the *d*C cluster. Resistance to ribonuclease action and thermal stability data indicated that the lengths of such runs are around 10–50 residues, probably with a few other bases interspersed within them. Shearing data suggested about one cluster for every few million molecular weight, that is for every few cistrons. Szybalski has proposed that such clusters may be the initiation points for messengers, possibly because they might have a unique secondary structure with a high affinity for the polymerase. Indeed, X-ray crystallographic studies suggest that homopolymer sequence runs in DNA may have a structure different from that of the normal duplex (Richardson, 1969). It is not necessary to postulate that the recognition regions are transcribed, so no restriction is placed upon the initial sequence of the messenger.

Burgess and coworkers (1969) reported that it is possible to separate the RNA polymerase of *E. coli* into two components by chromatography on a phosphocellulose column. One component comprised a 'purified' enzyme which was able to transcribe calf thymus DNA almost normally, but was much less active when T4 DNA was provided as template. However, the activity of this *minimal enzyme* fraction was greatly enhanced by the addition of the second component—the *sigma* (σ) *factor*. This lacked enzyme activity itself, but somehow stimulated the minimal enzyme possessing the catalytic activity of synthesizing phosphodiester bonds. The factor must be capable of existing in a complex with RNA polymerase since it was purified with it through several fractionation steps prior to the phosphocellulose chromatography; as observed above, the *complex enzyme* consists of an association between a minimal enzyme ($\alpha_2\beta\beta'$) and a sigma factor to give the complete unit of $\alpha_2\beta\beta'\sigma$.

The sigma factor stimulates the synthesis of RNA not by exerting an effect upon the activity of the minimal enzyme in chain elongation, but rather by assisting it to initiate new chains. Travers and Burgess (1969) found that when they followed the effect of the factor on initiation by measuring the incorporation of a P^{32} label into RNA, the stimulation of total RNA synthesis was paralleled by a similar stimulation of initiation. Each sigma factor appears to be capable of re-use several times to assist initiation by more than one minimal enzyme. When the complex enzyme (minimal enzyme plus sigma factor) is added to a reaction mixture at low ionic strength, initiation on T4 DNA ceases within 8–10 minutes. At this point, one RNA chain has been started for each complex enzyme molecule added (the DNA–complex enzyme–nascent RNA complex is termed the *preinitiated complex*). If a large excess of minimal enzyme is added to this, there is renewed initiation, corresponding to the re-use of each sigma factor at least five times, and probably more.

This conclusion has been confirmed by experiments utilizing the antibiotic rifampicin, which inhibits RNA synthesis by interfering with the action of the minimal enzyme. The antibiotic inhibits RNA synthesis *in vitro*, and a rifampicin-resistant mutant has an altered RNA polymerase (Di Mauro and coworkers, 1969). The use of C^{14} derivatives of the antibiotic showed that it binds to the polymerase, but is unable to do so with the enzyme derived from the resistant strain. *In vitro* experiments showed that the rifampicin resistance of RNA synthesis depends only on the source of the minimal enzyme and not on the sigma factor. Rifampicin must act on one of the early steps in RNA synthesis since it does not inhibit RNA synthesis already under way when it is added; one possibility is that it may prevent stabilization of the enzyme–DNA initiation complex by the first purine nucleotide through competing for its binding site on the minimal enzyme.

Travers and Burgess (1969) reported that when rifampicin and minimal enzyme, derived from a resistant mutant, were added together to a preinitiated complex, there was a renewal of initiation. This did not occur when the minimal enzyme was derived from a strain sensitive to rifampicin. Since the enzyme present in the preinitiated complex was sensitive to the antibiotic, the sigma factor present in the complex must be able to stimulate the action of the added rifampicin-resistant minimal enzyme; this implies re-use of the factor.

There are two possible ways in which the sigma factor might act to stimulate initiation. One possibility is that it remains attached to DNA at the initiation site and assists further minimal enzyme molecules to bind at this site and commence synthesis of more RNA chains. Alternatively, the factor may be released from the DNA after initiation and then be free to combine with a minimal enzyme to yield a complex enzyme able to undertake initiation elsewhere. These possibilities were distinguished by initiating synthesis on phage ϕ80 DNA and then by adding an excess of T4 DNA together with further minimal enzyme. 83 per cent. of the RNA synthesized proved to be T4 specific; since RNA synthesis on T4 DNA is almost completely dependent on the sigma factor, this implies that the factor must have been released from the ϕ80 DNA and re-used at an initiation site on the T4 DNA. The lack of much factor-dependent RNA synthesis on the ϕ80 template suggests that the sigma factor *must* be released after one initiation before it can be used for another—otherwise, rather more synthesis would occur on the template to which it was first attached.

The equilibrium of association between sigma and minimal enzyme is probably highly favourable to complex formation under most conditions. The cycle of activity of the factor can thus be represented as shown in Figure 6.1. Its exact function in initiation is not known although it has been suggested that it may cause the start of a local unwinding of DNA necessary for RNA synthesis. Another function which it might exert would be to provide a specific recognition of certain initiation sites; in this case, there might be several factors available, the different species having specific affinities for different initiation sites. This could provide a mode of control of transcription, and is discussed more fully later (page 261). (And see also page 153.)

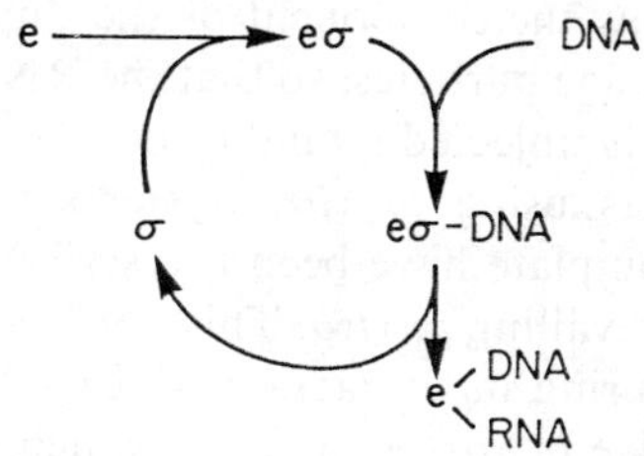

Figure 6.1. The sigma factor cycle. A free sigma factor (σ) associates with a minimal enzyme (e) to yield a complex enzyme (eσ) which initiates the transcription of RNA from DNA. After initiation, the sigma factor is released and the minimal enzyme continues to transcribe RNA. Because the sigma factor has a high affinity for the minimal enzyme, it probably remains free for only a short time before it associates with another minimal enzyme molecule. (After Travers and Burgess, 1969)

Models for transcription

Only one strand of the DNA duplex is employed for transcription into messenger in any particular region of the genome, but which of the complements is transcribed may vary with the gene concerned. Indeed, in T4 phage it is known that for some genes the strand transcribed lies on one chain of the duplex, whilst for other genes messenger synthesis is directed by its complement. McCarthy and Bolton (1964) found that the messenger species produced by *E. coli* could only anneal with about half of the denatured DNA, suggesting that transcription corresponds to only a single strand overall. Direct experiments have been performed on certain phage systems where it has proved possible to identify each of the two separate strands of the DNA duplex. Marmur and Greenspan (1963) denatured DNA from phage SP8 and separated the components by CsCl density gradient centrifugation into a denser strand (H) containing more pyrimidines and fewer purines than a lighter strand (L). Only the H strand hybridized with the RNA synthesized in the infected host. Hayashi, Hayashi and Spiegeleman (1963) used phage ϕX-174; the mature phage is a single strand of DNA, but is converted to a duplex replicating form (RF) by synthesis of its complement during infection. Hybridization tests and the base composition of the RNA synthesized in infected cells revealed only RNA complementary to one of the RF strands; this was the component of the duplex complementary to the strand of mature phage particles, so that the RNA which is synthesized bears the sequence of the injected strand only.

In many experiments using *in vitro* systems for transcription, both strands of the DNA template have been transcribed instead of the single-strand transcription prevailing *in vivo*. This appears to depend on damage caused to the DNA during its extraction, since sufficiently gentle extraction procedures can give *in vitro* systems in which only a single strand is transcribed. When ϕX-174 replicative form DNA was allowed to synthesize messenger *in vitro*, Hayashi, Hayashi and Spiegelman (1964) found that with the intact circular DNA duplex as template, only one strand was transcribed. But as soon as the circles were broken, RNA complementary to both strands was synthesized. It appears to be not circularity *per se* which is the critical feature, but the introduction of breaks in the DNA; even one single strand break seems sufficient to cause loss of asymmetric transcription. Thus the mechanism of strand selection cannot be by some control mechanism separable from the DNA and polymerase, but must be an inherent feature of the intact double-stranded species. Similarly, it has been found with transcription of phage T4 DNA *in vitro* that if it is prepared by gentle procedures, only messenger RNA corresponding to the early genes is produced. Rougher extractions result in transcription of both early and late genes. (The most recent paper on T4 transcription is by Milanesi, Brody and Geiduschek, 1969.) Here also, therefore, there is a control over gene transcription which is intrinsic to native DNA which has not lost its integrity. (But see also page 153.)

Two types of scheme for messenger production have been proposed. Stent (1958) suggested that there need not be unwinding of the DNA duplex, but a third polynucleotide chain could grow within its wide groove. This would lead to a transient three-chain structure, as shown in Figure 6.2. The specific sequence of this third chain would be governed by the base pairs of the DNA, each base pair being capable of forming another pair of H-bonds with only one other base if this third base is to be in the right orientation for formation of the messenger chain. Thus G–C would fit only with C, C–G only with G, T–A with A, and A–T with U, as shown in Figure 6.3. The third chain would unwind at some subsequent stage and be released to function as messenger. Zubay (1962) suggested a similar scheme on the basis of model building.

However, the model which is now commonly accepted proposes that transcription proceeds by a process analogous to that involved in DNA-polymerase catalysed DNA replication, when unwound single-stranded DNA is used as the template to direct formation of the

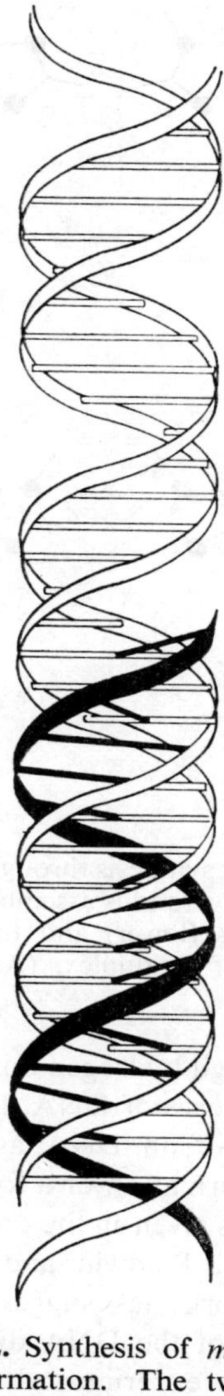

Figure 6.2. Synthesis of *m*RNA by triplex formation. The two DNA strands do not unwind and the messenger RNA grows within the wide groove of the double helix. (After Stent, 1958)

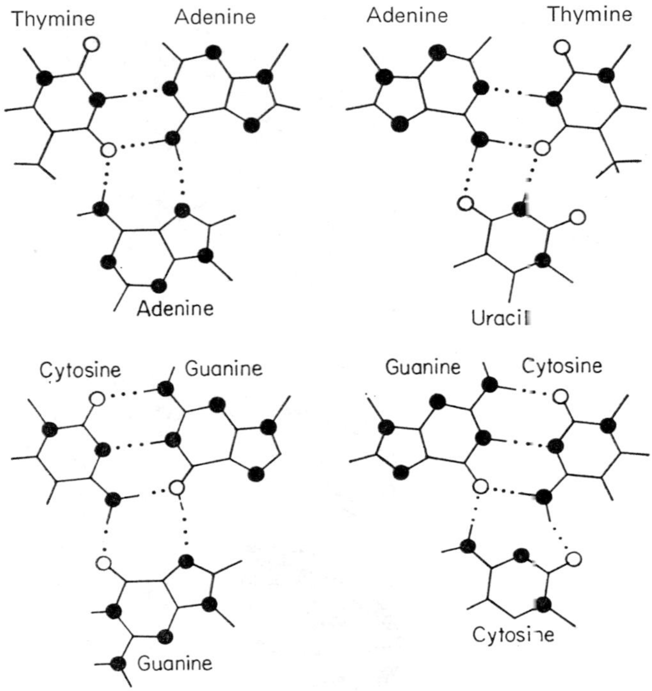

Figure 6.3. Messenger synthesis through base triplet hydrogen bonding. The *m*RNA strand is assembled by hydrogen bonding between its bases (lower) and the Watson–Crick base pairs (upper) of the duplex. (After Stent, 1958)

complementary strand through hydrogen bonds to the new bases. Various workers have isolated complexes of DNA, RNA and polymerase enzyme involved in synthesis of RNA from DNA, and it appears that the reaction does not proceed through formation of a long hybrid one-strand DNA—one-strand RNA complex as given in the annealing reaction (Bremer and Konrad, 1964; Chamberlin, Baldwin and Berg, 1963; Schulman and Bonner, 1962). Current work has suggested rather that the process involves a *local* unwinding of the DNA duplex just at the head of the RNA under synthesis, and experimental results on the behaviour of DNA–polymerase–RNA complexes can be interpreted in terms of the movement of this region along the DNA as synthesis proceeds.

Novack (1967) found that some 1·5 to 2 per cent. of the DNA in the complex formed by an *in vitro* system using *Bacillus subtilis* DNA and

Micrococcus lysodeikticus polymerase was resistant to DNAase, probably because the polymerase surrounds a portion of DNA in such a manner as to make it inaccessible to enzyme attack. Tongur and coworkers (1968) obtained a complex from *E. coli* (and other bacteria) either by phenol or by detergent extractions. Addition of actinomycin (which inhibits transcription) reduced its RNA content from the usual 10 to 3 per cent, and the intensity of incorporation of a short pulse dose (15 sec) showed that the specific activity of the bound RNA was very much higher than that of the total RNA; that is, the complex comprises DNA and nascent RNA. The RNA in the complex was resistant to RNAase, presumably reflecting its binding to DNA. The DNA was susceptible to DNAase attack, and after its degradation the RNA became sensitive to RNAase. Preparations of this complex contained some 5 to 7 per cent. of protein resistant to phenol and chloroform deproteinization. The possible role of protein as a stabilizing factor was investigated by incubation with trypsin or pronase, followed by treatment with RNAase; sensitivity to the latter was used as the criterion of whether the earlier treatment had destroyed the complex. Half the RNA was rendered sensitive, but the remainder was resistant even after complete protein hydrolysis. This would suggest that half the RNA is bound through protein and the other half directly, presumably through hydrogen bonds—these could represent RNA in different stages of its synthesis.

Hayashi (1965) and Hayashi and Hayashi (1968) examined the nature of the RNAase resistant part of the RNA bound in the DNA–RNA complex resulting from *in vitro* synthesis with ϕX-174 RF DNA. The sedimentation properties of the complex suggested that it comprised double-stranded DNA and single-stranded RNA. Although there was a gradual increase in the size of the RNA synthesized as transcription proceeded (shown by release of the RNA with formamide after various periods of incubation), the size of the resistant region remained constant. This region might represent either the tail of the growing RNA (at the initiation site on the DNA) or its head (the current growth point). In the former case, radioactively labelled RNA should not be chased out during growth, whereas in the latter situation the former resistant region should become sensitive to RNAase attack. The latter was found, with a resistant region about fifty nucleotides long moving along the DNA as the head of the growing chain.

Fuchs and coworkers (1967) proposed a model for RNA polymerase action in which the two strands are separated over only a small segment of the DNA at any time. The polymerase enzyme is envisaged as possessing two sites involved in transcription, the strand separation site and the strand exchange site, as shown in Figure 6.4. The first site separates the two strands of the template, and this would be necessary either with duplex DNA or with single-stranded DNA (in the latter situation the first round of transcription would produce a DNA–RNA duplex hybrid which would act as the template in subsequent rounds). However, with a double-stranded DNA template, the assembly of RNA by complementary base pairing in the unwound region is succeeded by an exchange of the newly synthesized RNA strand for the original complementary DNA strand. This exchange to restore the original duplex is not necessary with single-stranded DNA as the template. Overall, therefore, on duplex DNA

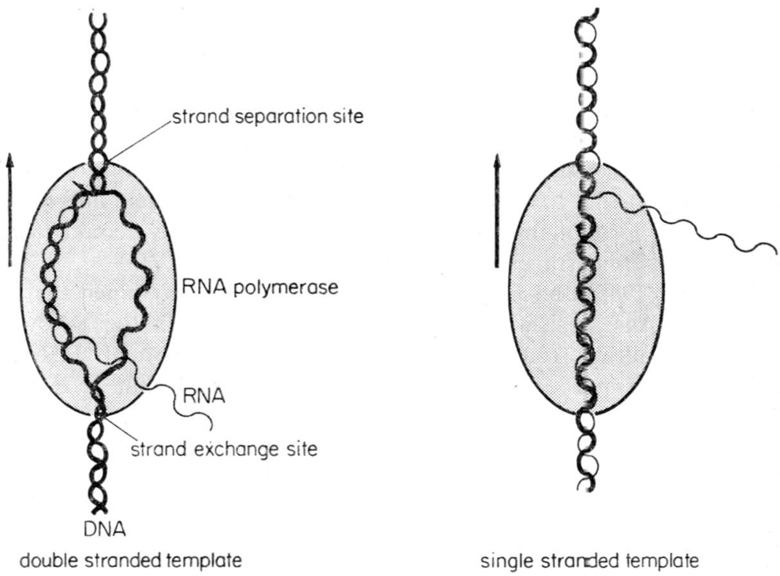

Figure 6.4. A model for transcription by local unwinding. The two strands of DNA are separated from each other at the strand separation site where they enter the polymerase enzyme (or for a single-stranded template the RNA made by the previous round of transcription is dissociated from its template). A complementary strand of RNA is synthesized on one of the unwound strands within the polymerase. The DNA leaves the polymerase and is reformed into a duplex at the strand exchange site, displacing the RNA if the template is a double helix; this site is not used if the template is single stranded. (After Fuchs and coworkers, 1967)

templates, transcription occurs at a locally unwound segment which can be visualized as progressing along the template during RNA synthesis.

This model was based upon a study of the effects of ionic species on transcription catalysed by *E. coli* polymerase. They found that ions appear to exercise two effects. First, divalent cations are required as cofactors; Mg^{2+} usually fulfills this role, but can be replaced by other ions with similar properties, such as Co^{2+} or Mn^{2+}. Secondly, ions are required for a function exercised when RNA is synthesized using double-stranded DNA as the template, but not when a single-stranded template is provided. They observed that the suppression of RNA synthesis at low ionic strength followed by its reappearance at higher levels, shown by duplex templates, was not found with single-stranded templates. If the effect of salt concentration is exerted on the strand separation step, its effect should be the reverse of that observed, because ionic strength increases the T_m. Thus they proposed that strand exchange is the reaction affected by ionic strength; this also accounts for its effect on duplex templates only. Inactivation at low ionic strength could be caused by a competitive displacement of the DNA strand by the RNA strand at the exchange site, but this would be reversed at a higher salt concentration. This is also consistent with the results of kinetic studies, which suggested that the inactivation and reactivation is due to arrested enzyme molecules restarting without being released from the template and having to reinitiate new chains.

Walter and coworkers (1967) found that with a duplex template there was a pronounced lag phase at low temperature or high ionic strength before transcription began, but this was not shown by single-stranded DNA. The lag phase did not appear to correlate with an effect of the enzyme binding reaction, and increased with a fall in temperature, bearing a close relation to the DNA melting curve. They suggested that this could result from the melting of a small segment of the double strand, that is the strand separation reaction. The lag phase increased sharply above a certain ionic strength, implying that a high energy of activation is required for this reaction. Its length could be shortened by a preincubation of the enzyme with the DNA at a low ionic strength, independent of the presence of the substrate nucleoside triphosphates, which would be consistent with involvement of a physical process. Transcription can thus be described in terms of the three sequential reactions:

enzyme+DNA	⟶	enzyme–DNA	(binding)
enzyme–DNA	⟶	enzyme–DNA*	(melting)
enzyme–DNA*+NTP	⟶	enzyme–DNA–RNA	(polymerization)

Release and transport of the messenger

When RNA is synthesized *in vitro*, it is not released from its template, but remains associated with the synthetic complex. Since polynucleotides inhibit the action of the polymerase, the RNA product itself causes synthesis to be restricted to comparatively small polynucleotide lengths. Both these effects can be explained if some component missing from the *in vitro* system is responsible for actively removing nascent RNA from its template *in vivo*. It has been suggested that the ribosome may fulfil this role, and there is some evidence that ribosomes may assist transcription *in vivo*, although their action is not exerted through translation of the messenger.

The sedimentation properties of the complex formed when RNA synthesis is directed *in vitro* by phage T4 DNA suggest that it may be associated with ribosomes (Byrne and coworkers, 1964). In studies using electron microscopy, this complex was shown to consist of long fibres 15–20Å across, with strands branching off; these strands had small particles attached to them (Bladen and coworkers, 1965). The data are consistent with the suggestion that the long fibres represent DNA, the branching strands RNA and the particles 70S ribosomes.

The addition of ribosomes to *in vitro* systems for synthesizing RNA alters their behaviour so that they more resemble the *in vivo* process. Shin and Moldave (1966) isolated and partially purified a deoxyribonucleoprotein (DNP) preparation from *E. coli*, which possessed both the DNA template and RNA polymerase activity, when the four nucleoside triphosphates were added. However, the nascent RNA produced was of only low molecular weight, remained attached to the complex during the polymerization reaction, and synthesis proceeded only for a short time at a slow rate. But after the addition of ribosomes, RNA synthesis continued for a longer time to yield a greater overall production; the species produced had a higher molecular weight; and a significant proportion was released from the complex in the form of ribosome–RNA complexes. Similarly, Jones and coworkers (1968) found that when ribosomes were added to a T7-*E. coli in vitro* system, the newly synthesized RNA sedimented in a separate peak with the ribosomes, instead of with its DNA template as hitherto. The effect of the ribosomes depended upon the stage of transcription at which they were added; when early (after 2 minutes), only 10 per cent. of the nascent RNA was released from its template, but once the RNA had reached its maximum length (at 20 minutes), the proportion rose to 50 per cent. This suggests that the messenger may have to reach some minimum length before attachment of the ribosomes can take place.

Although ribosomes do seem to assist the release of messengers, there is evidence that this does not demand that they translate the messenger (see also pages 275–276). When Shin and Moldave (1966) added the components required for protein synthesis to their system, no incorporation of amino acids could be detected, and the addition of chloramphenicol did not inhibit the stimulatory effect of the ribosomes. Revel and coworkers (1968b), using a T4-*E. coli* system, found that ribosomes exerted a marked effect on transcription when added under conditions enabling them to bind to RNA—the RNA molecules synthesized were longer, and more were released from the template. However, the same stimulation was exhibited by 30S particles in the absence of 50S subunits, excluding the possibility that translation might be involved. Similarly, Brown and Doty (1968) found that ribosomes assisted transcription when the initiation factor responsible for *m*RNA–ribosome binding was present, but not in its absence. It has been proposed that ribosome movement of some nature, even if not involving translation, might be required for transcription, but it seems unlikely that this could be achieved by 30S subunits alone. One possibility is that their attachment to the messenger stimulates transcription by preventing the RNA product from inhibiting the polymerase. Some scheme such as this would be supported by the finding of Shin and Moldave that the stimulation could be exerted by ribosomes from any system, and was not dependent upon their species of origin.

A problem which arises in higher organisms, although not in bacteria, is how messenger RNA is transported to the cytoplasm after it has been synthesized in the nucleus. Although newly synthesized *m*RNA can be detected in polysomes, it is not clear how and when it becomes attached to ribosomes. 'DNA-like' RNA (presumably messenger species) can be detected in both the nucleus and the cytoplasm, where it is characterized by a sedimentation rate which suggests that it is included in a ribonucleoprotein (RNP) complex. When the newly synthesized *m*RNA of mammalian cells is released from polysomes (for example, by treatment with EDTA), it sediments as an array of RNP complexes between 12S and 60S, mostly around 40S (Henshaw, 1968; Perry and Kelly, 1968). Their buoyant density values suggested a ratio of about 40 per cent. RNA, 60 per cent. protein. The protein components involved in these complexes have been shown to have electrophoretic properties indicating a lower basicity than histones or ribosomal proteins (Parsons and McCarty, 1968). Perry and Kelly (1968) found that when protein synthesis was blocked by puromycin, *m*RNA continued to emerge from the nucleus and form an RNP complex as usual; thus complex formation does not require the polysomes to participate in protein synthesis. So far, however, it has not proved possible to eliminate the possibility that the RNA structures may be artifacts of isolation procedures, and in any case, their exact role in transport through the nuclear membrane and in polysome function is not clear.

The visualization of transcription

We cannot do better to conclude this chapter than discuss the impressive electron micrographs of DNA in the throes of transcription taken by Miller and Beatty (1969a, 1969c). This work made use of a system in which extrachromosomal DNA coding only for ribosomal RNA (so far as is known at present) is engaged in very intensive transcription. Amphibian oocytes (eggs) pass through a lengthy growth period which in *Xenopus laevis* may be some six months and in *Trituris viridescens* may be up to two years. During this period, the nucleolar organizer region of the genome (chromosomal DNA associated with the nucleolus and coding for the synthesis of ribosomal RNA—see pages 171–173) is multiplied to produce about 1000 nucleoli within each nucleus. These nucleoli comprise a compact fibrous core (containing DNA, RNA and protein), surrounded by a granular cortex (containing only RNA and protein). By using a solution of very low ionic strength, the core and cortex can be separated, and the core dispersed for electron microscopy.

When the cores are maximally unwound, each consists of a thin axial fibre some 100–300Å in diameter, periodically coated with a matrix [see Figure 6.5 (Plate III)]. The axial fibre of each core forms a circle; treatment with DNAase breaks the core axis, and treatment with trypsin reduces its diameter to about 30Å. This suggests that the axis is a single duplex of DNA (the diameter of a DNA double helix is 20Å) coated with protein.

The length of the core axis covered by one matrix segment (2–3μ) is in good agreement with estimates for the length of DNA required to code for the expected ribosomal precursor molecule. Each matrix consists of about a hundred thin fibrils connected by one end to the core axis, and increasing in length from the 'thin' to the 'thick' end of the unit. After administration of a tritiated ribonucleoside, the label is found in the matrix. Each fibril terminates upon the core in a spherical granule of about 125Å in diameter; this almost certainly must represent an RNA polymerase molecule engaged in transcription. (This means that about one-third of the total length of the gene must be covered in enzyme molecules—a high concentration.) Treatment with ribonuclease or trypsin removes the fibrils from the core, implying that they are ribonucleoprotein (RNP) in composition. Protein-specific staining shows that as the RNA molecules are synthesized, newly made portions are immediately coated with protein. An RNA chain corresponding to a gene 3μ long would be about 6μ in length; since the maximum fibril length observed (at the 'thick' end of the matrix) is $0{\cdot}5\mu$, the chain must be coiled so as to give a roughly twelve-fold reduction in its length. The pattern of a hundred fibrils steadily increasing in length along the matrix indicates that successive RNA chains must be sequentially initiated before the completion of earlier chains.

Note added in proof

Since this book went to press, it has become apparent that the sigma factor plays an important role in the control of transcription. The activity shown by the core enzyme in transcribing DNA *in vitro* (from calf thymus or from phages T4, λ, or *fd*) appears to represent the transcription of random regions of the DNA template. The addition of the sigma factor directs the attention of the core enzyme to the transcription of only certain genes. Bautz, Bautz and Dunn (*Nature*, **223**, 1022, 1970) have shown that, with T4 DNA *in vitro*, the sigma factor stimulates the *E. coli* core enzyme into transcribing specifically those genes of the phage which are usually transcribed *in vivo* during the first minute after infection (the early genes). Sugiura, Okamoto and Takanami (*Nature*, **225**, 598, 1970) reported that when the double-stranded RF form of the single-stranded DNA phage *fd* is used *in vitro*, the core enzyme transcribes RNA species randomly from either of the two strands. When the sigma factor is added to the incubation medium, transcription is restricted to that strand of the phage which is transcribed *in vivo*. The relative properties of the RNA species synthesized under these two conditions suggest that the effect of the sigma factor is to restrict the activity of the core enzyme so that it commences transcription at only certain, specific sites on the DNA template; presumably, these correspond to the promotors at which transcription is initiated *in vivo*.

CHAPTER 7

The ribosome

Ribosomal nucleic acid

Three types of RNA are found in all ribosomes. Each subunit possesses a major species; RNA from the bacterial 30S particle is characterized by a sedimentation value of 16S, and that of the 50S subunit by a value of 23S. This would correspond to an amount of RNA in the large subunit of about twice that present in the smaller. In higher organisms, the RNA from the small subunit sediments at 18S and that from the larger at 28S. In addition to the two major species, a very much smaller RNA which sediments at 5S is present in the larger subunits of both bacterial and mammalian ribosomes; this species comprises some 120 residues in a definite structure whose nucleotide sequence has been determined for *E. coli* and KB cells.

Many workers have suggested that each of the major RNA components comprises a single covalently linked chain; the sedimentation characteristics of bacterial ribosomal RNA molecules would correspond to a chain length of about 1500 nucleotides for the smaller ribosomal RNA, and about twice that for the larger. For example, Bruening and Bock (1967) subjected yeast ribosomal RNA to heating at 80°C for one minute in order to rupture any non-covalent bonds. After this treatment, the RNA from each subunit sedimented as would a single polynucleotide chain. However, chemical estimations of the number of 3′-hydroxyl termini have been made by their oxidation with periodate and subsequent reaction with C^{14}-isonicotinic hydrazide—INH (McIlreavy and Midgley, 1967; Midgley and McIlreavy, 1967a, 1967b). The radioactivity bound suggested that both the 16S and 23S *r*RNA molecules from *E. coli* have the same single chain length of 1500–1600 nucleotides. Two possibilities were proposed to account for this. Both species might indeed be the same length but have different features in their secondary structures which result in their different sedimentation properties. Alternatively, there might be two chains associated together in the large subunit, although their dimerization does not involve the usual form of covalent linkage. In yeast *r*RNA, determination of the proportions of terminal nucleotides has suggested that the 26S species consists of two covalently linked 16S-like species (Van den Boas and Planta, 1968).

One feature of *r*RNA is that it has an unusually high methyl group content. Brown and Attardi (1965) studied the incorporation of radioactively labelled methyl groups into Hela cell *t*RNA and *r*RNA. Analysis after hydrolysis showed that only 20–24 per cent. of these were attached to the bases, with the major portion present on the ribose moiety of the nucleotide, probably esterified as 2′-0-methylribose. The two major components were not equally methylated; the 18S contained some 1·5 times more methyl groups than the 28S. In *t*RNA, by contrast, the major portion of the methyl groups (80 per cent.) was attached to the bases, and only a minor amount to the ribose.

There are several cistrons representing *r*RNA in bacterial species and a high multiplicity of these in higher organisms. This has raised the problem of whether all the *r*RNA molecules are identical or whether there is heterogeneity between ribosomes. Gould (1967) argued that, if there is a multiplicity of *r*RNA types, there should be a large number of fragments resulting from digestion with nuclease. However, with reticulocyte 28S RNA, there was a high homogeneity amongst such fragments, imposing an upper limit of only three different *r*RNA species. Fellner and Sanger (1968) studied the nucleotide sequences around the methylated sites in 23S and 16S RNA in *E. coli*. This showed that methylation must be highly specific and occur at only a small number of sites. The sequences around these methylated sites suggested that both the 16S and 23S *r*RNA species are largely homogeneous (although distinct from each other). All the major methylated oligonucleotides occurred twice in the 23S, suggesting that it is composed of two sections which, if not identical, display considerable homology. It is possible that it comprises two identical or similar polynucleotide chains, or a single chain with a duplicated sequence—this could have arisen during evolution by gene duplication. Certain of the methylated sequences of the 16S were also repeated, although the degree of repetition was less. Since some repetition occurs in 5S RNA also, it may be a general property of ribosomal nucleic acid. The finding by Sugiura and Takanami (1967) that the two major ribosomal RNAs of *E. coli* have unique sequences at their 5′ end would also argue against heterogeneity between ribosomes. An examination of various other species showed that these too possess considerable homogeneity at this terminus.

Some evidence has been cited for general similarities in the ribosomal RNA of several species, irrespective of the degree of difference between their genomes. Miura (1962) and Midgley (1962) examined the nucleotide composition of RNA from a variety of bacteria with genomes with a range of GC/AT ratios varying from 1·75 to 0·6. Only a rapidly labelled species (*m*RNA) showed the same overall base composition as the DNA. Whilst *t*RNA did not show the full range of GC/AT ratios, it did exhibit some dependence upon the nucleotide composition of the DNA. However, all the *r*RNA species showed a similar base composition, and there was no obvious variation in this which could be correlated with the GC/AT ratio of the genome. (This implies that the segments of the genome which code for *r*RNA must be atypical in GC/AT ratio compared with the remaining DNA). Cross hybridization experiments have suggested that there is a fair degree of homology between *r*RNA molecules from closely related organisms, both in bacteria and in plants (Moore and McCarthy, 1967). The unique 5′ terminal sequences of bacterial *r*RNA molecules obtained by Sugiura and Takanami (1967) also tended to be fairly similar between related species. Presumably, the retention of these common characteristics in *r*RNA, despite gross differences in genome, reflects some common evolution of the protein biosynthetic machinery.

5S RNA was first reported as a distinct ribosomal species by Rosset and Monier (1963). Subsequently, Comb and Sarkar (1967) showed that it is bound to a specific site on the larger ribosomal subunit, and Kaempfer and Meselson (1968) that the attachment is permanent: once 5S RNA is linked to the subunit it remains so and does not undergo exchange with other ribosomes. There is one 5S RNA for every 28S RNA in Hela cell extracts, that is one for every 50S subunit, and none free in the cytoplasm (Knight and Darnell, 1967). The 5S RNA is very tightly associated with the ribosomal subunits since treatment with LiCl, NH_4Cl, CsCl, or deoxycholate in the presence of Mg^{2+} ions, does not dissociate it (although such treatment removes a considerable amount of protein from the ribosome structure). However, Siddiqui and Hosokawa (1968) and Morell and Marmur (1968) found that treatment with NH_4Cl or CsCl under conditions of low Mg^{2+} ionic strength (which unfolds the structure of the ribosome) releases the 5S RNA. Treatment with EDTA, which also causes unfolding, produces similar results.

The nucleotide sequence of the 5S component from *E. coli* was determined by Brownlee, Sanger and Barell (1967, 1968). Unlike *t*RNA, it contains no minor bases. In both the strains of *E. coli* studied there were two 5S species, one of which was common to both strains and the other derived from it by a single base mutation. There must thus be more than one gene responsible for its synthesis, and since there are two 5S species and only one 5S molecule per ribosome, this implies the presence of two types of ribosome in the cell (although this is unlikely to have biological significance). Two base sequences—one of eight and one of ten residues—were repeated twice in the molecule, and if the structure is written so that these are aligned there is a fair degree of homology, suggesting that 5S RNA may have arisen by gene duplication. 5S RNA from human tumor KB cell ribosomes was sequenced by Forget and Weissman (1969), and proved to be either 120 or 121 nucleotides long. Two sequences of nine residues showed complementary pairing, and two other sequences (of six and seven nucleotides) were repeated—but these were in the same, not different halves as in *E. coli.* There was little homology between the mammalian and bacterial species. Williamson and Brownlee (1969) reported that 5SRNA from each of two mouse cell lines in culture shows the same T1 and pancreatic ribonuclease digest fingerprints as that from KB cells; this suggests that its sequence is probably the same both in man and mouse (although the presence of small differences cannot be excluded by this analysis).

Other secondary structures have been proposed for *E. coli* 5S RNA since the original determination of its sequence. Cantor (1967, 1968) reported that optical studies suggested extensive double-stranded regions, and proposed a model with 40–49 base pairs, that is involving 80–98 of the 120 nucleotides in complementary hydrogen bonding. This is illustrated in Figure 7.1. Data obtained by Cramer and Erdman (1968) on the extent of A–U base pairing, by studying the availability of adenine residues for oxidation, would support the Cantor model.

Aubert and coworkers (1968) showed that treatment of native 5S RNA from *E. coli* with urea or heat in the absence of Mg^{2+} could achieve a fractionation into two forms (A and B) which were distinguished by their different 3D conformations. The A form retained part of its binding activity for ribosomes, but the B form had no such activity; heat treatment of the B form gave a species with much the same activity as the A form, so that the $A \rightleftharpoons B$ transition seems to be reversible. Scott and coworkers (1968) studied the optical properties of these two forms by hyperchromocity, optical rotatory dispersion (ORD), and circular dichroism (CD). The A form proved to have some 38–39 base pairs, which is similar to the native form, but although the extent of hydrogen bonding is similar the actual arrangement is less stable. The B form proved rather different, with only some thirty-one base pairs.

Ribosomal proteins

There is now considerable evidence that the protein component of ribosomes consists of a large number of protein molecules; assuming an

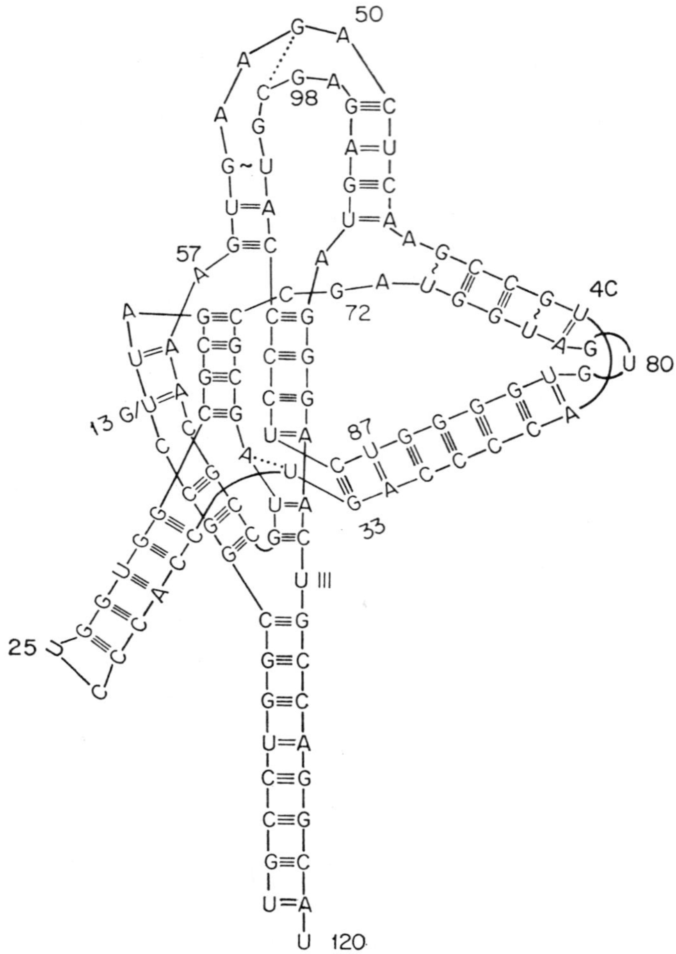

Figure 7.1. The two-dimensional projection of a highly base-paired model for 5S RNA. 80–98 of the 120 nucleotides are involved in complementary hydrogen bonding. Conventional base pairs are indicated as (G≡C, U=A) and unusual combinations as (G~U). (After Cantor, 1967)

average molecular weight of 25,000 would imply a total of about forty proteins in the *E. coli* ribosome, fourteen in the 30S and twenty-six in the 50S subunits. The proteins comprising the two subunits are different, and it appears that ribosomes possess one copy of each of a number of different proteins rather than many copies of a few protein species. Some of these proteins have recently been identified as playing specific roles in the functioning of the ribosome. In addition to these species, there are also the various protein factors involved in protein synthesis, which may become associated with the ribosome.

Waller (1964) isolated proteins from purified 70S *E. coli* ribosomes and identified twenty-four bands by starch gel electrophoresis. The band patterns from each subunit were quite different, although the majority of proteins in each showed similar basic behaviour. Otaka, Itoh and Osawa (1968) labelled 50S subunits with H^3-lysine and the 30S with C^{14}-lysine and then analysed them simultaneously on carboxymethylcellulose (CMC) columns. Fractions obtained on the columns were checked against acrylamide gel electrophoresis and shown to be homogeneous. The results demonstrated that *E. coli* (strain Q13) subunits have no common components (except possibly one or two very minor species). Traut (1966) found twenty-one proteins in the larger *E. coli* subunit (seventeen were basic and four acidic) and thirteen in the smaller subunit (eleven basic and two acidic). It has been suggested that this electrophoretic heterogeneity might arise as an artifact through interactions amongst a smaller number of polypeptide chains, but this possibility was excluded by Traut's results, and also by Traub, Nomura and Tu (1966). Similar results showing heterogeneity of ribosomal proteins have been reported from work on *Neurospora crassa* (Alberghina and Suskind, 1967).

Recent work has resulted in purification and analysis of some of the ribosomal proteins. Traut and coworkers (1967) obtained amino acid compositions and tryptic fingerprints from several purified *E. coli* ribosomal proteins; both criteria suggested that the different species were chemically distinct. Fogel and Sypherd (1968) separated proteins from the smaller subunit by preparative polyacrylamide gel electrophoresis and checked their homogeneity by molecular sieve chromatography. Amino acid analyses demonstrated that each of the fifteen proteins found had a distinct composition, and an analysis of the data excluded the possibility that some of the larger proteins might have arisen by aggregation of smaller species. Eleven out of these fifteen were also shown to have unique tryptic fingerprints. Moore and coworkers (1968) purified thirteen proteins from the *E. coli* 30S subunit by CMC followed by Sephadex column chromatography. Molecular weights were determined by sedimentation equilibria, and the proteins fell into the three groups, 4500–5500 (three), 9000–14,500 (seven), and 20,000–27,500 (three). Their average molecular weight proved to be 14,100; that of the 50S subunit proteins was somewhat higher at 15,200. Amino acid compositions confirmed that they were distinct species. The estimation of molecular weights by this method produces a somewhat lower value than the previous estimate (25,000) based upon end group analyses; this might be because some ribosomal proteins have blocked ends. A quantitative estimation of each of the proteins by counting the radioactivity associated with bands on polyacrylamide gel electrophoresis suggested that there is one molecule of each protein for every 30S subunit.

Leboy, Cox and Flaks (1964) confirmed that this heterogeneity is under genetic control by electrophoresing ribosomal proteins from various *E. coli* strains on polyacrylamide gels. They found that a single protein from the 30S subunit of strain K12 has a lower electrophoretic mobility than its counterpart in other strains. The locus governing its synthesis was mapped by genetic conjugation and by transduction with phage P1, and proved to be located very close to another locus (the streptomycin sensitivity site) which is also thought to elaborate a ribosomal protein (see page 215). Further work (Birge and coworkers, 1969) has shown that the corresponding proteins from *E. coli* B and K both have molecular weights of about 22,000–25,000. This suggests that a difference in the charge composition of their amino acids is likely to be responsive for their different electrophoretic mobilities rather than a difference in size. An amino acid analysis showed significant differences in their composition with respect to eleven amino acids; the B strain protein has three more arginines and one more lysine than that from the K strain; this could account for the change in mobility.

It is also worth noting that, in contrast to the similarities retained between the ribosomal nucleic acids of different bacterial strains, the *E. coli* ribosomal proteins show little similarity with any of the *Bacillus* strains, although there is a somewhat greater similarity with *Salmonella abony* (Otaka, Itoh and Osawa, 1968).

The structure of the ribosome

Although there has been controversy as to whether all the ribosomes within a cell are identical, or whether there are various types, models have been proposed for ribosome structure in generalized terms of its behaviour as a ribonucleoprotein particle. Gavrilova, Ivanov and Spirin (1966) found that when *E. coli* ribosomes were washed in magnesium-free NH_4Cl, they dissociated into their subunits with the loss of a considerable part of their endogenous magnesium ions. Reduction of the ionic strength of the NH_4Cl then caused conformational changes, as shown by decreased sedimentation coefficients. These are shown in Figure 7.2. There was no fragmentation of *r*RNA or loss of protein, so these changes appear to represent an unfolding of the particles into a less compact form, possibly because the NH_4^+ ions initially substitute for the Mg^{2+} so that with their removal the internal ionic bonds break down. Discontinuous changes occur from 50S to 35S and thence to 22S, and these are presumably due to specific unfoldings of the structure. The smooth changes within these unfolded classes are probably due to interactions depending upon the polyelectrolyte nature of the nucleic acid. All the changes except the initial one are reversible. These changes can be interpreted as being due to the disruption of the structure of a specifically folded single ribonucleoprotein strand.

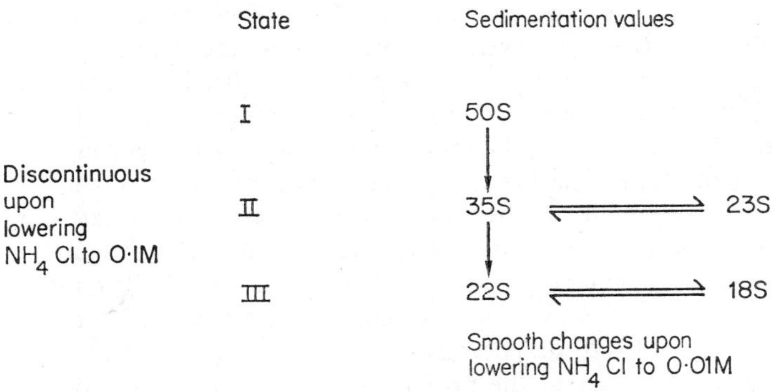

Figure 7.2. Structural changes in the 50S subunit. (After Gavrilova, Ivanhov and Spirin, 1966)

The dye acridine orange can be bound to RNA, and the change in its spectrum with an increasing polynucleotide/dye ratio is related to the amount of double helical content of the nucleic acid. Furano, Bradley and Childers (1966) performed such studies on ribosomal RNA free in solution and bound in ribosomes. Their results suggested that there is little double helical content in the ribosome, and this is less than that present when the *r*RNA is free in solution. They found that some 90 per cent. of the RNA phosphates could bind to the dye, and this, together with other data on the interaction of the nucleic acid and dye, suggests that the nucleotide phosphates are located at or near the ribosomal surface. This would imply a model for the ribosome in which the RNA is maintained in a single-stranded configuration with its phosphates exposed to the solvent. This, and the strong dependence of subunit association on Mg^{2+} concentration, suggests that the subunit particles may be linked to each other by Mg^{2+} salt bridges between their respective RNA strands. A similar interaction could be responsible for binding *m*RNA, so that some relatively non-specific interaction of this nature might provide the basic mechanism for its affinity, with other factors influencing and controlling the exact nature of the interaction.

Cotter, McPhie and Gratzer (1967), working with yeast ribosomes, found that the temperature melting curves of free extracted *r*RNA and ribosomal subunits virtually coincided. This implies that the RNA in the ribosome has a conformation very similar to that exhibited when it is free in solution. Temperature difference spectra showed that extracted *r*RNA consists of a heterogeneous population of helices of varying composition which melt over a wide temperature range; the regions melting out become progressively richer in G–C content. On the basis of data obtained from synthetic polynucleotides, the total double helical content was estimated as some 60 per cent. (this is probably a lower limit). This conclusion was confirmed with infrared studies (Cotter and Gratzer, 1969). Various factors were proposed to account for the difference in this estimation from that of Furano, Bradley and Childers, (1966); one possibility is that the conditions used by the latter workers might have partially or wholly disrupted the ribosome structure.

When basic polymers, such as ribosomal proteins, bind to nucleic acid, its double helical state is strongly stabilized, and the melting temperature raised by some 15–20°C. The absence of any such stabilization in these experiments showed that the proteins of the ribosome are not bound to the helical regions of its RNA. Dye binding experiments with acridine orange confirmed the conclusion of Furano, Bradley and Childers that a large proportion of the RNA was available for reaction, again suggesting that the surface of the ribosome must consist substantially of nucleic acid rather than protein. The greater part of the protein must be bound to the non-helical regions, so that the exposed RNA should be largely double helical. This concept is supported by the resistance of the intact ribosome to ribonuclease (which is known to act preferentially on single-stranded RNA). The model proposed on this basis is illustrated in Figure 7.3.

Dissociation and reconstitution of the ribosome

Much recent work has concentrated on removing certain specific proteins from the ribosome and then testing for biological activity both

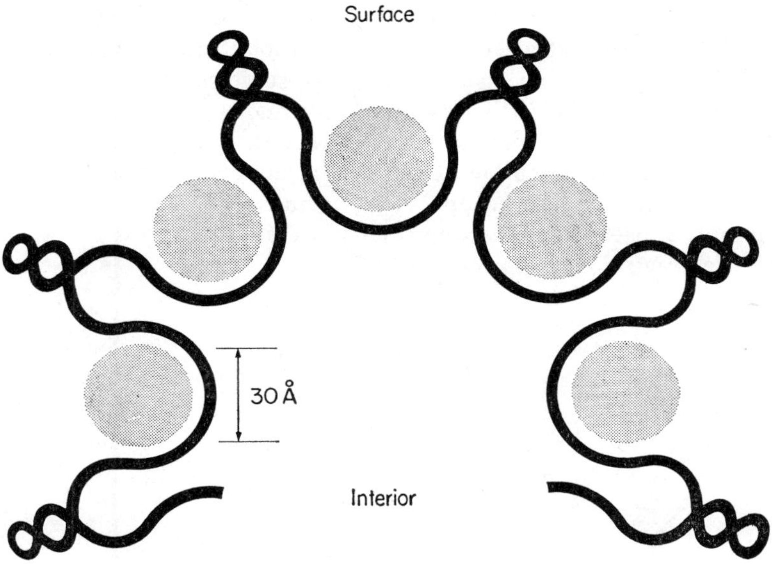

Figure 7.3. A model for the ribosome nucleoprotein structure. The RNA exposed at the surface of the ribosome is double-stranded; single-stranded regions of RNA are associated with internal proteins (shaded areas). (After Cotter, McPhie and Gratzer, 1967)

the remaining particle and that reconstituted by addition of the protein(s) removed. This has shown that, under appropriate conditions, functional ribosomes may be assembled directly from their macromolecular components; some of the details of how this assembly occurs have been worked out. The formation of *subribosomal* particles in reconstitution mixtures which lack some protein component of the ribosome has enabled the role played by the missing protein to be identified; it is indicated by the aspect(s) of protein synthesis in which the reconstituted particle is defective. In addition, the use of heterologous reconstitution mixtures (in which the ribosomal components may be derived from different organisms) has shown the extent to which the properties of these various components have been conserved in evolution between different bacterial species.

The initial work which made this approach possible was the finding of Meselson and coworkers (1964) that density gradient centrifugation of 50S (or 30S) subunits in CsCl yields 42S (or 23S) particulate *cores* by the loss of a discrete amount of protein (the *split proteins*). More detailed studies (Itoh, Otaka and Lerman, 1968; Osawa and coworkers, 1966) have since shown that when ribosomes are subjected to high concentrations of CsCl or LiCl, they suffer a series of successive disruptions, each involving the loss of a further group of proteins, as shown in Figure 7.4. The discrete release of protein suggests that the dissociation does not represent a gradual liberation of various protein components, but rather that the ribosome has groups of cooperatively organized proteins; disruption of each ordered group results in the more or less cooperative loss of large groups of molecules together.

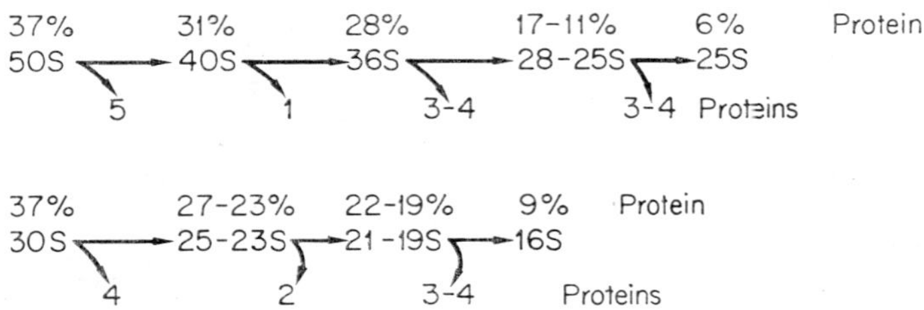

Figure 7.4. Discrete dissociations of ribosomal proteins by addition of high LiCl concentrations. (After Itoh, Otaka and Osawa, 1968)

The stepwise dissociation can easily be reversed in the absence of CsCl, provided that Mg^{2+} ions are present. Ribosomes so reconstituted possess appreciable biological activity (Spirin and Belitsina, 1966). Staehelin and Meselson (1966a) reported that cores could be reconstituted by addition of split proteins to yield ribosomes which are active in cell-free systems for protein synthesis, and Nomura and Traub (1968) demonstrated that each reconstituted particle consists of one equivalent of each of the core and split proteins. The reconstitution is specific for the proteins split-off; the addition of 50S split proteins would not restore activity to 30S cores and vice versa (Hosokawa, Fujimura and Nomura, 1966). The proteins split-off from the 50S subunit by CsCl centrifugation have been termed the SP50 fraction, and the corresponding species from the 30S subunit the SP30 fraction.

In further work, Traub and Nomura (1968b) dissociated the 23S cores of the 30S subunit into their component core proteins (the CP30 fraction) and the 16S *r*RNA. Subunits were then reconstituted by mixing the 16S *r*RNA with the CP30 proteins at 37°c and subsequently adding the SP30 fraction in the cold. After dialysis and sedimentation by high speed centrifugation, the particles recovered had a biological activity virtually equivalent to that of native 30S subunits when tested for protein synthesis in either the poly-U system or under direction from phage f_2 RNA. Gel electrophoresis showed that the reconstituted particles had the same protein composition as the native 30S subunits. None of the 30S proteins could be substituted by other species at any stage of the process. The ability of mature ribosomal components to undergo this autonomous reassociation to yield a functional unit suggests that the information required for assembly of the ribosome is contained in the structure of its macromolecular components.

The kinetics of the reconstitution process were followed by incubating standard reconstitution mixtures at various temperatures, recovering the particles, and assaying them for activity in an *in vitro* system at 37°C (Traub and Nomura, 1969). Few active particles resulted from the 10 and 20°C low temperature incubations, and centrifugation of these reaction mixtures produced inactive species termed RI (reconstitution intermediate) particles. These lack several of the 30S subunit proteins, which can then be recovered from the supernatant (the S proteins). The proteins present in the RI particles are similar to those of the 23S core particles, and the S proteins resemble those of the SP30 fraction, although there are minor differences. The S fraction comprises seven to ten proteins, and although these are not capable of binding to free 16S *r*RNA, when heated together with RI particles at 40°C for 20 minutes, almost fully active 30S subunits resulted. This suggests that the reconstitution process takes place through discrete steps, the addition of S proteins occurring later in the sequence.

After heating the RI particles alone, cooling to 0°C and adding the S proteins at the low temperature also gave active 30S subunits. Although the overall assembly of the 30S subunit involves some twenty components, the rate-limiting reaction appears to be unimolecular; this probably comprises a structural rearrangement of the RI particle, and it is this step which is permitted to take place by the increase of temperature from 0 to 40°C. Thus the reconstitution process can be represented as:

$$16S\ r\text{RNA} + \text{RI proteins} \longrightarrow \text{RI particles} \xrightarrow{\text{heat}} \text{RI*} \xrightarrow{\text{S proteins}} \text{30S subunits}$$

where RI* is the active intermediate particle and its formation is the temperature-dependent rate-limiting step. It seems likely that the stepwise reconstitution of the 30S subunit described above mimics an *in vivo* process for self-assembly.

Dialysis of 30S subunits against a solution containing mercaptoethanol and the chelating agent EDTA for 12 hours in the cold removes magnesium ions from ribosomes. The resultant particles sediment at 13S; although they are intact in that they retain all their components, they have been unfolded. When these unfolded particles were subjected to the same procedure used for the reconstitution of 30S subunits from 16S *r*RNA and proteins, they were converted into active 30S subunits. However, upon incubation at low temperature, the conversion was prevented and instead yielded particles—termed RI(u)—lacking some of the 30S proteins and only weakly active in protein synthesis. The missing protein components were present in the supernatant, and heating these S(u) proteins with the RI(u) particles gave active 30S subunits in a manner similar to their formation from S proteins and RI particles. In fact, the RI and RI(u) particles proved to be functionally identical, and the S and S(u) fractions capable of substituting for each other in the reconstitution process. This suggests that reconstitution—and thus probably self-assembly *in vivo*—involves a step in which a structural conversion is effected by a folding of the particle structure.

The functional activities of reconstituted subunits

Traub and Nomura (1968a, 1968b) obtained split proteins from both subunit types by centrifugation on CsCl density gradients, and separated each of the SP50 and SP30 fractions on DEAE-cellulose into an acidic fraction (SP50A or SP30A) and a basic fraction (SP50B or SP30B). Gel electrophoresis confirmed that the pattern of proteins in each fraction was distinct; the SP50A and SP50S each comprised about four proteins, the SP30A two, and the SP30B some five protein components. All four split protein fractions were soluble in buffers without requiring addition of urea or high salt concentration.

Subribosomal particles were reconstituted from the core particles by addition of the various split protein fractions. For each of the subunits, each of its split protein fractions was added both separately and in the presence of the other. As a control, the unfractionated split proteins were added to the cores. Each reconstituted subunit was added to normal subunits of the other type and tested for various biological activities. Both reconstituted subunit types were tested for their activity in the *in vitro* poly-U system, and for their ability to bind *t*RNA (for the 50S in the absence of *m*RNA, for the 30S in its presence). In addition, the 30S were tested for their ability to bind to the poly-U messenger.

The results obtained with the reconstituted 50S are shown in Table 7.1.

*t*RNA can normally undergo a reversible binding to the 50S subunit without addition of *m*RNA—this is probably at the P site. SP50B is clearly required for this activity, although since it is not necessary for peptide bond formation, full activity in *t*RNA binding cannot be required for protein synthesis. SP50A is not concerned with *t*RNA binding, but appears to be essential for peptide bond synthesis.

Table 7.1.

Particle	Poly-U system	*t*RNA binding
40S core	Almost completely inactive	Almost no activity
40,A	Half activity of 40,AB	Almost no activity
40,B	Inactive	Half 40,AB activity
40,AB	Half activity of native 50S	30% activity of native 50S

For the reconstituted 30S particles, the specific binding of *t*RNA closely paralleled the incorporation of phenylalanine in the poly-U system, and the ability to bind poly-U showed a similar behaviour (see Table 7.2). For all activities, SP30B proteins appear to be indispensable and the SP30A not required, although in the presence of SP30B they cause stimulation of *t*RNA binding and poly-U directed incorporation ability.

Table 7.2

Particle	*t*RNA binding/poly-U activity	Poly-U binding
23S core	Inactive	Inactive
23,A	Inactive	Only weakly active
23,B	Half of 23,AB activity	Comparable to 23,AB
23,AB	Up to 85% of native 30S activity	Highly active

Traub, Soll and Nomura (1968) investigated the functions of the apparently dispensable SP30A fraction by examining in detail the differences between 23,AB and 23,B reconstituted particles. One possibility is that SP30A might be necessary for initiation (not required in the poly-U system). However, using f_2 *m*RNA it appeared that the relative activities of 23,B and 23,AB were the same as with synthetic messenger. The system responded to the initiation factors required for binding fmet-*t*RNA so that the ribosome structure necessary for formation of the initiation complex must be intact, and again, 23,B showed the same activity relative to 23,AB in binding fmet-*t*RNA to AUG or random poly-(A, U, G). The possibility that SP30A might be required to maintain correct reading of the code was excluded by the finding that the degree of error in reading the code was the same for both 23,B and 23,AB. They found that addition of the SP30A fraction about doubled the number of sites binding aminoacyl-*t*RNA; this is unlikely to be due to creation of a new site, since it would have to be one which the ribosome could function without. Thus this protein fraction probably affects the functioning of a site already present. The data are consistent with the idea that 23,B particles can assume any of several structures, all of which have the binding site although some are active and others inactive. In the presence of SP30A only the active form is structurally permitted in the 23,AB particle.

The SP30B fraction appears indispensable for all the activities of the 30S subunit, and this was investigated further by its fractionation into five proteins on a phosphocellulose column. The five types of reconstituted particle each lacking one of these five proteins were then tested for functional activity. As a control, the activity of 23,AB (1,2,3,4,5) particles was comparable to the 23,AB reconstituted particles or native 30S subunits. The deficiencies gave the results shown in Table 7.3.

Table 7.3.

Deficiency	Activity
B_1	Amino acid incorporation reduced to 10–40%
B_2	Retain partial activity for both *t*RNA binding and amino acid incorporation
B_3 or B_5	Almost completely inactive in both specific *t*RNA binding and amino acid incorporation, although normal capacity to bind poly-U messenger
B_4	Full activity in the poly-U system, but reduced (24–44%) activity in f_2-dependent binding of fmet-*t*RNA_f; this fraction may be required only for initiation and not elongation

23S core proteins show some polyacrylamide gel bands of the same mobility as B_1 and B_2, so it is not possible to exclude the possibility that the residual activities in these deficient particles are due to the presence of small quantities of these fractions. It is clear that all active ribosomes must possess proteins B_3 and B_5, so there can be no heterogeneity in the ribosome population with regard to these, at least.

The reconstitution process is highly specific for all components of the ribosome. Traub and Nomura (1968a) examined the specificity of the split protein fractions in restoring synthetic activity by assaying each of the sixteen reconstituted subribosomal particle types obtained by all possible combinations between the four split protein fractions and each of the four core particle types 23,A; 23,B; 40,A; 40,B. Stimulation was restricted to the appropriate combinations and was highly specific, confirming that each protein fraction has distinct functions which cannot be substituted by any of the other protein fractions.

There is a high specificity for the nucleic acid component of the ribosome as well as for its proteins; 30S particles reconstituted after treatment of their RNA with heat or alkali had no biological activity at all. Nomura, Traub and Bechmann (1968) constructed hybrid 30S particles by using heterologous systems in which the various ribosomal components were derived from different organisms. They found that none of 16S *r*RNA from yeast, 18S *r*RNA from rat liver, or a '16S' RNA derived by degradation of *E. coli* 23S RNA could substitute for the endogenous *E. coli* 16S *r*RNA in reconstituting active 30S particles. Indeed, the particles formed were physically distinct from the normal subunit, so the requirement for RNA appears to be more specific than merely for size. The normal *E. coli* 16S *r*RNA itself is highly susceptible to loss of activity upon chemical modification; only six to eight alterations induced by nitrous acid are sufficient to destroy the activity of reconstituted particles. This may be due to a loss of ability to fold into its normal compact structure since it still seems to be able to bind ribosomal proteins.

They also achieved reconstitution with 16S *r*RNA from various other bacterial species with a wide range of G–C content in their DNA and ribosomal RNA (*E. coli* has 50 per cent. G–C in the genome and 53 per cent. in *r*RNA). Both *Azobacter vinelandii* (66 per cent. G–C in its DNA) and *Bacillus stearothermophilus* (47 per cent. in its DNA, 61 per cent. in its *r*RNA) were as efficient in giving active reconstituted 30S particles as a homologous *E. coli* system. *Micrococcus lysodeikticus* (72 per cent. G–C in genome, 54 per cent. in *r*RNA) was not nearly so efficient, and gave particles which sedimented more slowly.

The properties of the two active hybrid systems were investigated in more detail. Homologous 30S subunits of each type were reconstituted and, in addition, the two hybrid types containing *r*RNA from one species and ribosomal proteins from the other. With *A. vinelandii* the homologous 30S particles (AA) were as efficient in their reconstitution as the *E. coli* (EE). The hybrid particles with 16S *r*RNA from *A. vinelandii* and ribosomal proteins from *E. coli* (AE particles) were as active as the two homologous (AA and EE) types, but the reverse heterologous combination with RNA from *E. coli* and proteins from *A. vinelandii* (EA) showed only 40 per cent. of this activity. Reconstitution for *B. stearothermophilus* was not very effective at the 40°C optimum temperature used for *E. coli*, but was efficient at around 55°C. BE particles (RNA from the *Bacillus*) were very active, but the reverse EB particles (proteins from the *Bacillus*) were only partially active, as, indeed were the BB homologous particles.

Thus some functional property of 16S *r*RNA has been conserved in evolution, in that it has the ability to interact successfully with sets of 30S proteins derived from a fairly wide range of bacterial species. Hybridization experiments showed that these *r*RNAs are distinct, so the requirement for a specific base sequence of the *r*RNA is not absolute. Of course, it is possible that only certain sequences interact with the proteins and these have similar base sequences, with the remainder of the RNA not conserved. Similarly, functionally equivalent ribosomal proteins from different species should have certain structural features in common.

Experiments with *B. stearothermophilus* suggest that non-conserved portions of the RNA are also important. This bacteria grows at the high temperature of 60–65°C, and its ribosomes are much more thermostable than those of *E. coli*. The thermostability of both types of ribosome, and of the hybrids EB and BE, was examined by heating reconstituted particles at 65°C for five minutes. The *E. coli* (EE) were completely inactivated, but the *Bacillus* (BB) retained 60–100 per cent. of their activity. Both the hybrids proved to be heat labile, although the EB had a thermostability slightly greater than that of EE or BE. Thus the thermostability seems to depend on both components of the ribosome, and, since whatever is responsible for this in the *r*RNA is clearly not present in *E. coli* *r*RNA, it must depend upon non-conserved sequences. It seems possible that only small conserved regions might be directly involved in interaction with protein, and whilst non-conserved regions may not be involved in this interaction directly, they may have important functions of some other nature in the mature particle.

The cistrons coding for ribosomal RNA

The amount of DNA in the genome corresponding to *r*RNA has been measured by hybridization techniques for a variety of organisms. In bacteria, the level of hybridization would correspond to about ten genes for each of the major species (Oishi and Sueoka, 1965; Yanofsky and Spiegelman, 1962). Hybridization experiments with *t*RNA show that it is represented by a total of some forty cistrons (Goodman and Rich, 1962). In higher organisms, there is a great multiplicity of cistrons representing ribosomal nucleic acid; in *Drosophila* (Ritossa and Spiegelman, 1965;

Vermeulan and Atwood, 1965) there appear to be some 150 to 200 cistrons for each of the two major *r*RNA species. In *Xenopus laevis* (Brown and Weber, 1968a) and in Hela cells (McConkey and Hopkins, 1964) there are slightly more, with between 400 and 450 cistrons for each type.

The cistrons coding for *r*RNA are fairly well localized in the starting region of the chromosome in *Bacillus subtilis* (Oishi and Sueoka, 1965), and the genes for *t*RNA are located in the same region (Dubnau, Smith and Marmur, 1965). The pattern of arrangement of the two types of cistron was determined by Smith and coworkers (1968), using a technique in which *B. subtilis* cells were grown to the stationary phase in D_2O and then transferred to normal light medium. DNA was isolated at various times during the first synchronous cycle of chromosome replication, and showed a density corresponding to the time for which it had been able to replicate. The times of appearance of transforming activity were followed for various genetic markers, and correlated with the hybridizing affinity for various RNA species in the hybrid density (newly replicated) band. This mapped the 5S, 16S and 23S *r*RNA and *t*RNA genes as all near the origin, but showed that they do not occur in blocks of each type, but rather are interspersed amongst each other.

There has been evidence for some time of a relationship between *r*RNA biosynthesis and the nucleolus in higher organisms (a recent review is by Perry, 1967). Indirect evidence is that rapid protein synthesis is correlated with an increase in the size of the nucleolus, and the composition of nucleolar RNA resembles that of *r*RNA. Further, the nucleolus contains particles which resemble ribosomes under the electron microscope. Several workers (Birnsteil, Chipchase and Hyde, 1966; Vaughan, Warner and Darell, 1967) have isolated these particles and have shown that their sedimentation properties correspond to those expected of ribosomes. Their RNA shows the same base composition as ribosomal nucleic acid, and sediments similarly. Kinetic studies have suggested that these particles are precursors to the cytoplasmic ribosomes. An analysis of their protein content on polyacrylamide gels shows similarities with ribosomal proteins, although some differences, such as in the overall RNA/protein ratio, would be consistent with a role as precursors.

McConkey and Hopkins (1964) showed that there is a concentration of DNA complementary to *r*RNA (termed *r*DNA) in the chromatin associated with the nucleolus of Hela cells. Huberman and Attardi (1967) developed a method for fractionating isolated Hela cell metaphase chromosomes by their sedimentation velocity in low speed centrifugation through glycerol–sucrose density gradients. DNA was purified from the various chromosomal fractions and hybridized with *r*RNA and *m*RNA. Whereas the messenger species were responsive to chromosomal sites distributed more or less uniformly through chromosomes of all classes, the 18S and 28S *r*RNA hybridized only with the smaller chromosomes; cytological evidence has shown that the nucleolar organizer is associated with this class.

Brown and Gurdon (1964) used a mutant in *Xenopus laevis* (*anucleolate*) which is a recessive lethal responsible for lack of the nucleolus; heterozygotes have only one nucleolus as opposed to the two of the normal diploid. This mutation also prevents synthesis of 18S and 28S *r*RNA, although DNA, *t*RNA and *m*RNA are synthesized as normal. Wallace and Birnsteil (1966) showed that DNA from homozygotes anneals very poorly to *r*RNA whilst the DNA from heterozygotes anneals at levels intermediate between the homozygous mutant and wild type. Thus this mutant appears to represent a deletion of the nucleolar region, and the cistrons representing both ribosomal species must be located within it. Ritossa and Speigelman (1965) used mutants of *Drosophila melanogaster* containing an inversion involving the *No* (nucleolus organiser) region on the chromosome to breed strains containing duplications and deletions. Hybridization experiments on these strains showed that the amount of ribosomal RNA which would hybridize was proportional to the dose of *No* in the genome.

Brown and Weber (1968a) developed a technique for fractionating DNA in a CsCl density gradient according to its buoyant density and subsequently subjecting the fractions to hybridization with various RNA species. In *Xenopus laevis* the *r*RNA has a higher G–C content than the average of the DNA, so that DNA corresponding to *r*RNA shows as a distinct satellite band apart from the remaining nuclear DNA on the gradient. This *r*DNA band is absent in anucleolate mutants. The *r*RNA hybridizes with this band of high buoyant density, *t*RNA hybridizes with DNA in the average region (that is the main peak on the sedimentation profile), and 5S RNA tends to hybridize with the lighter regions (although not entirely so).

They then applied this technique to determine the mode of clustering of the ribosomal RNA species (Brown and Weber, 1968b). DNA preparations were fragmented to a range of molecular weights and tested to show whether regions homologous to 18S and 28S *r*RNA were present on the same molecule. The results of such linkage experiments depend upon their arrangement along the genome. If cistrons representing the two species lie in two contiguous blocks, one for each type, there should be no molecules which hybridize to both 18S and 28S *r*RNA, irrespective of the molecular weight of the *r*DNA preparation (blocks of 400 genes would give many fragments under the conditions used). If the two types of cistron are in strict alternation, any DNA stretch longer than the size of the 28S *r*RNA gene should hybridize to both. It also is possible to distinguish the situation in which regions of DNA non-homologous to either species are interspersed in the region coding for them. The results indicated that there could not be more than two sequences for either 28S or 18S *r*RNA before a gene of the other type followed. Further, non-homologous sequences were intermingled with these genes. The most likely model for their arrangement would be alternating sequences of 18S and 28S, but with each pair of genes separated from the next by a non-homologous sequence of about the same or smaller size than the ribosomal cistrons themselves. A drawback of this analysis is that it gives only the average arrangement of genes, and does not exclude heterogeneity of linkage within small clusters.

Birnsteil and coworkers (1968) also examined the arrangement of *r*RNA genes in this satellite *r*DNA, but used sedimentation characteristics instead of radioactive labelling to distinguish the species formed by hybridization. The 18S and 28S *r*RNA molecules have different G–C ratios (although both are distinct from the average of the genome), so that the two types of *r*DNA–*r*RNA hybrid should have different buoyant densities. Since such blocks with different buoyant densities are not found, the two types of cistron must be interspersed in some manner. Fractionation to the sub-cistronic level was necessary to release DNA of one *r*RNA type uncontaminated by DNA of the other type, confirming that these must be closely integrated in the genome. However, the buoyant density of *r*DNA fragments is too high to be accounted for solely by cistrons coding for *r*RNA, so these must be separated by DNA of a higher G–C content. The data would be consistent with an alternating model in which stretches of higher G–C content separate the pairs of cistrons coding for ribosomal RNA. The annealing reaction itself proved to be very rapid, and this would suggest homogeneity in the *r*RNA cistrons, that is any 18S (or 28S) *r*RNA molecule could anneal to any 18S (or 28S) coding segment.

The biosynthesis of ribosomal RNA

Biochemical studies on the synthesis of ribosomal RNA have been performed largely in the mammalian Hela cell system (recent reviews are by Perry, 1967, and Maden, 1968). The first stage is synthesis at the nucleolus of a large precursor RNA molecule, sedimenting at about 45S, which is subsequently cleaved to yield the mature ribosomal RNA species. Support for the accuracy of such biochemical studies has been provided by the work of Miller and Beatty (1969b) on the visualization of RNA synthesis by nucleolar extrachromosomal DNA (see pages 152–153). This showed that the DNA comprises units some 2–3μ long which are active in transcription (that is covered by the RNA-protein matrix), separated from one another by inactive (matrix-free) lengths of DNA. Most of the inactive segments are about one-third the length of the adjacent matrix-covered units, but may be up to ten times the length of the active segment. There does not appear to be any pattern to the distribution of the larger matrix-free regions along the axis. The average length of the inactive inter-gene segments is about two-thirds the length of the gene, indicating that about 60 per cent of the nucleolar DNA codes for the precursor *r*RNA, with the remainder inactive. This corresponds well with the results of hybridizations between *r*RNA and the isolated DNA. The RNA chain synthesized by a gene of 2·5μ should have a molecular weight of about $2{\cdot}5 \times 10^6$, and this is about the lower of the estimates for the size of amphibian precursor *r*RNA ($2{\cdot}5$–$3{\cdot}5 \times 10^6$ as opposed to the larger $4{\cdot}5 \times 10^6$ in mammals). As expected, the temporal incorporation of a radioactive label into the matrix corresponded well with the results obtained by biochemical studies.

The process by which ribosomal RNA is synthesized has been followed by observing the fate of a radioactive pulse label during its conversion through the various stages of synthesis; Greenberg and Penman (1966) used a C^{14}-methionine label (this is incorporated into methyl groups), and Zimmerman and Holler (1967) used labelled 6-methylaminopurine and 2′-0-methylcytidine. Ten minutes after incorporation, only the 45S nucleolar species was labelled. Its specific activity declined rapidly after 40 minutes, when a 32S nucleolar RNA reached its maximum labelled level. This in turn then declined, and the 28S *r*RNA reached a maximum label at 80 minutes; the 18S *r*RNA reached the cytoplasm rather more rapidly than this.

This suggests two possible schemes for the biosynthesis of *r*RNA. The one which seems most likely at present proposes that both the 28S and the 18S *r*RNA mature species are derived from the same precursor molecule. According to this, the RNA sedimenting at 45S must comprise a single homogeneous species which is cleaved to give one molecule of each of the two mature species; since the 45S precursor is about twice the size of the 28S and 18S combined, its surplus material must be lost during the maturation process. This model is illustrated in Figure 7.5. It has the advantage that it can readily account for the arrangement of the cistrons representing *r*RNA in the genome which has been suggested by hybridization experiments. That is, a series of cistrons would code for the 45S precursor, probably in much the same way as shown in the electron micrograph of Figure 6.5, and, in effect, these would represent a series of pairs of *r*RNA cistrons (one 28S and one 18S) which are separated from each other by a stretch of non-ribosomal RNA. An alternative scheme, which has proved difficult to exclude conclusively, would suggest that the two *r*RNA species are derived from different precursor molecules, although these happen to be of the same size, and both are characterized by a sedimentation value of 45S.

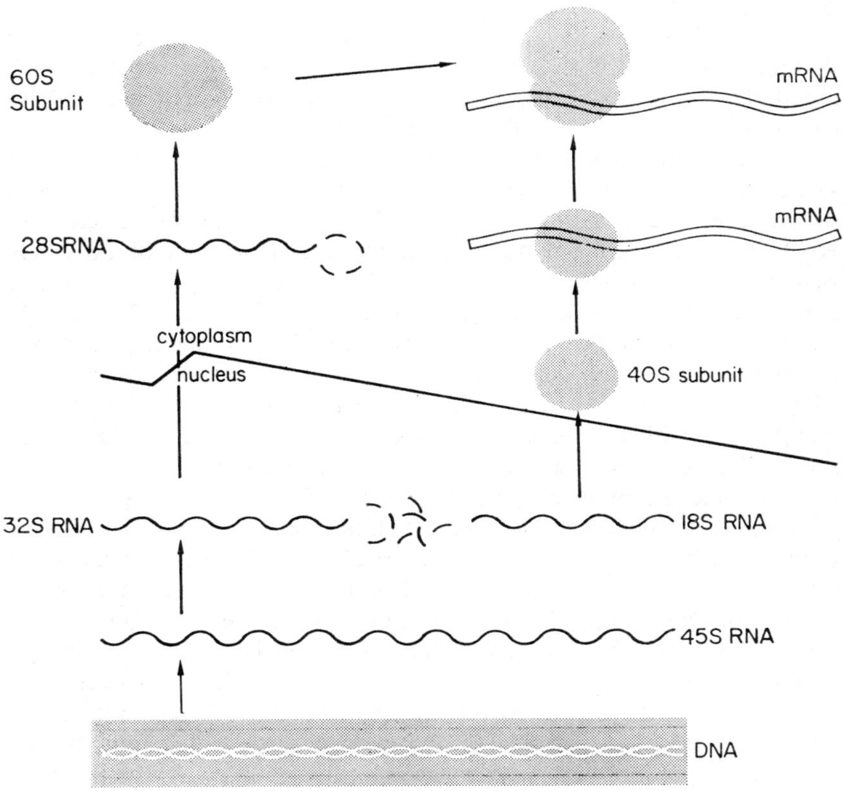

Figure 7.5. The biosynthesis of the ribosome. The 45S RNA molecule is broken to a 32S and 18S, and the 32S then loses further material to yield a 28S. It is probable that some ribosomal proteins associate with the *r*RNA in the nucleolus and the remainder are added in the cytoplasm

Jeanteur, Amaldi and Attardi (1968) confirmed by heat denaturation experiments that the 45S and 32S precursors are each continuous polynucleotide chains, with molecular weights corresponding to lengths of about 14,000 and 7000 nucleotide residues, respectively. (The mature 28S and 18S molecules have chain lengths of about 5000 and 3000 nucleotides each.) Digestion with ribonuclease was used to obtain oligonucleotide patterns of both the precursor and the mature species. The 45S and 32S were similar, but very different from either the 28S or the 18S (or the expected pattern for the weighted sum of the two mature species). This demonstrates that the maturation of ribosomal RNA involves the loss of precursor segments which have a different nucleotide composition and sequence from those of the mature species.

Chemical analyses have been performed on both the precursor and mature ribosomal RNA species to find whether both 18S and 28S *r*RNA are derived from a single precursor molecule. A comparison of base compositions (Amaldi and Attardi, 1968; Willems and coworkers, 1968) has shown that the 45S and 32S precursors each have a rather higher G–C content (70 per cent.) than the mature species (the 28S is 67 per cent, the 18S 56–59 per cent.). This implies that the fraction of RNA discarded during maturation of the precursor must be extremely rich in G–C content—of the order of 75–77 per cent. Jeanteur and Attardi (1968) performed hybridization experiments with 18S and 28S *r*RNA, and with the purified 32S and 45S precursors. The competition between the mature species and the 45S precursor was quantitatively in agreement with the theoretical levels expected if the latter comprises one of each *r*RNA type, with the remaining half of the molecule non-ribosomal in nature. The 32S species would correspond to one 28S, with the remaining third of its length non-ribosomal.

Mature ribosomal RNA is methylated, and the pattern of methylation of the 2′ position of ribose (which comprises some 80 per cent. of the total methylation) shows that the base methylation frequencies do not correspond to the base composition frequencies (Wagner, Penman and Ingram, 1967). This suggests that methylation is a specific and not merely a random process. By pulse-labelling with C^{14}-methionine, Greenberg and Penman (1966) demonstrated that the methylation of Hela cell *r*RNA takes place on the nucleolar 45S species. A large proportion of this methylation probably occurs even before completion of synthesis of the molecule. Zimmerman (1968) has observed that, in addition to the major methylation taking place during or soon after transcription of the 45S precursor, there is a subsequent secondary methylation which probably takes place at cleavage of the 45S molecule. This is concerned with forming one base of dimethylaminopurine per 18S molecule. A function for the methylation has been suggested by the finding of McConkey and Hopkins (1969) that the segments of the 45S RNA which are lost during maturation are the non-methylated ones; thus methylation may indicate to the enzymes responsible for cleavage of the 45S precursor which sections are to be retained for conversion to *r*RNA and which are to be degraded.

The biosynthetic mechanisms involved in the production of 5S RNA appear to be parallel to those for the larger ribosomal RNA species. Hecht, Bleyman and Woese (1968) studied 5S RNA synthesis in *B. subtilis* where the mature forms of 16S and 23S *r*RNA arise from precursors some 5–20 per cent. larger; thus post-transcriptional cleavage is involved, as in mammalian systems. Kinetic experiments and the response to inhibitors of transcription or protein synthesis suggested that 5S RNA is also synthesized via a precursor molecule. Although the precise nature of this precursor is not known, it does not seem to be a species of similar low size. One possibility is that the 5S RNA is derived as a segment split off in formation of the larger species. Knight and Darnell (1967) examined the rate of flow of 5S RNA into newly synthesized Hela cell 50S subunits. The label appeared rather more slowly than expected, suggesting that the cell maintains a pool through which 5S RNA must pass en route to the ribosome.

Assembly of the ribosome

The precise mode of assembly of the ribosome is not known, but the association of its nucleic acid and proteins probably occurs in the nucleolus (for higher organisms), and may take place at the precursor RNA stage before its maturation to *r*RNA. Hela cells appear to contain pools of preformed functional ribosomal proteins, since, when protein synthesis is stopped completely by cycloheximide, cells can still initiate and complete synthesis of new *r*RNA which reaches the cytoplasm in ribosomes (Warner and coworkers, 1966). By contrast, different results have been obtained with puromycin as the inhibitor; Soeiro, Vaughan and Darnell (1968) found that it permitted synthesis of 45S precursor, but there must be a block at some subsequent stage in maturation since none of this RNA appears in cytoplasmic ribosomes. However, this does not seem to reflect on the biological assembly process, but rather is due to the particular mode of inhibition exercised by puromycin on protein synthesis. Probably, released nascent polypeptidyl-puromycin combines with the newly synthesized *r*RNA in competition with the pool of normal ribosomal proteins.

Warner (1966) subjected Hela cells to a pulse dose of C^{14}-amino acids, followed by a chase with normal species. Ribosomes were prepared from the cytoplasm at various times after the chase and examined for labelled proteins. Analysis on polyacrylamide gels showed that most of the protein bands were subject to a continuous increase in radioactive label; the kinetics of appearance of the radioactivity suggested that they are sequentially withdrawn from their pools and added to the nucleic acid. Some proteins did not show this increase; instead of entering cytoplasmic ribosomes only by transport in association with *r*RNA, they are probably in rapid equilibrium between the cytoplasm and ribosomes. This suggests that ribosome assembly is a sequential process, in agreement with the results of experiments on the reconstitution of the ribosome.

Yoshikawa-Fukada (1967) observed that the 45S precursor of FL (mammalian) cells in tissue culture does not exist as a free molecule, but is present in the nucleus as a ribonucleoprotein particle (presome) of between 70S and 90S. This suggests that the ribosomal proteins associate with their nucleic acid before its maturation, and implies that cleavage of 45S RNA takes place when it is in the form of a ribonucleoprotein complex. Presomes could be distinguished from ribosomes by their content of precursor RNA, a lack of dissociation into subunits in EDTA, and a greater susceptibility to ribonuclease. The latter would suggest that part of the ribosomal protein is lacking in the presome, and presomes do indeed appear to possess less protein, especially the more basic, than ribosomes. The inhibition of protein synthesis with puromycin or cycloheximide resulted in an accumulation of presomes in the nucleus, suggesting that their protein components are derived via a pool. Maturation of presomes to ribosomes must require protein synthesis as well as cleavage of the RNA. The use of actinomycin demonstrated that maturation could be achieved without RNA synthesis, but in this case, the ribosomes formed remain in the nucleus and require RNA synthesis for their transport to the cytoplasm.

Most is known about the actual association of ribosomal proteins with nucleic acid to form a ribonucleoprotein complex in *E. coli* (for review see Osawa, 1968). The process of ribosomal RNA synthesis is rather different from that prevailing in higher organisms. Mangiarotti and coworkers (1968) found that the same kinetics hold for the formation of both 16S and 23S *r*RNA. It takes about 2 minutes to synthesize either type of chain (although the latter is twice as large). A cell in exponential growth synthesizes some 20,000 ribosomes every generation, and this would imply synthesis of about thirty *r*RNA copies per minute from each *r*RNA cistron. There does not appear to be any large precursor RNA molecule analogous to that found in mammalian systems, and assembly of the ribosomes appears to take place by successive addition of proteins to 16S and 23S *r*RNA.

Various means have been used to identify the intermediate ribonucleoprotein complexes through which ribosome synthesis must pass. Addition of chloramphenicol to *E. coli* cells prevents the synthesis of ribosomes, and Kurland, Nomura and Watson (1962) showed that after its addition, ribonucleoprotein particles of 18S and 25S can be extracted from the cell. These 'CM' particles contain 75 per cent, RNA, that is, they are deficient in protein. Subsequent work (for review see Vasquez, 1966) has shown that they contain some proteins with the electrophoretic mobilities characteristic of ribosomal proteins but that they lack others. When the antibiotic is removed, the accumulated CM particles are transformed into normal 30S and 50S subunits. This can be explained if the proteins in the CM particles are derived from preexisting pools so that protein synthesis is not required for their attachment to *r*RNA. This led to the suggestion that they represent a stage in ribosome biosynthesis and removal of chloramphenicol permits completion of the process.

However, Yoshida and Osawa (1968) observed that their presence could also be due to formation from nascent *r*RNA and protein present during the extraction procedures, rather than a reflection of their presence in the intact cell. These possibilities were distinguished by radioactive labelling experiments in which CM particles were prepared under two sets of conditions, as shown in Figure 7.6. Under condition A, half the cells were grown under non-radioactive conditions, and the other half grown in C^{14}-lysine, after which protein synthesis was inhibited by the addition of chloramphenicol and subsequent RNA synthesis labelled with H^3-uridine. Under condition B, one set of cells was grown in C^{14}-lysine, and the other on normal amino acid; protein synthesis was blocked in the second set by chloramphenicol, and H^3-uridine was added simultaneously. The H^3 radioactive label indentifies the *r*RNA synthesized after ribosome biosynthesis has been halted, so there is no confusion with preexisting ribosomes. In each case, the two sets of cells were mixed together before the extraction of CM particles. If the CM particles are formed within the cell, those from A should be labelled with C^{14}, but those from B should be unlabelled. On the other hand, if they are formed during the extraction procedure, those from both A and B should be equally labelled. The latter was found, indicating that the particles are artifacts of the preparation procedure and are not due to an intracellular process. Schlief (1968)

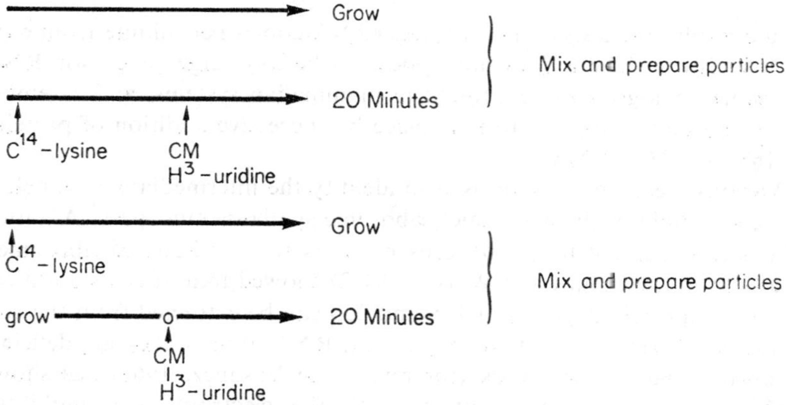

Figure 7.6. Preparation of CM particles under two conditions to distinguish between the possibilities that they exist as such *in vivo* or are artifacts of preparation. (After Yoshida and Osawa, 1968)

has demonstrated that the proteins of CM particles are derived from pre-existing pools of non-ribosomal basic protein and not from ribosomal proteins.

The 'relaxed' *E. coli* mutant RC^{rel} does not require protein synthesis in order to achieve transcription, unlike the wild strain (RC^{str}) in which blocking protein synthesis prevents transcription (see pages 289–292.) Thus, using this mutant, starvation for an essential amino acid blocks protein synthesis but does not interfere with RNA synthesis. Nakada, Anderson and Magasanik (1964) found that methionine starvation of a methionine requiring RC^{rel} mutant resulted in the production of ribonucleoprotein 'relaxed particles' of 18S and 28S. Acrylamide gel electrophoresis (Nakada, 1967b) demonstrated that the proteins of RC particles were of the ribosomal type, but that some of the normal ribosomal proteins were missing. The total quantity of ribosomal proteins synthesized during amino acid starvation is certainly insufficient to account for their protein content, so this is probably derived from pools of preformed species.

Nakada (1967a) observed that the accumulated relaxed particles were converted to normal ribosomes after restoration of the essential amino acid to a previously starved culture. Nakada and Unowsky (1966) reported that relaxed particles could combine with solubilized ribosomal proteins *in vitro* to form functional ribosomes. If the conversion of RC particle to ribosome is to be accounted for by the addition of complementary proteins lacking in the former, ribosomal proteins can be classified into two types: old proteins, already present and conserved in RC particles, and new proteins, which are added to RC precursors to complete ribosome formation. Experiments (Nakada, 1967b) in which relaxed cells were grown in labelled medium containing methionine, starved in normal medium and allowed to recover in the presence of methionine and 5-fluorouracil (the latter prevents *de novo* synthesis of further ribosomes), showed that about 60 per cent. of the ribosomal proteins can be classified as old. This suggests that ribosome formation proceeds through a stage of addition of ribosomal proteins from preformed pools to give relaxed particles, after which newly synthesized proteins are added.

It is difficult to prepare ribosomal precursor particles directly by following the fate of a radioactive pulse label in *r*RNA because of contamination with *m*RNA. One approach (see Osawa, 1968) avoided this by using 'shift-up' culture conditions where *m*RNA synthesis is reduced to a low level relative to that of *r*RNA. This identified 22S and 26S precursors to the small subunit, and 30S and 40S precursors to the large subunit. Mangiarotti and coworkers (1968) used a 'fragile' mutant of *E. coli* in which *m*RNA can be removed as polysomes after a gentle lysis of the cells. Precursor particles containing 16S *r*RNA sedimented at 26S, and peaks sedimenting at 32S and 43S contained 23S *r*RNA. The 32S and 43S precursors probably correspond to the 30S and 40S found by Osawa, but it is not clear why a 22S particle should be present in one case and not the other.

Unlike mature subunits, the precursor particles were highly sensitive to degradation by ribonuclease—this would suggest that their RNA is not yet fully covered by protein. Osawa and coworkers (1969) observed that ribonucleoprotein particles sedimenting at 30–32S and 40–43S are found in extracts from *E. coli* cells treated with low concentrations of chloramphenicol; these proved to be indistinguishable from the precursors to the 50S subunit by criteria of sedimentation behaviour, the properties of their RNA, protein composition, and 5S RNA content. Experiments to check the possibility that these LCM (low-chloramphenicol) particles are artifacts suggested that they are genuine precursors (unlike the CM particles extracted after treatment of cells with much higher levels of the antibiotic). Upon removal of the chloramphenicol, the RNA of LCM particles was converted to mature *r*RNA, and the protein content of the particles adjusted to that of mature subunits.

The 30S particle-precursors contain only three of the nineteen protein components recognized for the 50S subunit by CMC column chromatography, while 40S particles contain twelve of the nineteen (including those of the 30S). By comparison, the particles obtained by treatment of 50S subunits with LiCl contain thirteen of the nineteen, only two of these being different from those in the 40S precursor. It seems likely that the addition of proteins to the *r*RNA proceeds in three steps through these two precursors. The 23S *r*RNA from both the 30S and 40S precursors is about 60 per cent. methylated, and the remaining 40 per cent. of the methyl groups is inserted during the conversion to the 50S subunit. It appears that 5S RNA is also inserted at this stage since the 40S particles have only 30 per cent. of the 5S RNA content which would be associated with 50S subunits. The maturation of the 50S subunit can thus be represented as:

$$\text{23S } r\text{RNA} \xrightarrow[\text{group I proteins}]{\text{60 per cent. methylation}} \text{30S} \xrightarrow[\text{group II proteins}]{} \text{40S}$$

$$\xrightarrow[\text{group III proteins and 5S RNA}]{\text{40 per cent. methylation}} \text{50S}$$

RNA prepared from relaxed or precursor particles (nascent *r*RNA) can be distinguished from the mature species by slight chemical and physical differences (Osawa, 1968). Nascent 16S *r*RNA has only 10–20 per cent of the mature methyl group content, and 23S nascent *r*RNA is about 60 per cent. methylated. The content of psuedo-uridine is also lower in the nascent species. Nascent *r*RNA from the smaller subunit is also distinguished by a slightly increased sedimentation coefficient (17S); this is probably due to features of its secondary structure which differ from the mature species. This is not merely due to a difference in methylation, since Sypherd (1968) found that the different kinetics of the structural conversion and methylation of nascent *r*RNA to mature *r*RNA would exclude methylation as a cause of structural change.

Kinetic experiments (Mangiarotti and coworkers, 1968) suggest that *r*RNA chains begin to associate with ribosomal proteins during their transcription, so that 16S and 23S molecules appear as 26S and 32S precursor particles as soon as they are complete. For cells with a doubling time of 120 minutes, the 26S precursor takes an average of 18 minutes to mature to a 30S particle. The completed 23S *r*RNA spends about 1 minute in a 32S precursor pool before conversion to a 43S particle, where it remains for 17 minutes before maturation to a 50S subunit. Control of ribosome synthesis might be exercised by ribosomal protein through its interaction with nascent *r*RNA chains during their transcription, and Dalgarno and Gros (1968) investigated this experimentally. After 150 minutes of amino acid starvation, relaxed *E. coli* showed no net synthesis of *r*RNA. Relaxed particles accumulated during this period, but if the cells were then starved for glucose, these broke down. Restoration of glucose after 60 minutes resulted in a resumption of *r*RNA synthesis. This would be consistent with a model in which ribosomal nucleic acid is synthesized until the cell runs out of some ribosomal protein(s). The breakdown of relaxed particles during glucose starvation would release these proteins, thus permitting RNA synthesis upon restoration of glucose.

CHAPTER 8

Transfer RNA

Primary structure

Transfer RNA is the smallest cellular RNA, sedimenting at 4S, with between seventy-five and eighty-five nucleotides in a molecule. It contains a large number of unusual bases which are not normally present in RNA, as well as the usual U, C, A and G; amongst the more common unusual nucleosides are pseudo-uridine (ψ), dihydrouridine (UH_2), and dimethylguanine (Me_2G). The unusual bases are probably produced by an enzymic modification of the *t*RNA molecule after its transcription. The methylation of *t*RNA is known to be catalysed by methylases which use *S*-adenosylmethionine as the methyl group donor, and *in vitro* studies of methylase action using as substrate methyl-deficient *t*RNA from methionine-starved *E. coli* have suggested that there is a high degree of enzyme specificity (Baguley and Staehelin, 1968). Eight fairly distinct enzyme fractions with *t*RNA methylase activity have been obtained from yeast (Bjork and Svensson, 1969). *E. coli* extracts also contain an enzyme which transfers sulphur from cysteine to *t*RNA (Hayward, Elicieri and Weiss, 1966); the major reaction product is 4-thiouridine.

Primary structures have now been established for several transfer species from yeast, *E. coli*, rat liver and wheat germ, and also for the 5S RNA from *E. coli* and the human tumor KB cells. The analytical procedure is analogous to that involved in the amino acid sequence determination of proteins, although this tends to restrict its application to only small RNA species (for review see Holley, 1968). Digestion with T1 ribonuclease, which attacks internucleotide linkages selectively on the 3′ side of guanosine, gives large oligonucleotide fragments. Smaller size fragments are given by digestion with pancreatic ribonuclease, which attacks more generally on the 3′ side of pyrimidine residues. The whole structure is then obtained by the conventional technique of examining the overlaps between the various sequenced fragments.

Although determination of the nucleotide content of small oligonucleotides is straightforward, sequence determination is more difficult.

Partial digestion with snake venom phosphodiesterase, which degrades polynucleotides stepwise from the 3′ end, can establish most of the sequence, but is not applicable for the last two or three nucleotides at the 5′ end. Micrococcal nuclease attack can be used to release a dinucleotide from the 5′ end. Small fragments, such as tri- and tetranucleotide sequences, were at first separated and identified on DEAE-cellulose columns (Holley and coworkers, 1965). Sanger, Brownlee and Barrell (1965) and Sanger and Brownlee (1967) reported that the ion exchange chromatography could be replaced by a two-dimensional paper electrophoresis with a high degree of resolution. A critical factor in sequencing *t*RNA is the difficulty of obtaining sufficient pure material, and this method has the advantage that it can be used with small amounts of material. The technique is used in conjunction with a highly radioactive (P^{32}) label so that the nucleotides are detected on the chromatograms by autoradiography and quantitated by counting.

The first dimension of fingerprinting is by electrophoresis, usually at pH 3·5 on a cellulose acetate membrane—this gives less streaking and sharper spots than paper. The second dimension uses DEAE paper, also at low pH (1·9), and this is comparable to the DEAE-cellulose columns. The fractionation is performed by electrophoresis so that the resulting separation is partly due to this and partly due to ion-exchange chromatography; the effects complement each other to produce an effective separation.

It is possible to prepare a fingerprint for the chromatogram in the form of a graticule showing the relationship between the composition of a nucleotide fragment and its position on the two-dimensional chromatogram. A line is drawn through the positions of related spots, for example with the compositions UG, (CU)G, (C_2 U)G, (C_3 U)G, etc. Similarly, lines are drawn through the spots representing other fragments which differ in only the number of A residues, or the C residue number. This gives three graticules, one with no U, one with one U, and one with two U, since uridine is the mononucleotide which has the greatest effect on mobility in this system. The exact separation of the graticules depends on the solvent systems. Graticules may overlap in certain areas, so it is not possible to make an unequivocal assignment for the nucleotide composition of spots in these regions. One such system is shown in Figure 8.1.

Most of the transfer species so far sequenced have been from yeast; the first was a *t*RNA accepting alanine (Holley and coworkers, 1965), and since then the sequences of two serine species, a valine, tyrosine and phenylalanine have been determined (for review see Madison, 1968). Both methionine $t\text{RNA}_m$ and $t\text{RNA}_f$ from *E. coli* have been sequenced

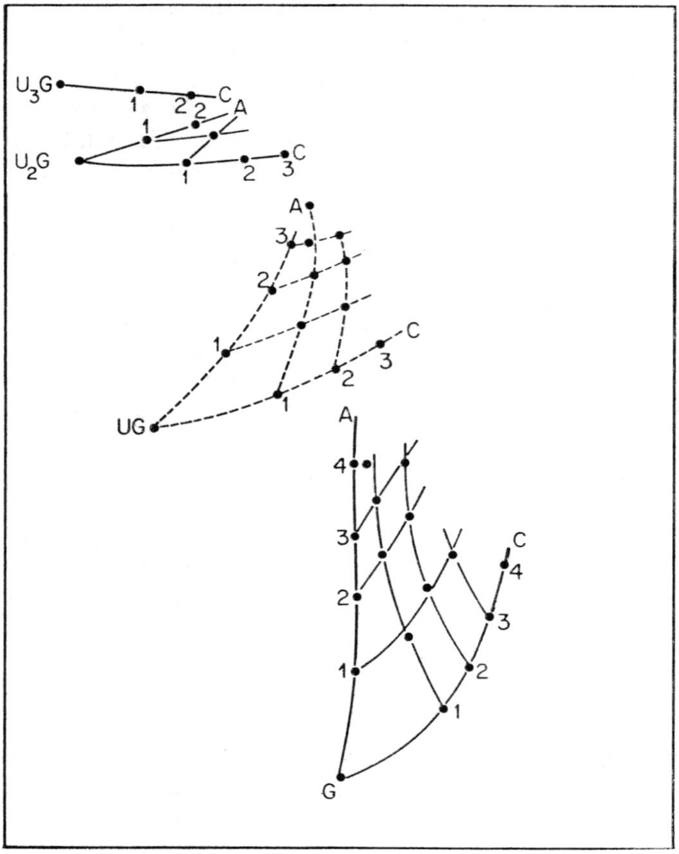

Figure 8.1. A two-dimensional fingerprint for separation of nucleotides. (After Sanger, Brownlee and Barrell, 1965)

(Cory and coworkers, 1968; Dube and coworkers, 1968), and a tyrosine *t*RNA and a suppressor mutant derived from it (Goodman and coworkers, 1968). Serine *t*RNA from rat liver is the only mammalian species so far to have been sequenced (Staehelin and coworkers, 1968).

There are certain similarities between all these species. One feature is that there is no concentration of unusual bases in any one part of the molecule; they are scattered throughout its length. The sequence G-*r*T-ψ-C-G is present in all the species (in one of the yeast serine *t*RNAs and in met-*t*RNA_f its final G is replaced by an A), located some twenty nucleotide residues from the amino acid accepting end of the molecule (the 3′). Most have some dihydrouridine residues in the region sixteen to twenty-one nucleotides from the 5′ end. Although there may be several triplet sequences which could represent the anticodon, all transfer RNA molecules have a suitable sequence in about the same position of the primary (and secondary) structure. This is about half-way along the molecule. There is always a U on the 5′ side of the anticodon and an A or modified A on its 3′ side. It is significant that there is a striking degree of homology between the serine *t*RNA from yeast and that from rat lever; there are long stretches of identical nucleotide sequences corresponding to over half the total length of the molecule.

Secondary structure

A feature of the primary structure is that no transfer species shows long complementary sequences which could allow G–C, A–U pairing—the longest is usually about five nucleotides. This means that either the double-stranded regions must be short, or if pairing is more extensive, other base pairs, such as G–U, must be present. It is possible to form secondary structures by considering the possibilities for hydrogen bonding in the primary structure. Maximizing the number of (normal) base pairs gives various possible models; the two most probable for alanine *t*RNA (Holley and coworkers, 1965) have been termed the clover-leaf and hairpin models, and are shown in Figures 8.2 and 8.3. The sequences of further *t*RNA molecules have suggested that the clover-leaf model is likely to hold generally, and Figure 8.3 shows the transfer RNAs sequenced so far in terms of this model.

All these species isolate the anticodon in the arm directly opposed to the amino acid acceptor end. The common pentanucleotide sequence G-*r*T-*ψ*-C-G lies in the right-hand arm, and the left-hand arm is characterized by possession of dihydrouridine residues. Another arm is of variable length, sometimes representing a full arm, and sometimes only a 'lump' in between the anticodon and the pentanucleotide arms. Base pairing starts five nucleotides along from the terminal adenine, except for the *E. coli* initiator species which starts at six residues. Another difference between the initiator and other transfer molecules lies in the anticodon loop, in which the base following the last letter of the anticodon is an unmodified A instead of the modified A or alkylated purine present in most *t*RNAs. This may be concerned with permitting an unusual first position wobble of the triplet so that it can recognize either AUG or GUG. The generalized clover-leaf model is shown in Figure 8.4.

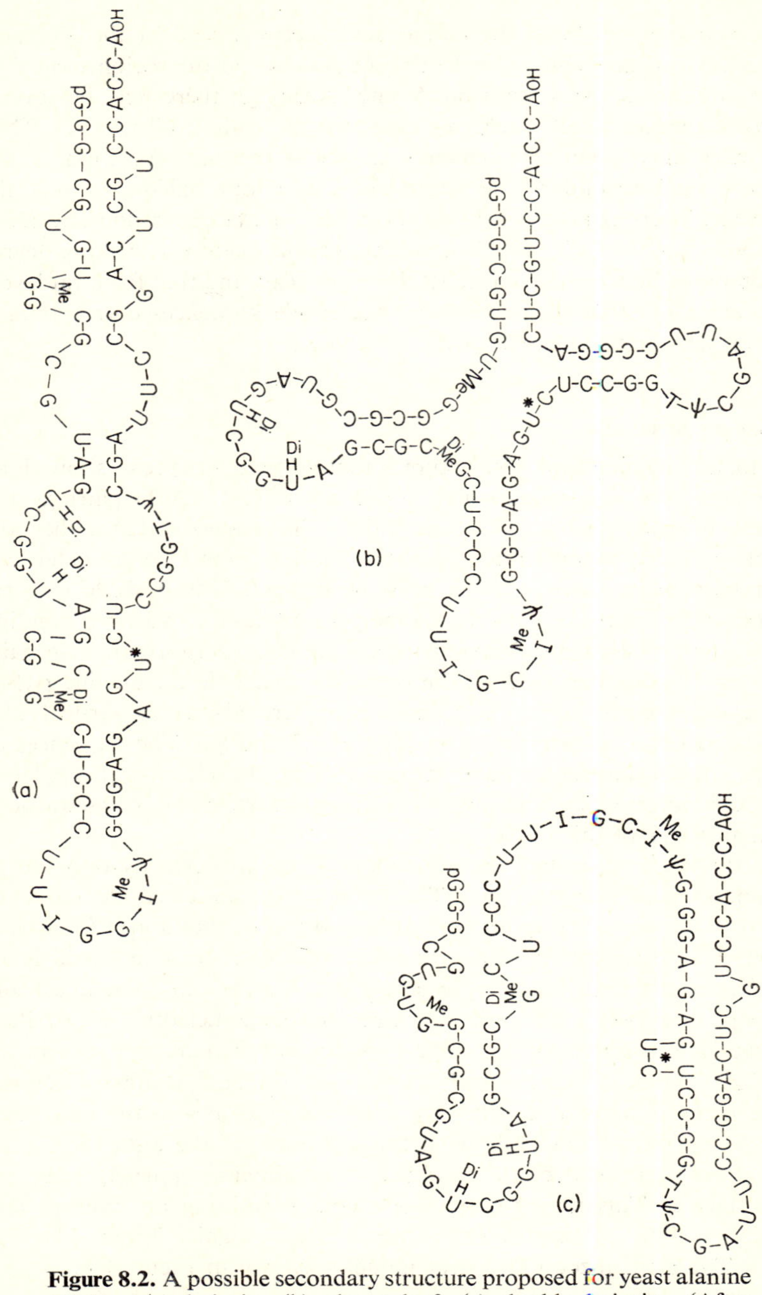

Figure 8.2. A possible secondary structure proposed for yeast alanine *t*RNA: (a) hairpin, (b) clover leaf, (c) double hairpin. (After Holley and coworkers, 1965. Copyright by the American Association for the Advancement of Science, 1969)

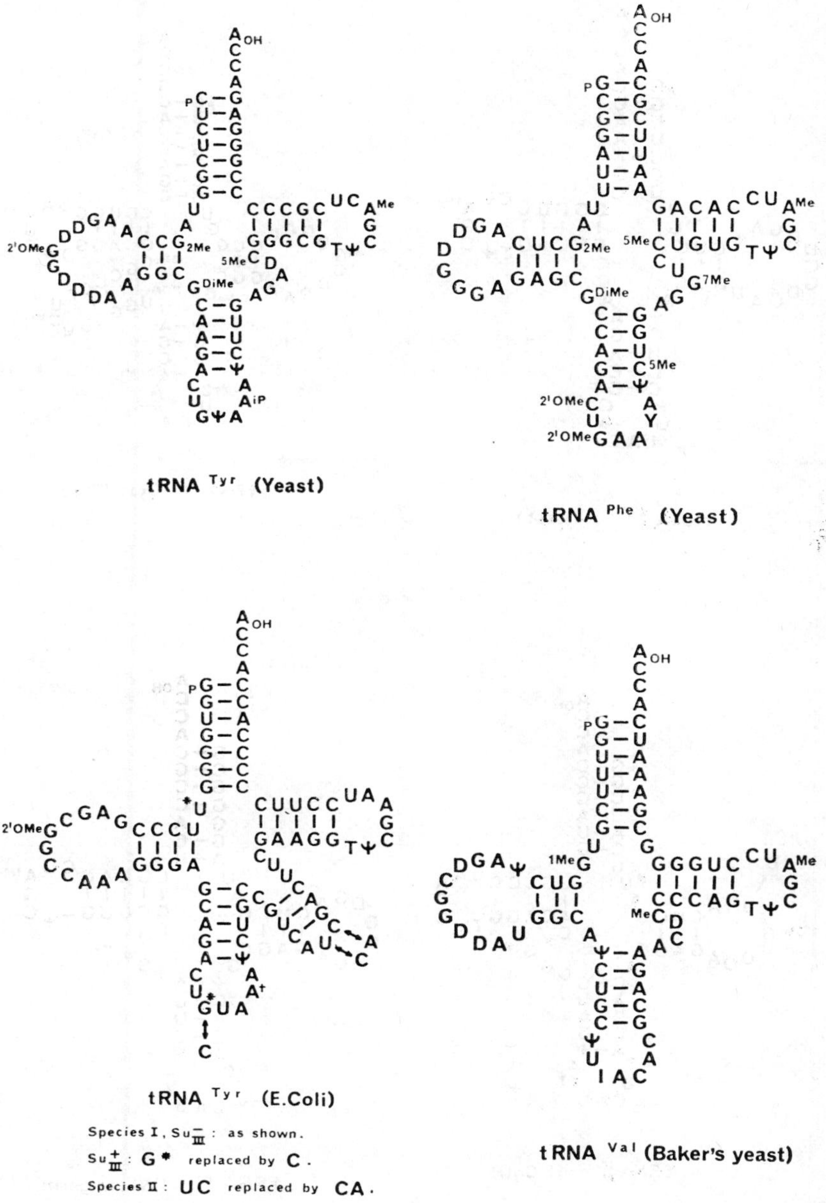
tRNA Tyr (Yeast)
tRNA Phe (Yeast)
tRNA Tyr (E.Coli)
Species I, Su$_{III}^{-}$: as shown.
Su$_{III}^{+}$: G* replaced by C.
Species II: UC replaced by CA.
tRNA Val (Baker's yeast)

Figure 8.3.

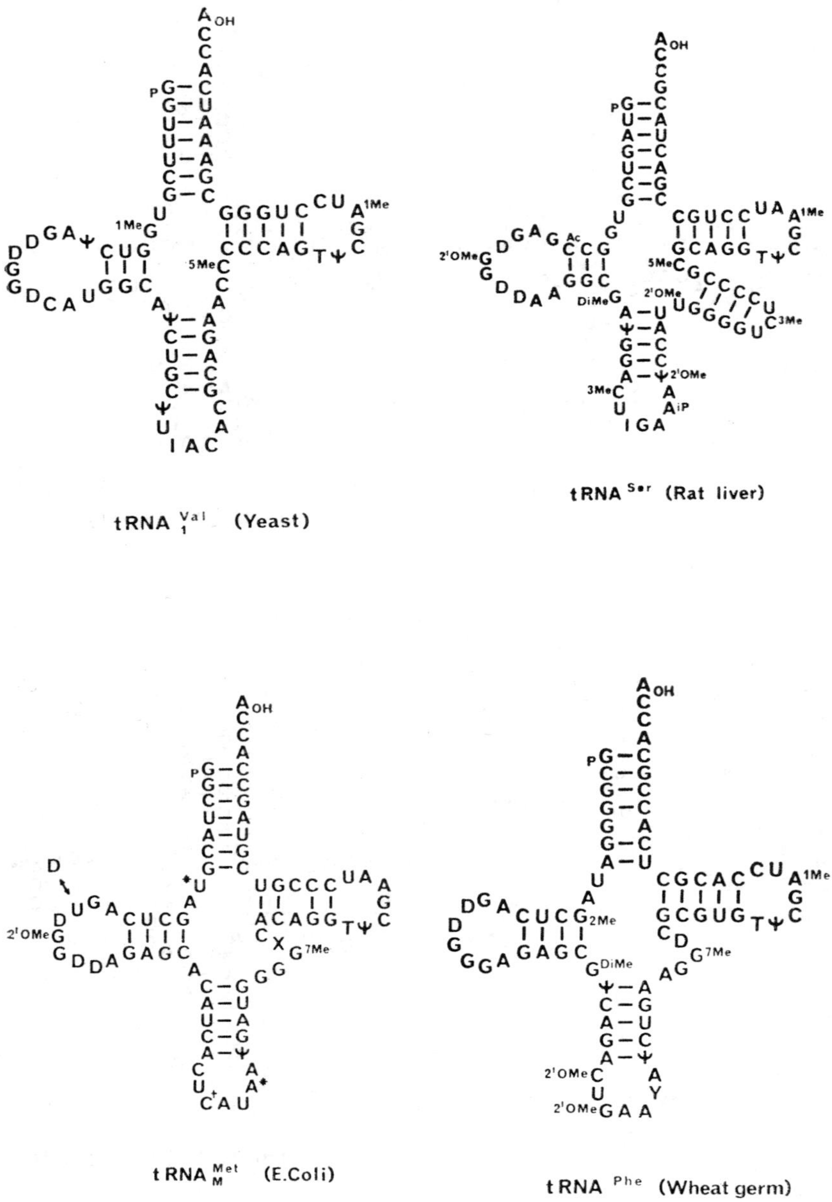
tRNA Val 1 (Yeast)
tRNA Ser (Rat liver)
tRNA Met M (E.Coli)
tRNA Phe (Wheat germ)

Figure 8.3.

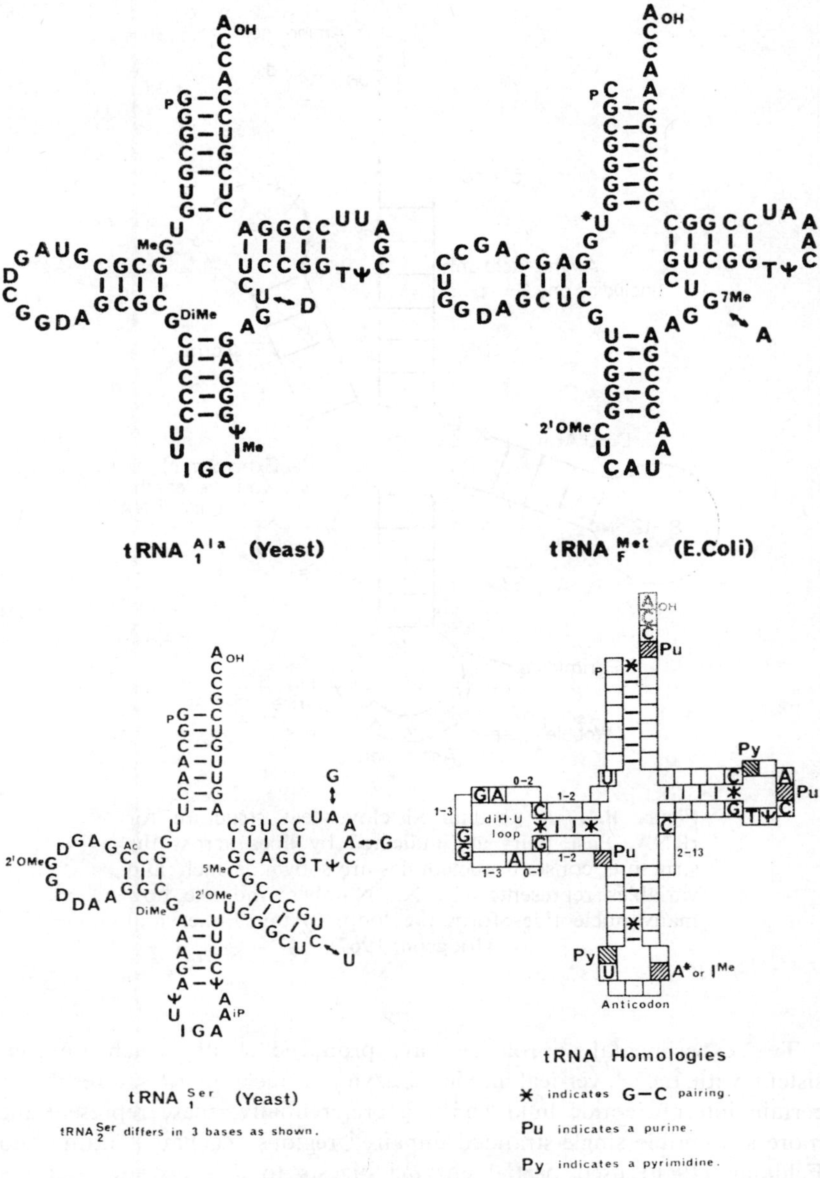

Figure 8.3. Clover-leaf models for the *t*RNA species so far sequenced from *E. coli*, yeast, rat liver, and wheat germ (D represents dihydrouridine. The homologies between these are also shown. (Courtesy of M. Waring and N. Webb)

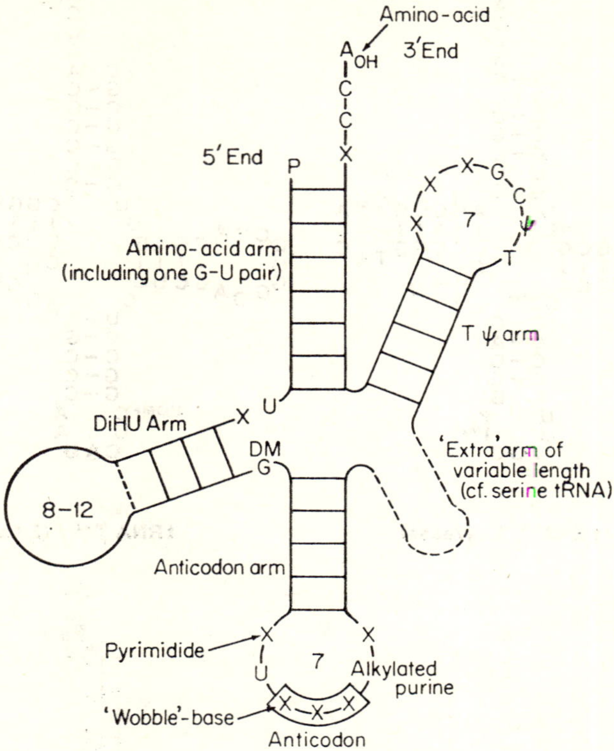

Figure 8.4. A generalized clover-leaf structure for *t*RNA. Base pairs are indicated by lines across the structure, constant nucleotides are shown as such, and variables represented by X. Numbers indicate how many nucleotides form the loop. (After Fuller and Hodgson, 1967)

Two experimental approaches have produced results which are consistent with the clover-leaf model. Enzymic attack on *t*RNA results in certain internucleotide links splitting preferentially; these represent the more susceptible single-stranded unpaired regions. Zachau, Dutting and Feldman (1966) used partial enzyme digests to demonstrate that the susceptible regions were those which would be predicted from a clover-leaf model. Also, the various possible secondary structures can be distinguished on the basis of their sensitivity to single-strand breaks; hairpins would be subject to rapid disruption, and clover-leafs would be fairly resistant. Bell and Russell (1967) found that 35–40 per cent. of the nucleotide content of *t*RNA could be removed and still leave a residue with molecular parameters closely resembling the original undegraded molecule. This seems most likely to result from a clover-leaf model.

Another approach is to study the susceptibility of the different parts of the molecule to mutagenic treatment; bases in the loops of a clover leaf should be more sensitive than those in the central regions. Nelson, Ristow and Holley (1967) used bromination and deamination. The susceptible parts of the molecule proved to be those expected in a clover-leaf structure, with the proviso that for some reason the pentanucleotide loop is less reactive than the others, This could be caused by some protection due to a tertiary structure.

Fuller and Hodgson (1967) built scale models utilizing chemical information about nucleotide sequence and X-ray data on the conformation of base paired regions in double helical RNA. When stereochemical restraints were maintained so that the negatively charged phosphate groups do not come too close to each other, hydrogen bonding groups are available and not buried, and single bond orientations are kept within known limits, the generalized structure of Figure 8.4 was obtained. This suggests that there is very little freedom in the anticodon configuration. Its position as nucelotides 8, 9 and 10 of its arm would allow *m*RNA to pair to it without steric hindrance between the rest of the arm and adjacent codons on the messenger, and would exclude recognition of any triplet other than the proper anticodon. It appears to be possible to arrange the clover-leaf so that there is no steric hindrance between two aminoacyl-*t*RNA molecules being read at adjacent codons on the *m*RNA. The wobble required in codon–anticodon recognition could be accommodated by a distortion of the anticodon conformation—it is not necessary to distort the codon.

It has been suggested that other bonds as well as hydrogen bonds may be involved in maintaining the structure of *t*RNA. Lipsett (1966) found that in *E. coli* there is sulphur in *t*RNA corresponding to about 1·4 atoms per *t*RNA molecule. Most of this is in the form of 4-thiouridine (the oxygen at position 4 is replaced by a sulphur). If a *t*RNA possessed two such molecules, disulphide bridge formation would be possible. The level of sulphur in nucleotides differs greatly between organisms, and so does its base distribution; yeast contains very much less sulphur, and in rabbit liver, which possesses appreciable sulphur, there is no 4-thiouridine. However, no disulphide bridges have been identified in the transfer molecules sequenced as yet.

Tertiary structure

The clover-leaf secondary structure of *t*RNA raises the possibility that there might be a distinct tertiary structure. Completely helical macromolecules, such as DNA, exist in a statistical distribution of all possible conformations, but if the helical regions of *t*RNA were fixed in definite conformations in three dimensions, there could be a specific tertiary structure. This would accord with the concept that transfer RNA represents an attempt to make a nucleic acid fulfil the type of function normally exerted by protein (Crick, 1966), as shown by its ability to recognize amino acids through its activating enzyme and the codon on *m*RNA through a ribosomal interaction.

Fresco and coworkers (1966) found evidence for a definite tertiary structure through direct physico–chemical methods. Denaturation of *t*RNA was caused by temperature increase or lowering of counterion concentration, and was observed by hyperchromicity (sensitive to loss of secondary structure) or the hydrodynamic parameters of viscosity and sedimentation characteristics (sensitive to gross conformation). The form of thermal denaturation depended upon which salt form the *t*RNA was in. With the sodium salt, a non-cooperative process (that is different physical properties are affected at different stages) occurred in two steps. The first involved a slight shape change with only slight loss of secondary structure, and the second a major disruption of the secondary structure. The first step represents a loss of tertiary structure. However, the magnesium salt—presumably the *in vivo* form—did not show the same indications of tertiary structure. It was stable to higher temperatures, when a cooperative change occurred in all its physical properties. This suggests that magnesium ions may serve to maintain the tertiary structure of *t*RNA, so that when it does finally collapse its loss is simultaneous with that of secondary structure, which obscures it. The viscosity of *t*RNA decreased when ionic conditions were changed from Na^+ to Mg^{2+}, which would suggest that the tertiary structure is more complete under this latter condition. Biological activity was associated with the lower viscosity form.

It proved possible to undertake reversible denaturation and renaturation of yeast and *E. coli* *t*RNA. Certain *t*RNAs were found to be inactive after preparation by methods involving denaturing steps, such as phenol extraction or dialysis against the chelating agent EDTA. The denatured form was stable and had to be 'destabilized' before it could be converted to the native (biologically active) state; its direct transfer into dilute Mg^{2+} effectively 'froze' it in a thermodynamically unstable state which took some two days to convert into the native. By contrast, renaturation was rapid upon incubation with 1 mM Mg^{2+} at 60°C—this procedure overcomes the stability of the denatured form to allow a rapid transition into the native state. The renatured form could be denatured within about 1 minute without any comparable need for destabilization. These findings were interpreted as implying the existence of denatured, destabilized and native energy states, separated by the activation energy barriers shown in Figure 8.5. In the absence of Mg^{2+}, the native state rapidly collapses into the denatured form, and this can be retained for some time, even in the presence of magnesium ions, because of its high energy activation barrier. Renaturation is assisted by destabilization because this gives the energy necessary to overcome the activation energy barrier.

The activation energy required for the conversion of denatured *t*RNA

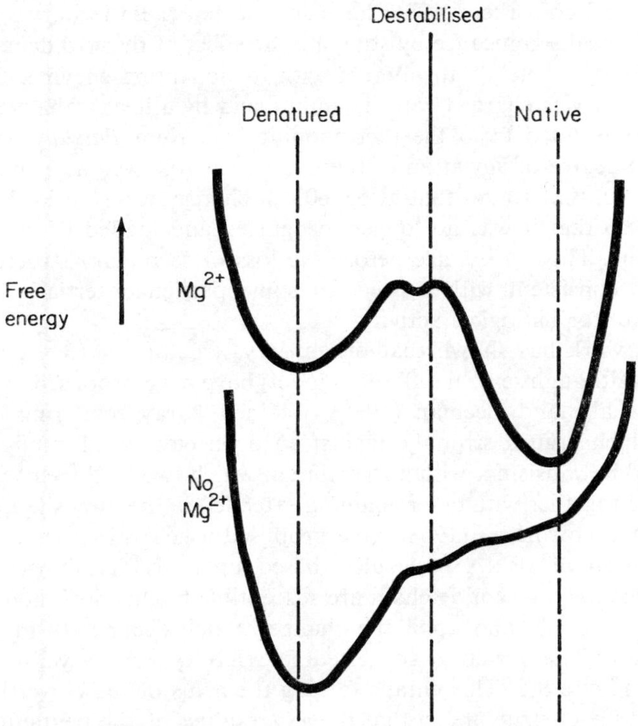

Figure 8.5. Energy states of transfer RNA. (After Fresco and coworkers, 1966)

to the native state is much too high to be accounted for merely by the imposition of a correct tertiary structure on the secondary. Rather, it must first involve disruption of a definite alternative tertiary structure before the biologically active native state is formed. Adams, Lindahl and Fresco (1967) compared the conformational features of native and denatured *t*RNA using *t*RNA$_3^{leu}$ from yeast. It was not possible to determine the qualitative changes in secondary structure, although quantitatively there was a reduction of about 10 per cent. in helical content with denaturation. Although both states had a similar extent of secondary structure, the denatured was less compact and more accessible to ribonuclease action. This would be consistent with a major difference in their tertiary structures. The tertiary structure may depend upon bond formation between unpaired bases in the loops, and in this case, the denatured form could be such a 'wrong' association.

Fresco and coworkers (1966) have shown that these tertiary structures have biological significance by studying the effect of thermal denaturation on the ability of *t*RNA to interact with its activating enzyme. For this, valine *t*RNA was charged with its amino acid by a heat stable val-*t*RNA synthetase isolated from the thermophilic bacterium *Bacillus coagulans*. When the degree of acylation of the *t*RNA was followed as a function of temperature, they found that at 55–60°C a change occurred in the *t*RNA structure so that it was no longer recognized and loaded with valine by the enzyme. This took place before the loss of secondary structure, and would be consistent with a need for some particular tertiary structure (the native) for biological activity.

Recent work has shown that the tertiary structure of *t*RNA is rather compact, although several different models have been proposed to account for this. Lake and Beeman (1968) obtained X-ray scattering data on *t*RNA which argued strongly against an open clover-leaf configuration, but would be consistent with a structure in which two or three of the arms are folded together with one remaining extended in the opposite direction. Cramer and coworkers (1968) have proposed a general structure for the conformation of *t*RNA molecules, based upon results showing which parts of the polynucleotide chain are susceptible to chemical modification or enzymic attack, and upon the changes which occur with increase in temperature. The structure suggested for *t*RNA$_{phe}$ from yeast is illustrated in Figure 8.6. This entails folding the arms of the clover-leaf into a more compact structure so that the ψC residues of the pentanucleotide arm can hydrogen bond to the AG bases in the dihydrouridine loop, and the GG sequence in the latter loop can hydrogen-bond to the CC of the —CCA stem. Its particular structure would be unique for each *t*RNA

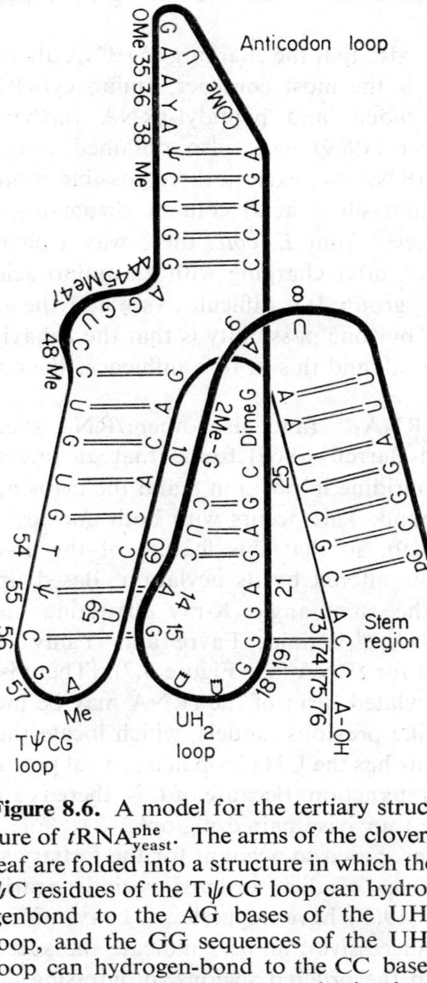

Figure 8.6. A model for the tertiary structure of $t\text{RNA}^{\text{phe}}_{\text{yeast}}$. The arms of the cloverleaf are folded into a structure in which the ψC residues of the TψCG loop can hydrogenbond to the AG bases of the UH_2 loop, and the GG sequences of the UH_2 loop can hydrogen-bond to the CC bases of the stem region. This results in these three loops pointing in one direction, with the anticodon loop exposed at the opposite end of the molecule. The overall length of the structure is about 70Å if the anticodon arm is extended, and the width is about 44Å. (From Cramer and coworkers, 1968)

species, although all would follow the same general pattern of a compact folded clover-leaf.

Cramer suggests also that the charging of *t*RNA alters its conformation; uncharged *t*RNA is the most compact, aminoacyl-*t*RNA appears to be slightly more unfolded, and peptidyl-*t*RNA further unfolded. Stern, Zutra and Littauer (1969) have also obtained evidence which would suggest that each *t*RNA can exist in three possible configurations. MASA (methylated albumin–silicic acid) column chromatography was used to resolve *t*RNA species from *E. coli*; there was a change in the elution properties of *t*RNA after charging with an amino acid, and again after blocking the $-NH_2$ group. It is difficult to see how the acylation influences the conformation, but one possibility is that the behaviour of the —CCA stem region is altered, and this in turn influences the conformation of the whole molecule.

When *E. coli* $t\text{RNA}_1^{\text{val}}$ (and also other *t*RNA species) is irradiated, Yaniv, Favre and Barrell (1969) found that an interaction takes place between the 4-thiouridine in position 8 and the cytosine of position 13 to form a covalent bond. This occurs with both uncharged and aminoacyl-*t*RNA (Ninio, 1969), so that this feature of the conformation of the molecule cannot be altered by its acylation. Based upon this evidence, and also upon the small-angle X-ray scattering curve obtained for phenoxyacetyl-$t\text{RNA}_1^{\text{val}}$, Ninio, Favre and Yaniv (1969) constructed a molecular model for *t*RNA (see Figure 8.7). (There is evidence that the curve for the unacylated form of the *t*RNA may be more complicated—Ninio, 1969). Unlike previous models, which locate the loops at the end of the molecule, this has the UH_2 loop in a central position, to account for the position 8–13 interaction. Because of this, there is a highly asymmetric distribution of the four base paired regions.

All the models proposed to account for the tertiary structure of *t*RNA have the helical axes of their ordered regions nearly parallel. Ninio, Favre and Yaniv (1969) have suggested a convention to describe these models in which successive numbers indicate the successive numbers of strands involved in the ordered regions, progressing along the long axis of the molecule. This is shown in Figure 8.8. The 6–2 and 4–4 models fit the data of Lake and Beeman (1968), the 6–2 represents that of Cramer and coworkers (1968), and the 4–2–2 that of Ninio, Favre and Yaniv.

We should note also that several workers have recently reported the formation of crystals of *t*RNA; this is a necessary prerequisite to studying its structure by X-ray crystallography (see, for example, Kim and Rich, 1968). Crystal formation indicates that the *t*RNA molecules must have a regular size and shape and can be organized in a lattice array.

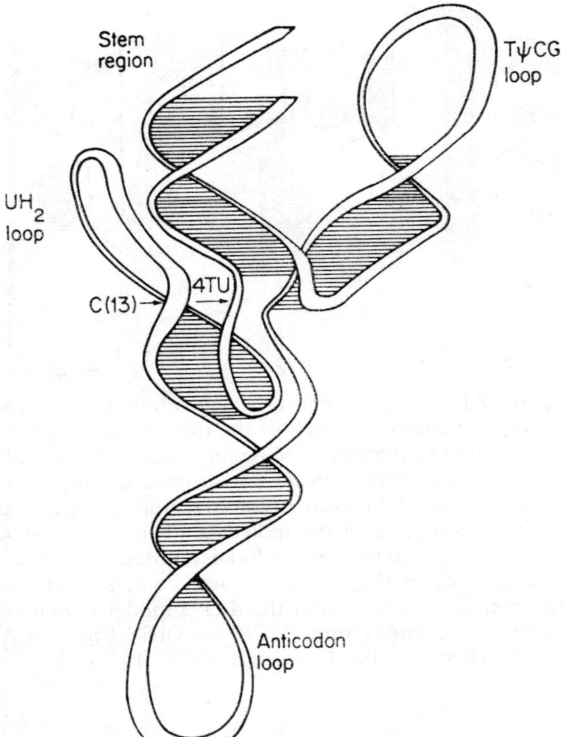

Figure 8.7. A model for the tertiary structure of *E. coli* $tRNA_1^{val}$. The UH_2 loop is located in a central position so that the 4-thiouridine of residue 8 at the base of the stem region can interact with the cytosine of position 13 in the UH_2 loop. The anticodon arm is extended at one end of the molecule, and the stem region and UH_2 loop at the other. The structure has a length of about 100Å, and a width of around 40Å. (From Ninio, Favre and Yaniv, 1969)

Biological interactions of *t*RNA

More than one factor appears to contribute to recognition of only the correct *t*RNA and no other species by its activating enzyme. Although it has been proposed that the anticodon arm possesses a certain element of enzyme recognition, recent evidence has suggested that it is not critically involved in the interaction with synthetase. Integrity of the amino acid

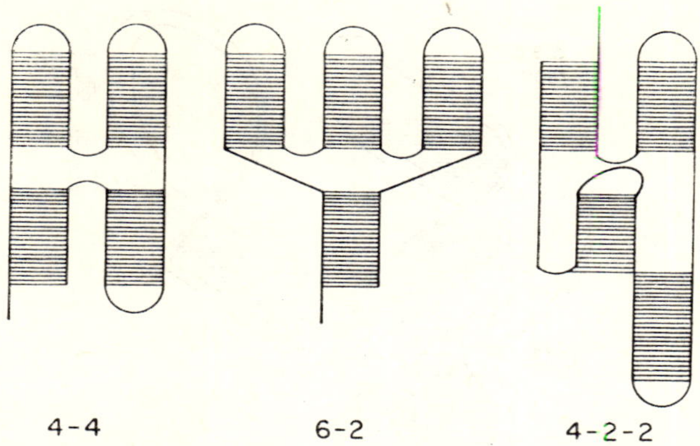

Figure 8.8. A numbering convention for *t*RNA tertiary structures. In each of the three models shown, the helical axes of the ordered regions are nearly parallel. Successive numbers indicate the successive number of strands involved in each ordered region, progressing along the long axis of the molecule. The 6–2 and 4–4 models agree with the data of Lake and Beeman (1968); the 6–2 model is that of Cramer and coworkers (1968) shown in Figure 8.6; and the 4–2–2 model is that of Ninio, Favre and Yaniv (1969) shown in Figure 8.7. (From Ninio, Favre and Yaniv, 1969)

acceptor arm appears necessary if the *t*RNA is to be charged with its amino acid, and it is possible that the dihydrouridine arm (the left arm) may possess an enzyme recognition site. Whilst it is necessary for the anticodon to remain intact for binding to the ribosome, its modification does not prevent the interaction with synthetase. The common pentanucleotide arm is probably required for binding to the ribosome.

Ultraviolet irradiation can cause loss of amino acid accepting ability in *t*RNA by modifying its nucleotide components, as shown by the altered patterns of oligonucleotides given by enzyme digestion of irradiated *t*RNA. Schulman and Chambers (1968) reported that it is possible to produce extensive changes in the primary structure, which result in large changes in conformation, without destroying amino acid acceptor ability. This would indicate that, although the integrity of some particular region(s) may be required for acceptor ability, it is not necessary to retain intact the overall conformation of the whole molecule. They found that photomodification of nucleotides located five, six or seven residues from the acceptor end of yeast alanine *t*RNA destroyed its ability to accept alanine. This suggests that the first three base pairs of *t*RNA play some specific role in recognizing the synthetase. Imura, Weiss and Chambers (1969) split alanine *t*RNA_2 from yeast into large fragments (quarter molecules); they found that the combination of the fifteen 5′-terminal and nineteen 3′-terminal oligonucleotides could be charged with alanine, but not with any other amino acid. The UH_2 region, the anticodon, and the pentanucleotide regions were not required for this recognition. This again suggests that it is the stem region which is responsible for synthetase recognition.

Stulberg and Isham (1967) obtained 'limit digests' of *t*RNA by digestion with venom phosphodiesterase at varying temperatures. A cooperative effect of temperature and nuclease action results in a progressive loss of the ordered structure of the molecule; temperature causes a stepwise disruption of its structure, probably by melting out of the various double-stranded regions, and the phosphodiesterase attacks the resultant stable structures to varying degrees. Although the limit digest is incapable of accepting phenylalanine, it can inhibit formation of phe-*t*RNA by competing with undegraded *t*RNA_{phe} for the enzyme. The degree of hydrolysis can be varied considerably without altering the extent of the inhibition; this suggests that *t*RNA possesses some specific sequence which is responsible for enzyme recognition, and limit digests can compete for the enzyme binding site so long as they retain this sequence. Up to forty-two nucleotides of the *t*RNA molecule can be digested away from the 3′ end without loss of the inhibition ability, but it is lost by the time sixty-five nucleotides have been removed. This would suggest that there is an enzyme recognition site somewhere between these limits, that is, in the dihydrouridine arm.

Thiebe and Zachau (1968a) reported that an unknown base Y is located adjacent to the 3′ nucleotide of the anticodon of yeast phenylalanine *t*RNA. Mild treatment of the anticodon of the *t*RNA with acid resulted in excision of this base, but without breaking the phosphate–sugar backbone, so that the polynucleotide chain retained its integrity. The treated *t*RNA did not bind to poly-U in the ribosome binding test and did not transfer its amino acid onto the growing peptide chain; this loss of activity can probably be attributed to the change in conformation of the anticodon loop which might be expected to result from the base excision. However, despite this, *t*RNA_{phe}^{HCl} could be charged with its amino acid to the same extent as the untreated species, and this suggests that Y, and probably the whole anticodon loop, is not an essential part of the synthetase recognition site of the *t*RNA. Subsequent work (Thiebe and Zachau, 1968b) used a heterologous synthetase (from *E. coli*) instead of the homologous enzyme from yeast to charge the *t*RNA; although the normal untreated *t*RNA could be loaded with phenylalanine, the treated species was unreactive. The requirements for heterologous charging thus seem to be more stringent, raising the possibility that such a system might be useful as a sensitive test for detecting conformational changes in *t*RNA. Certainly, in this case, the anticodon loop appears to play a role in recognition of the heterologous enzyme, although it does not do so in the homologous system.

Gefter and Russell (1969) obtained results which implicate the anticodon arm in ribosomal binding but not in synthetase recognition. They isolated three forms of the *amber* suppressor tyrosine *t*RNA of *E. coli* (see pages 213–214), which differed only in their degree of modification of the base adjacent to the 3′ nucleotide of the anticodon. (The modifications represent successive reactions in the biosynthesis of the *t*RNA.) These differences did not affect the ability of the *t*RNA to accept its amino acid, but they did alter the binding activity to the ribosome in response to the UAG triplet. The fully modified form (biosynthetically complete) was the most active; the partially modified form showed 54 per cent. of this activity; and the unmodified form only 14 per cent.

Ofengand and Henes (1969) found that the TψGC tetranucleotide, which is common to all *t*RNA molecules so far sequenced, by itself inhibits the binding of phe-*t*RNA to ribosomes. This suggests that this sequence may be part of the ribosomal binding recognition site(s) of *t*RNA.

The charging of *t*RNA

The multiplicity of *t*RNA molecules raises the question of whether there is a similar repetition of synthetases, with one enzyme for each *t*RNA, or whether one synthetase can charge more than one of the transfer species representing its amino acid. In the latter case, this might apply to recognition of *t*RNAs with the same anticodon but differing elsewhere in the molecule, or to multiple transfer molecules with different anticodons but representing the same amino acid. At present, evidence is somewhat limited, but it does appear that one synthetase can charge *t*RNA molecules which have different anticodons. In *E. coli*, two isoleucine *t*RNA species which respond to different codons appear to be charged by one synthetase, and a similar situation holds also for leucine and for serine (Baldwin and Berg, 1966a; Bennett, 1969; Sundharadas and coworkers, 1968). The ability of one enzyme to charge *t*RNA molecules with different anticodons with the same amino acid implies that the anticodon does not serve as a specific recognition site for the enzyme, but this must be located elsewhere on the *t*RNA.

There is some evidence that the characteristic of the activating enzyme may change during complex formation. Many of the aminoacyl-*t*RNA synthetases are fairly unstable, but Chuang, Atherly and Bell (1967) found that, for proline and valine, the enzyme–*t*RNA complex is more resistant to heat inactivation than is the enzyme alone. Ohta, Shimada and Imahori (1967) found that the binding of *t*RNA to the synthetase–amino acid–AMP complex caused a conformation change in the protein structure. This could not be followed by the usual technique of optical rotatory dispersion because the ORDs of the enzyme and *t*RNA overlap, but their effects were clearly distinguished by using circular dichroism. When tyr-*t*RNA synthetase was incubated with its specific *t*RNA there was a CD change indicating a slight loss of helical structure in the enzyme—no other *t*RNA caused this. Inactivation of the *t*RNA with periodate (this attacks only the 3′ terminal ribose) caused loss of the ability to induce the conformational change.

Yarus and Berg (1969) found that the binding of isoleucine to its synthetase promotes an increase in both the rates of dissociation and association of its *t*RNA—since both are raised about six-fold, the equilibrium constant remains unaltered. This suggests that the availability of the site to *t*RNA—both for entry and for exit—is increased by the binding of the amino acid. The maximum velocity of the acylation reaction itself is higher than the maximum rate at which the ile-*t*RNA can leave its site, so that the release step may be the rate-limiting one. Measurements on the affinity of association between valine *t*RNA and its synthetase also suggest that dissociation of the enzyme–aminoacyl-*t*RNA complex is likely to be the rate-limiting step in the production of aminoacyl-*t*RNA (Yaniv and Gros, 1969).

Although their data only require two conformational states for the synthetase (one with slow entry and exit, and one with fast), Yarus and Berg (1969) have suggested a model in which the enzyme passes through four states during its catalytic cycle, as shown in Figure 8.9. Each moiety which binds to the enzyme causes a conformational change. The binding of isoleucine (reaction 1) 'opens' the *t*RNA binding site, which is then filled by *t*RNA (reaction 3). Binding of the *t*RNA in turn affects the amino acid binding site. When the amino acid is transferred onto the *t*RNA and then leaves its site (reaction 5), the binding of another isoleucine and AMP is necessary to reconvert the *t*RNA site into its 'open' form (reaction 6). A second isoleucyl–adenylate complex is formed before the first ile-*t*RNA is released, assisting the rate-limiting step

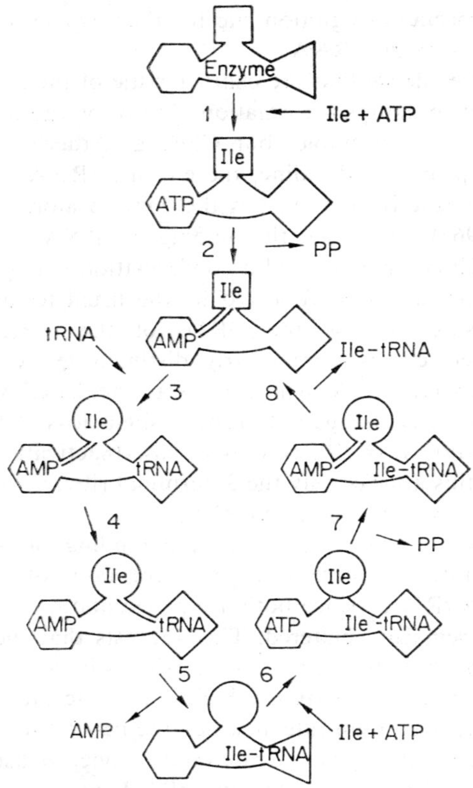

Figure 8.9. The catalytic cycle of isoleucyl *t*RNA synthetase. This model suggests that the enzyme can exist in any of four conformations, and the binding of its amino acid or *t*RNA alters its conformation. When isoleucine is bound, the *t*RNA binding state is in an 'open' state, permitting rapid entry and exit of the molecule. Because the model permits a second isoleucyl –AMP complex to be formed before the first ile-*t*RNA has been released, the amino acid binding site is occupied at the time of ile-*t*RNA release, thus increasing the rate of its dissociation from the enzyme, which is probably the rate-limiting step in the production of aminoacyl-*t*RNA. The detailed reaction cycle postulated is:

1. Isoleucine and ATP bind. This converts the *t*RNA binding site into its 'open' form.
2. Isoleucyl–AMP is formed.

(*continued opposite*)

(reaction 8) in which the ile-*t*RNA is lost. The cycle is then repeated. Yaniv and Gros (1969) found that the valine enzyme has *two* distinct and relatively independent sites, one which is responsible for the specific recognition of the acceptor *t*RNA, and one which activates the amino acid and transfers it to the *t*RNA. Since the Yarus and Berg model does not require the AMP binding site to change its configuration, the model is consistent with these data.

Baldwin and Berg (1966b) examined in some detail the mechanism responsible for excluding transfer of the wrong amino acid to *t*RNA. They formed two complexes with the activating enzyme for isoleucine; in one the correct amino acid, isoleucine, was bound, and in the second the related amino acid, valine, was bound to the enzyme through the aminoacyl–adenylate complex. The complexes were isolated by gel filtration and then incubated with *t*RNA$_{ile}$; the ile–AMP–enz$_{ile}$ complex transferred its amino acid to the *t*RNA, whilst the val–AMP–enz$_{ile}$ complex broke down to release valine. Breakdown of the 'wrong' complex was specific for *t*RNA$_{ile}$ and no other transfer species could achieve this effect. Digestion of the isoleucine *t*RNA with ribonuclease caused loss of ability to produce complex breakdown, and there was a strict correlation between the abilities of the partially degraded transfer molecule to accept its amino acid and to cause complex breakdown. There seems to be a positive mechanism for achieving breakdown of the wrong complex—it is not likely to be due to just a conformational change caused in the protein by *t*RNA binding, because several chemically modified *t*RNA$_{ile}$ species which could interact with the enzyme did not cause breakdown. A requirement for free 3′ hydroxyl group on the *t*RNA$_{ile}$ suggests that the enzyme may start the procedure for transfer of valine to the *t*RNA$_{ile}$, but somewhere during or after its formation the complex val-*t*RNA$_{ile}$ is destroyed.

---

3. The *t*RNA binds. This changes the conformation of the amino acid binding site.
4. Isoleucine is transferred from AMP to *t*RNA.
5. The amino acid leaves its site. This changes the *t*RNA binding site into the 'closed' form.
6. Isoleucine and AMP bind, reconverting the *t*RNA binding site into the 'open' form.
7. Isoleucyl-AMP is formed.
8. Isoleucyl-*t*RNA is released, changing the conformation of the amino acid binding site.

(From Yarus and Berg, 1969)

Codon–anticodon recognition

A feature of the genetic code is that whilst a change in the first or second base of a triplet generally results in a codon representing a different amino acid, a change in the third base very often gives another triplet representing the same amino acid. Thus the third base is not as rigidly defined as are the first two. The patterns of third base degeneracy are:

in 8 cases	any of	U C A G	that is, 32 codons
1		U C A	3
7	either	U C	14
6		A G	12 (including 2 CT)
2	only	G	2
1		A	1 (CT)

This raises the question of whether one transfer molecule responds to all the codons related by third base degeneracy for some particular amino acid, or whether each triplet has a different *t*RNA. If only one molecule is required, it follows that the base pairing between the third base of the codon and its complementary base on the anticodon cannot always follow the normal pattern of A = U, G = C.

The pattern of third base degeneracy shows that it is not possible for the third base to be unique as a C alone or an A alone; any codon ending in C represents the same amino acid if U is substituted, and similarly, any amino acid codon ending in A is not affected by the substitution of a G (UGA represents nonsense and therefore does not have a *t*RNA). Crick (1966) proposed that there must be a certain amount of 'wobble' in the third base position so that the anticodon on any one *t*RNA responds to more than one codon. The degeneracy patterns can be accounted for if a certain amount of non-standard pairing takes place in the third position such that:

U in the anticodon recognizes A or G in the codon
C in the anticodon recognizes G only in the codon
A in the anticodon recognizes U only in the codon
G in the anticodon recognizes C or U in the codon
I in the anticodon recognizes U C or A in the codon

C and A in the anticodon will pair normally, but U and G also pair with each other. Inosine, one of the unusual bases in *t*RNA, has the base hypoxanthine which can pair equally well with any of the normal four except guanine.

Experimental work has concentrated on isolating single transfer species and determining which triplets they respond to in the ribosome binding test, or which synthetic polynucleotide sequences stimulate their activity in *in vitro* protein synthesis. Soll, Cherayil and Bock (1967) reported on the recognition of synonym codons (that is, different codons representing the same amino acid) in yeast and *E. coli* *t*RNA. They observed that *t*RNA species are capable of multiple codon recognition, and the patterns observed are consistent with the wobble hypothesis. Recognition of three codons by one *t*RNA always involved U, C and A as the terminal bases of the codon, recognition of two codons occurred when either both purines or both pyrimidines could occupy the terminal position, and when only one codon was recognized the third base was always guanine. The only predicted type of *t*RNA not found was one recognizing a codon terminating in only U, and this may be because adenosine in the anticodon could be subject to a ready deamination, which would change its coding properties.

Although the meaning of codons is the same in both *E. coli* and yeast, their patterns of codon–anticodon recognition differ. For example, in one a third base degeneracy of U/C/A/G may be provided by *t*RNAs responding to U/C and A/G, and in the other by species for U/C/A and G. There is no particular consistency in the type of recognition pattern shown. This finding also extends to mammalian systems. Caskey, Beaudet and Nirenberg (1968) compared the trinucleotide binding activity of *E. coli* and guinea pig aminoacyl *t*RNA preparations corresponding to twelve codon sets representing six amino acids. They found that only about half the codon–anticodon recognition arrangements were common to both, and the remainder were unique for each species. Figure 8.10 provides a summary of the patterns of recognition so far worked out for these three systems. (Marshall and Nirenberg, 1969, have since reported that thirty-seven codons in *Xenopus caevis* represent the same twelve amino acids as in *E. coli*, but again, although the overall correspondence is the same, there are differences in the precise pattern of responses.)

Direct proof of the validity of the wobble hypotheis can be provided by determining the coding properties of those *t*RNAs which have been sequenced. In this case, the sequences of the nucleotides comprising both codon and anticodon are known. As can be seen from Table 8.1, all show the expected codon recognitions. The tyr_{su} species is an *amber* suppressor which responds to UAG. Met-*t*RNA_f is not degenerate in the third position since wobble predicts recognition of only G, and its first-base degeneracy is presumably accounted for on another basis.

Wobble would enable considerably fewer than sixty-one transfer molecules to utilize all the code-words, but both yeast and *E. coli* contain

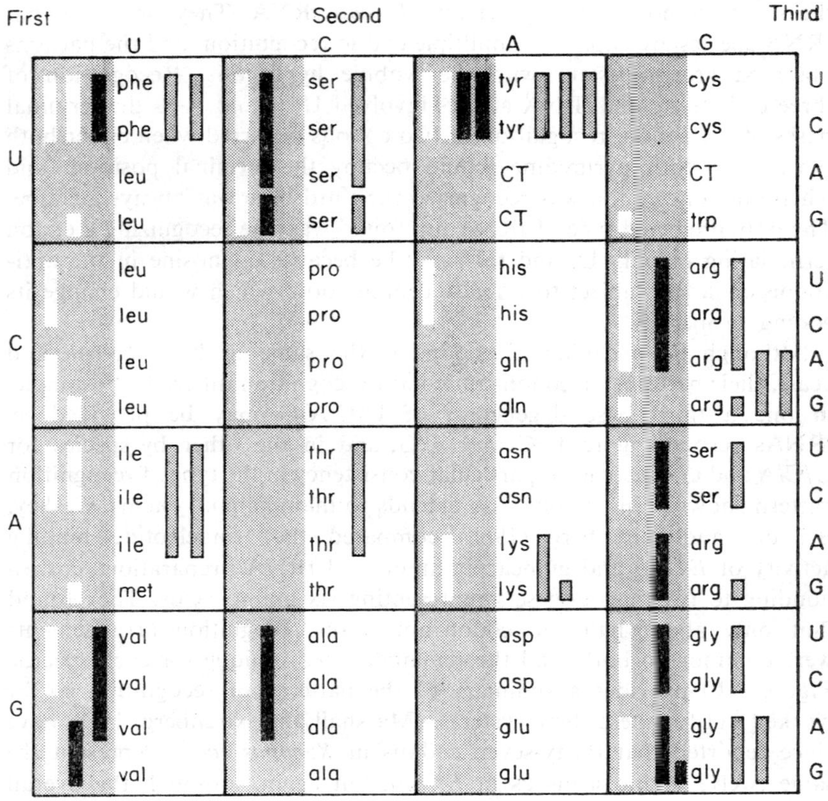

Figure 8.10. Patterns of *t*RNA codon recognition in *E. coli*, yeast and guinea pig liver. This data is based on binding of aminoacyl-*t*RNA fractions to trinucleotides. There does not appear to be any particular pattern of recognition used by any particular organism. Where more than one *t*RNA species responds to the same codons, this is indicated by more lines. (Based on data reported by Caskey, Beaudet and Nirenberg, 1968; Ishikura and Nishimura, 1968; Mirzabekov and coworkers, 1967; Nishimura and Weinstein, 1969; Soll, Cherayil and Bock, 1967; Soll and RajBhandary, 1967; Staehelin and coworkers, 1968)

Table 8.1.

*t*RNA	Anticodon	Codon Recognition	Organism
ala	IGC	GC U/C/A	yeast
phe	OMeGAA	UU U/C	yeast
tyr	GψA	UA U/C	yeast
ser	IGA	UC U/C/A	yeast
met_f	CAU	AU G	*E. coli*
met_m	CAU	AU G	*E. coli*
tyr	*GUA	UA C/U	*E. coli*
tyr_{su}	CUA	UA G	*E. coli*
ser	IGA	UC U/C/A	rat

more transfer RNA species than would be required to recognize all the meaningful codons. The multiplicity of *t*RNA molecules representing a single amino acid arises from two causes. As well as the degenerate species recognizing the synonym codons for a single amino acid, there is *redundancy* in that several *t*RNAs may respond to the same codon, but are structurally different in other parts of the molecule. According to the data of Muench and Safille (1968), there are (at least) some fifty-six transfer species in *E. coli* which can be separated by gradient partition chromatography. The number of *t*RNA molecules for each amino acid varies from two to five, and it is notable that each has more than one transfer type—this may have evolutionary advantages in resisting mutational damage. The number of *t*RNAs for a particular amino acid does not appear to have any correlation with its number of codons.

In *E. coli*, a particular *t*RNA may be represented by more than one cistron in the genome. Two *t*RNA molecules (with the same anticodon) accept tyrosine, and the first is represented by two genes, as shown by the production of an additional *t*RNA (a suppressor) when one of these genes is mutated. The three genes function in the proportions 25 : 15 : 60 (the first two synthesizing the same species), and the need for this is obscure since both *t*RNA types have almost identical sequences—they differ in only two nucleotides. Indeed, differences between transfer molecules responding to the same codon are usually rather limited, as could result if the *t*RNAs evolved by gene duplication. Zachau, Dutting and Feldman (1966) found that the two serine-accepting species of yeast are related by three base changes, and Dube and coworkers (1968) reported that met-*t*RNA of *E. coli* comprises a major (75 per cent.) species with a 7-Me-G where the minor has an A. In the serine species the base differences are located in the loops so that they both appear to have the same secondary structure.

CHAPTER 9

Suppression

A base change in the DNA of a cell will alter a codon in its RNA, resulting in the substitution of a different amino acid from wild type in the corresponding polypeptide chain. If this is acceptable to the protein there will be no phenotypic effect, but if unacceptable a missense mutation will be detected. Alternatively, a nonsense codon may arise, resulting in premature termination of the polypeptide chain. Mutants can be restored to the original phenotype either by reversion at the original mutation site, or by mutation at a different locus. The former returns the wild genotype of the organism; the latter phenomenon is suppression. Suppressors may be within the same cistron as the original mutation (although at a different site), or in a different cistron (either on the same or a different chromosome). Suppression may be considered in three main classes.

Indirect suppression is trivial; the primary mutation is circumvented rather than repaired. This could happen by opening another metabolic pathway, substituting the function of the product of the mutant gene by some other cell component, or changing cytoplasmic conditions to stabilize a previously unstable product.

Intragenic suppression falls into two categories. When there has been a mutation resulting in an amino acid substitution which causes inactivation of protein activity, a second substitution at a different point in the same polypeptide chain may restore the protein function. An example of this occurs in the *E. coli* tryptophan synthetase A protein; substitution of a glutamic acid for a glycine at one site causes inactivation, but a second substitution some thirty-six amino acids away of cysteine by tyrosine (also inactive by itself) restores activity to the protein.

When deletion or addition of a base has occurred, the correct code meaning can be restored by introduction of a second, compensating frame shift in the same cistron. The work of Terzhagi and coworkers (1966) showed that a (+−) combination of T4 *r*II acridine mutants which were themselves inactive resulted in activity in a protein differing from the wild type by only the five amino acids between the two sites of mutation (see page 82).

A major criterion for distinguishing between frameshift suppression and compensating substitutions is the extent of change in the amino acid composition of the pseudo-wild protein. It might be difficult to distinguish the extreme case when the acridine addition and deletion are so close together that only a couple of amino acid changes are involved.

In the case of informational (intergenic) suppression, the second mutation concerns one of the factors controlling the mechanism of transfer of genetic information from DNA to protein. The meaning of a codon is altered, and this may be due to mutation in the genome, and thus genetic, or due to an external effect, in which case the suppression is phenotypic. The suppression can take place at any one of several levels of action. The transcription process might be altered so that mistakes are made in the production of the messenger, or the messenger itself might be modified after transcription, for example, by a methylation causing recognition of the wrong anticodons. There might be alterations in components of the translation machinery, and these could comprise an effect on the specificity of an activating enzyme, alteration by mutation in the anticodon of a transfer molecule, or a modification of the ribosome affecting the accuracy of translation. These latter two phenomena are well known and have been studied in some detail.

*t*RNA translation errors

Benzer and Champe (1962) and Garen and Siddiqui (1962) first suggested that nonsense codons might be suppressed by production of a transfer species which recognized them as amino acid code-words, thus permitting protein synthesis to continue. Garen, Garen and Wilhelm (1965) studied three suppressor genes which act upon UAG mutations in the *E. coli* alkaline phosphatase gene. These mapped in different regions of the genome, and each could exist in an active (su^+) or inactive (su^-) state. The presence of any one active suppressor gene was sufficient to allow expression of nonsense mutants. Marked differences were observed in the enzymic behaviour of the phosphatase molecules produced by each suppression, and it was proposed that each suppressor results in a different structural alteration due to the insertion of a different amino acid at the position specified by the nonsense codon. Subsequent work (Weigert and Garen, 1965a, 1965b; Weigert, Lanka and Garen, 1965) showed that the three suppressors su_1, su_2, su_3 insert serine, glutamine, or tyrosine, respectively. Stretton and Brenner (1965) demonstrated that they insert the same amino acids at nonsense sites in T4 mutants, and allow propagation of chain synthesis with different but characteristic frequencies ($su_1 = 63$ per cent., $su_2 = 30$ per cent., $su_3 =$ 51 per cent.). Similar results have been obtained for *ochre* suppressors (see Brenner, Stretton and Kaplan, 1965), although these are less detailed, and the amino acid inserted is not always known, due to their lower efficiencies of suppression.

Capecchi and Gussin (1965) showed that the component active in suppression of the phosphatase gene of an su_1^+ *E. coli* is a serine-accepting transfer species not present in the non-permissive isogenic parent strain. Synthesis *in vitro* of coat protein of RNA phage R17 carrying a nonsense mutation was examined using extracts from both the suppressor and non-suppressor strains. No functional protein was produced in a system derived from the su_1^- parent strain, but when serine accepting *t*RNA from su_1^+ was added to this system, functional coat protein was produced.

Thus a strain which suppresses *amber* mutations possesses a *t*RNA which recognises UAG as a signal for an amino acid; this species is absent in the unsuppressed strain. The new *t*RNA in the su^+ could arise in either of two ways. The su^- might contain a *t*RNA corresponding to UAG which could be mutated in some way so that it carries an amino acid in the su^+—but this is rather unlikely since there is no evidence for the existence of such a chain-terminating transfer species. More likely, a normal transfer species in the su^- for the amino acid to be inserted in suppression might be modified in the su^+, so that it recognizes UAG instead of, or as well as, its normal recognition codon. In support of this, serine, glutamine and tyrosine all have codons (amongst their degeneracies) which are related to the UAG codon by a single base change. Also, *amber* suppressors can be converted to *ochre* by a single base mutation (Ohlsson, Strigini and Beckwith, 1968), and this can be accounted for by a further mutation, which converts the anticodon CUA (recognizes UAG) to UUA (recognizes UAA/G). This mechanism of suppression explains why *ochre* suppressors always suppress *amber* codons as well as *ochre*; according to the wobble hypothesis, the UUA anticodon must recognize both the codons UAA and UAG. The difference in suppression efficiencies (about five-fold) between these allelic *amber* and *ochre* suppressors implies that, in this instance, it must be a property of the codon–anticodon recognition.

Suppressor transfer molecules might be produced in either of two ways. Transfer RNA synthesis occurs by a process in which the immediate gene product (proto-*t*RNA) is modified enzymically to give the mature species. The *su* gene might code for one of the modifying enzymes (such as a methylase) to produce an altered enzyme activity resulting in a change in the coding properties of the *t*RNA. Alternatively, *su* might be the structural gene for *t*RNA, in which case the mutation would be a base change in its anticodon, resulting in recognition of UAG instead of the normal codon. The existence of allelic amber and ochre suppressors suggests that this is more likely.

Smith and coworkers (1966) reported that it was not possible to isolate the suppressor *t*RNA from su_3^+ cells because it is present in only very small amounts. Genetic mapping located the site of the suppressor gene as closely linked to the attachment site of phage $\phi80$; thus lysogenization followed by induction was used to obtain defective transducing phage in which a portion of the phage chromosome was exchanged for a fragment of the *E. coli* genome carrying the su_3 gene. Both su_3^+ and the wild-type su_3^- genes were placed on $\phi80$ to give $\phi80dsu_3^+$ and $\phi80dsu_3^-$, respectively. Using these derivatives, the level of the su_3 gene was raised in cells by infection and phage multiplication. The typical experimental procedure was to use phage defective in late functions so that cell lysis is delayed. After infection, up to 60 minutes are allowed for a pool of phage DNA to be established and chloramphenicol is then added; this blocks protein synthesis, thus preventing cell lysis, but DNA and RNA synthesis continue, allowing synthesis of increasing quantities of the su_3 gene product. After 180 minutes *t*RNA is extracted from the cells by phenol.

Comparison of the accepting level for tyrosine and other amino acids between cells infected with $\phi80dsu_3^+$ and cells infected with normal phage showed that the relative proportion of tyrosine *t*RNA was greatly increased (six to seven times) in the former, whereas the accepting capacity for other amino acids was unchanged. The coding properties of the tyrosine *t*RNA in these preparations was studied by the ribosome binding technique. Tyrosine is normally coded by UAU/C; in cells infected with phage $\phi80$ the tyrosine *t*RNA responded only to these species. However, in $\phi80dsu_3^+$ infected cells binding was stimulated by UAG as well as UAU and UAC, and there was a rough correlation between the proportion of tyrosine *t*RNA binding to UAG relative to UAU and the increase in tyrosine *t*RNA level upon infection with $\phi80dsu_3^+$. This suggests that the newly synthesized *t*RNA due to the su_3^+ gene recognizes UAG, but not the normal codons UAU and UAC. As expected, the suppressor transfer species did not recognize the UAA *ochre* triplet. When similar experiments were performed with $\phi80dsu_3^-$ there was an increase in tyrosine-accepting activity, but only in response to the usual codons.

These results suggest (although they do not formally prove) that the su_3 gene codes directly for a tyrosine-accepting transfer species rather than for a modifying enzyme. Landy and coworkers (1967) found that at least 35 per cent. of the P^{32}-labelled *t*RNA synthesized after infection with phage $\phi 80dsu_3^+$ was homologous to the DNA of the phage; the homology is specific for the su_3 gene since the *t*RNA did not bind to the DNA of $\phi 80$ itself. At saturation, there was one *t*RNA molecule bound to each $\phi 80dsu_3^+$ DNA. This confirms that the information carried by the defective phage is the structural gene for the *t*RNA.

Goodman and coworkers (1968) determined the nucleotide sequences of su_3^+ tyrosine *t*RNA and the *t*RNA specified by its su_3^- allele. The two species differ in only one fragment on their fingerprints. The expected anticodon for recognition of the tyrosine codons UAU/C is GUA, and this proved to be present from the su_3^-gene (yeast tyrosine *t*RNA has the anticodon GψA). In su_3^+ infected cells, this was replaced by the anticodon CUA, which would recognize the *amber* UAG triplet. Three genes in *E. coli* are responsible for synthesis of normal tyrosine *t*RNA; two specify the su_3^- species (*t*RNA_{tyr}I), and the other another closely related molecule (*t*RNA_{tyr}II). The su_3^+ mutation represents a change in the less active of the two genes specifying *t*RNA_{tyr}I, so that although this altered species can achieve suppression, the mutation does not deprive the cell of the *t*RNA_{tyr}I species.

Ribosomal mistranslation

The antibiotic streptomycin is an inhibitor of protein synthesis and also increases the level of misreading of messengers at translation; although both the effects result from its interaction with one of the ribosomal proteins, the inhibition is not caused by the misreading effect but has an independent basis. Davies (1964) and Cox, White and Flaks (1964) showed that the site of action of the antibiotic is the 30S subunit of the ribosome, for when 30S and 50S subunits prepared from *str-s* (streptomycin-sensitive) and *str-r* (streptomycin-resistant) strains of *E. coli* were reaggregated, only ribosomes possessing a 30S subunit derived from the sensitive strain suffered streptomycin-inhibition of protein synthesis *in vitro*. After dissociation of the 30S subunits into cores and split proteins, cross-reconstitution experiments with strains sensitive and resistant to the antibiotic showed that the cores are responsible (Staehelin and Meselson, 1966b). Traub and Nomura (1968c) showed further that the property of resistance is conferred by the core proteins and not the 16S *r*RNA; streptomycin was only able to exert its effect when the core proteins were derived from a sensitive strain—the origin of the *r*RNA had no effect. In fact, experiments in which the activity of cross-reconstituted

30S subunits was tested *in vitro* by poly-U or phage f_2 RNA directed translation showed that mutation in a single core protein is responsible for conferring resistance to both the inhibitory and misreading effects (Ozaki, Mizushima and Nomura, 1969). This protein was characterized as a single band (P_{10}) by migration on gel electrophoresis, and probably comprises a single molecular species. In a similar manner, Bollen and coworkers (1969) have been able to demonstrate that one of the split proteins of the 30S subunit confers the property of resistance to the antibiotic spectinomycin.

A single genetic locus on the *E. coli* chromosome governs resistance to streptomycin, and the genetic implication of this—now also confirmed by biochemical studies—is that it specifies an alteration in the 30S core protein which it synthesizes, such that any interaction with streptomycin, if it still occurs, no longer affects the functioning of the subunit. Indeed, several mutants which affect the 30S subunit of the ribosome all map in this region of the chromosome. Resistance to the antibiotic spectinomycin maps at a site closely linked to the classical streptomycin locus (Flaks and coworkers, 1966), and a ribosomal ambiguity mutation (*ram*) which affects the accuracy of ribosome function in translation (Rosset and Gorini, 1969) is closely linked to *spc*. The K12 ribosomal protein which is electrophoretically unique to the 30S subunit of this strain (see page 160) maps very closely indeed to the *str* locus. The clustering of these mutant activities suggests that this region of the genome may specify a number of its structural components. By contrast, Apirion (1967) isolated mutants in the response to the antibiotics erythromycin and lincomycin, which is expressed by the 50S subunit. Three genes proved to be concerned, and these are unlinked to each other or to the *str* region; thus there does not appear to be a similar clustering of the genes specifying the proteins of the larger subunit.

Gorini and Kataja (1964, 1965) discovered a class of *E. coli* mutants implying the existence of a suppression mechanism activated by streptomycin. Some 1–5 per cent. of the auxotrophic mutants obtained by mutagenic treatment were able to grow in the absence of their specific growth factor provided that streptomycin was present in the medium. This was termed the CSD (conditional streptomycin dependence) phenotype since these strains were dependent upon streptomycin for growth in minimal medium but did not require it in a medium supplemented with their specific growth factor. The distribution of CSD mutants amongst different auxotrophs was sufficiently random to exclude the possibility that this action could be due to its participation as substrate or cofactor in some reaction of intermediary metabolism. This implies that its presence must activate some mechanism which suppresses the mutation and thus obviates the need for the growth factor.

The suppression is exerted by an interaction of streptomycin with the ribosome to cause misreading of codons, with the consequence of amino acid misincorporation. Pestka, Marshall and Nirenberg (1965) and Kaji and Kaji (1965) found that there was an increase in the level of ambiguity in response to a poly-U template in the presence of streptomycin. The antibiotic inhibits the binding of phe-*t*RNA but promotes that of ile-*t*RNA to the ribosome messenger complex. It acts similarly on both aminoacyl *t*RNA and uncharged species, which suggests that its action is at the level of the interaction between *t*RNA and *m*RNA before the peptide bond synthesis step.

Davies, Jones and Khorana (1965) found that the effect of streptomycin on artificial polymer systems revealed definite patterns of misreading. With poly-U, the incorporation of isoleucine (first two nucleotides of codon AU) was stimulated, as was the incorporation of serine (UC). With poly-C, the incorporation of serine, histidine (CA), and threonine (AC) was stimulated in addition to the normal proline. (Possible effects on the 3′ terminal position of codons were ignored because wobble would make interpretation too difficult.) Since the triplets for the misincorporated amino acids are related to the messenger codons by a single base change it seems that only one base is misread at a time. The misincorporation results suggest that streptomycin encourages misreading of U as C or A, and of C as U or A in (at least) the 5′ and internal positions of the codon. Experiments with other polymers suggested that only the pyrimidine bases are misread, and most usually as a transition to the other pyrimidine, although there does seem to be a certain level of misreading as A. Davies (1966) performed experiments with chemically defined templates to test for possible neighbouring base effects. Poly-UC gave surprising results; its range of misreading was much more restricted than that of either poly-U or poly-C, so that the misreading of a particular base may be influenced by its neighbours.

Apirion (1966) and Apirion and Schlessinger (1967) demonstrated that mutations to the CSD phenotype, *str-d*, map at the same locus as the *str-s*/*r* alleles. Luzzato, Schlessinger and Apirion (1968) showed that, in fact, one allele (*str-r*) confers either resistance to streptomycin or dependence upon it; when *str-r E. coli* K12 were transduced into an *E. coli* C *str-s* strain, both streptomycin-resistant and -dependent transductants were found. These alternative expressions of the one gene are determined by the genetic background of the bacterium receiving the transducing fragment; another closely linked locus acts as a modifier to determine whether its phenotypic expression is as *str-R* or *str-D*. The two loci are closely linked since they can be cotransduced, but probably in different genes because they are frequently separated. It seems likely that the modifier locus specifies another ribosomal protein whose interaction with that determined by the *str-r* gene determines the phenotypic response of the latter to streptomycin.

Gorini, Jacoby and Breckenridge (1966) observed that the allele at the streptomycin locus, that is the structure of the ribosome, influences suppression mediated by altered *t*RNA molecules as well as affecting the interaction with streptomycin. A given mutant may impose a *restriction* on the efficiency of some other suppression, such as nonsense codon suppression by *t*RNA, and this can be so severe that the suppressed phenotype disappears. The *str-s/r* alleles affect suppression by different *t*RNA mutants independently, and the restriction of suppression is not necessarily associated with only streptomycin resistance, but can also be shown by a *str-s* allele.

Alterations in ribosome structure should be brought about not only by external agents, such as streptomycin, but also by mutation in the genes which code for ribosomal proteins. If the structure of the ribosome can control the degree of ambiguity with which the code is read, it should be possible to isolate ribosomal mutants which permit greater ambiguity. These would probably usually be lethal, and Rosset and Gorini (1969) used the phenomenon of restriction to assist in their isolation. Since restriction affects the translation machinery in a general manner, they argued that ambiguity and restriction might counteract each other, so that a double mutant would show the wild phenotype. They thus isolated mutants of a restrictive strain which are less restrictive than their parent, that is which allow the suppression restricted by the parent.

A mutant in another gene, *ram*, could neutralize the effect of a restrictive mutation in the *str* gene. The *ram* mutation alone produced a suppression of nonsense codons, that is increased ambiguity. The level of ambiguity was tested *in vitro* by measuring the incorporation of isoleucine and serine in response to poly-U, and the results obtained corresponded well with the effect of the mutant on suppression *in vivo*. The normal degree of misreading was about 7 per cent. incorporation of the first amino acid, and 2 per cent. of the second. A restrictive *str* allele caused a considerable decrease of misreading to about 1–2 per cent. and 0·5–1 per cent., but *ram* mutation in this strain restored the level of misreading to about the wild-type level. The *ram* mutation on its own about doubled the degree of misreading to 13 per cent. and 4 per cent. *In vivo*, the *ram* mutant could produce a suppressed phenotype similar to that resulting from the addition of streptomycin, and the effects of the mutation and the antibiotic proved to be additive (or possibly even to potentiate each other). Mixed reconstitution experiments with the *in vitro* system showed that *ram* is a genetic determinant of the 30S subunit of the ribosome, so that the effect of the mutation is to introduce an ambiguity in translation through an alteration in the structure of this subunit.

Both streptomycin-activated and genetic ribosomal suppression probably misread a number of codons to give a variety of amino acid replacements. Ribosomal suppressors should thus be far less specific than the codon-specific *t*RNA suppression. The effects of such mistaken reading should be wide ranging; it may cause suppression by reading an altered codon incorrectly so as to restore incorporation of the correct amino acid, but this codon will be used as sense elsewhere in the genome and thus will be translated there as missense. Effective missense suppressors would therefore probably be lethal to the cell, and, indeed, only weak missense suppressors have been found.

The streptomycin-induced misreading in bacteria results from the interaction of the drug with the P_{10} ribosomal 30S core protein; Ozaki, Mizushima and Nomura (1969) found that reconstituted subunits in which this protein was derived from sensitive strains showed a ten-fold increase in ambiguity over those with a resistant protein. It proved possible to reconstitute subunits lacking this protein altogether, and these sedimented in a manner identical to the control fully-reconstituted subunits. However, the P_{10}-deficient particles were completely resistant to streptomycin-induced ambiguity. Usually, ribosomes from strains resistant to streptomycin are sensitive to certain other antibiotics which also cause abiguity, such as neomycin and kanomycin. However, P_{10}-deficient particles were equally resistant to the ambiguity effects of both these agents. This suggests that the function of P_{10} in translational fidelity is to influence the ribosomal-error frequency, that is, the degree of misreading in translation is an inherent property of the bacterial ribosome.

Attempts have been made to induce misreading in a rabbit recticulocyte system with various factors, including streptomycin, ethanol and aliphatic polyamines such as spermine (Friedman, Berezney and Weinstein, 1968; Weinstein, Friedman and Ochoa, 1966). Although these enhance ambiguity considerably in bacterial systems, they had little or no effect upon systems containing the mammalian ribosomes. The difference in induced ambiguity between the two systems implies that the active role which the ribosome plays in codon–anticodon recognition has different characteristics in bacteria and mammals.

At first, it was thought that misreading might provide the basic mechanism for the antibiotic action of streptomycin through flooding the bacterial cell with non-functional proteins affecting all its activities. Misreading might also result in spurious chain initiation and termination. However, Davies (1966) found that streptomycin can exert an inhibitory effect on protein synthesis *in vitro* in the absence of misreading, such as when poly-AG is used as template. And further, it appears that at bacteriacidal concentrations, the antibiotic inhibits protein synthesis before there is time for misreading to occur.

Luzzato, Apirion and Schlessinger (1968, 1969) analysed the relative proportions of ribosomal subunits, monomers and polysomes present after the addition of streptomycin to cells. They found that a change in the distribution of ribosomes occurred concomitant with its inhibition of protein synthesis, such that the number of polysomes rapidly decreased and most became located as monomers attached to *m*RNA. The streptomycin-induced monomer formation is irreversible, and the very tight binding of the drug to the monosomes parallels the loss of ribosome function and cell viability. This suggests that the inhibition of ribosome function lies at or soon after chain initiation; this would allow ribosomes engaged in translation to complete polypeptide chain synthesis by running-off polysomes, but they would then accumulate as abortive initiation complexes with a new messenger. Although streptomycin inhibits chain elongation and promotes misreading when bound to ribosomes which are synthesizing proteins, its effect on initiation is probably the most relevant to its bacteriacidal action.

The idea that inhibition is an event concerned with initiation is lent further support by the finding that it is much greater *in vitro* when a natural messenger is used to direct translation, compared with an artificial polymer such as poly-U. Further, the rise in magnesium concentration which abolishes the need for proper initiation also overcomes the inhibitory effect. Experiments using P_{10}-deficient particles (Ozaki, Mizushima and Nomura, 1969) have shown that the interaction of streptomycin with P_{10} is responsible for its inhibitory effect as well as for misreading, and have also confirmed that the site of the inhibition lies at initiation. With poly-U as template, P_{10}-deficient particles possessed only half of the activity of control fully-reconstituted subunits, but when under direction from f_2 RNA, their relative activity was reduced to 16–20 per cent. of the control. The cause of this reduction can be attributed more precisely, since whereas the binding of normal aminoacyl-*t*RNA under direction of f_2 was only reduced to about 80 per cent. of its usual activity, the binding of fmet-*t*RNA$_f$ was only 25 per cent. of the control. The addition of P_{10} restored full activity. This suggests that particles lacking the P_{10} protein may be deficient in some activity concerned with initiation, and experiments on binding fmet-*t*RNA_f under direction from the AUG triplet support this conclusion; the P_{10}-deficient particles showed only 6–17 per cent. of the activity of fully reconstituted subunits. It thus appears that the P_{10} protein must play a role in (at least) two of the functions of the ribosome, initiation and translational fidelity, and the species present in strains sensitive to streptomycin interacts with the drug so as to inhibit initiation and enhance ambiguity.

III

The Control of Protein Synthesis

CHAPTER 10

The operon

Induction and repression

It was demonstrated at the beginning of the century that certain enzymes of yeast are formed only in the presence of their specific substrates; this effect was subsequently termed enzyme adaptation and is now known as *induction*. It is a considerable one; Jacob and Monod (1961) stated that *E. coli* grown in the absence of a β-galactoside contain only about five molecules of the enzyme β-galactosidase per cell, whereas when grown in the presence of the enzyme substrate there is an increase of up to a thousandfold to 5000 molecules per cell. Enzyme activity can be detected very rapidly (within 2 to 3 minutes) after addition of its substrate, and removal results in an equally rapid cessation of activity. The rate of production varies according to the particular substrate employed, reaching a different saturation value for different substrates.

The effect could be due to either the activation of an inactive precursor form of the enzyme already present in the cell, or to the *de novo* synthesis of new enzyme molecules. Immunological studies have shown that the induced enzyme is antigenically distinct from any protein species previously present, whilst isotopic studies have confirmed that it is not derived from a precursor molecule synthesized before induction. Nakada and Magasanik (1964) separated the phase of enzyme induction from the phase of enzyme production by removing the inducing substance at the end of the induction lag (the short period between addition of inducer and appearance of enzyme). The cells then produce enzyme for 5 to 10 minutes at an exponentially declining rate. The addition of chloramphenicol at the beginning of the production phase inhibited the synthesis of β-galactosidase, confirming that *de novo* synthesis is involved rather than activation.

The phenomenon of *repression*, that is specific inhibition of enzyme synthesis, was discovered by Monod and Cohen-Bazire (1953) when it was found that biosynthesis of the *E. coli* enzyme tryptophan synthetase is inhibited selectively by tryptophan. Yates and Pardee (1957) used isotope incorporation experiments to demonstrate that the effect involved repression of enzyme synthesis and not merely inhibition of activity.

The effect of induction or repression is not usually confined to a single enzyme. In bacteria, the genes governing the synthesis of the various enzymes sequentially linked in a metabolic pathway are often found to be extremely closely linked to form a cluster on the genome. In such cases, there is generally simultaneous induction or repression of all the enzymes coded by these genes. In the *lactose* (*lac*) system of *E. coli*, three adjacent genes, *z*, *y* and *a*, code for the enzymes β-galactosidase (involved in degradation of β-galactosides), β-galactoside permease (concerned with transport) and β-galactoside transacetylase (function unknown). All three are induced coordinately.

The induction effect is extremely specific, and is generally caused only by the substrate of an enzyme, or substances closely related to it. These are termed *co-inducers* (or *co-repressors* for systems of repression). For example, only molecules with an intact unsubstituted galactosidic residue can act as co-inducer for the *E. coli lac* enzymes (Jacob and Monod, 1961). The ability to induce is not dependent upon ability to act as a substrate, since thiogalactosides (which cannot act as a substrate), such as isopropylthiogalactoside (IPTG), are excellent co-inducers. There is no inherent correlation between the molecular structure of the co-inducer, and the structure of the active centre of the enzyme, as is shown by the lack of any quantitative correlation between the inducing capacity and substrate activity of various galactosides. For example, melibiose, an α-galactoside lacking affinity for the enzyme, can act as co-inducer. Conclusive confirmation of this was provided by certain mutants of the *z* gene which synthesize a protein with the same immunological response as the normal enzyme, but which is devoid of catalytic activity due to loss of the active centre. However, in diploids carrying both the normal and mutated gene, both normal and inactive enzyme are formed to the same extent in the presence of co-inducer.

Regulator genes

There is complete correlation between the induction of each of the three enzymes concerned with β-galactoside metabolism. Not only are the same compounds active or inactive as co-inducers of each enzyme, but the relative amounts of the enzymes synthesized in the presence of different co-inducers or various concentrations of one co-inducer are constant, even although the absolute amounts may vary considerably. This suggests that the co-inducer governs their synthesis by interacting with some common controlling element. That the specificity of induction is not related to the structural specificity of the enzymes, and that the rate of synthesis of the different enzymes is under common control, would suggest that this element is not elaborated by the genes coding for the enzymes themselves.

This led Jacob and Monod (1961) to propose the concept of two different types of gene, structural genes and regulator genes. Whilst structural genes are concerned solely with elaborating proteins required by the cell, regulator genes are responsible for controlling the synthetic activity of these structural genes (although this control is exercised through synthesis of a regulator protein). Mutants of the controlling system should not behave as alleles of the structural cistrons, and such genes in the *lac* system were detected by the discovery of *constitutive* mutants. Whilst mutations in the structural genes result in the loss of activity of their respective enzymes, mutation in two regions, termed *o* and *i*, could produce bacteria with the abnormal ability to synthesize large amounts of the *lac* system enzymes in the absence of a co-inducer. These were termed constitutive. The *o* region maps genetically immediately to the left of the structural genes, and the *i* region a little further along the genome:

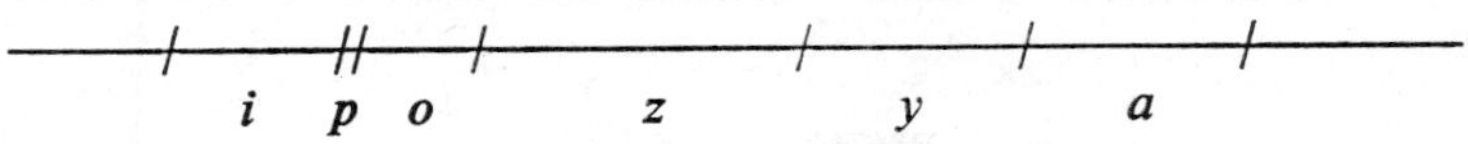

The normal alleles at these loci are termed o^+ and i^+ and the constitutive mutants i^- and o^c. Most constitutive mutants synthesize more enzyme than induced wild-type bacteria, but the relative proportions of the enzymes are the same as in normally induced cells, suggesting that these regulator loci provide the usual mechanism of control which interacts with the co-inducer.

The chromosomal location of the *lac* system is near the site of attachment of the F factor, and the use of F′ for sexduction gives diploids for these genes. Studies of the double heterozygotes

$$\frac{i^+z^-}{Fi^-z^+} \quad \text{and} \quad \frac{i^-z^+}{Fi^+z^-}$$

showed that the normal inducible allele i^+ is dominant over the constitutive i^-, and active in the *trans* position with respect to *z*. The same behaviour was found with regard to the other structural genes, *y* and *a*. Thus the *i* mutations must belong to an independent cistron which governs the expression of all three structural genes through some cytoplasmic component. The dominance of inducible over constitutive shows that the former is the active form of the gene; this was confirmed by strains carrying deletions of the *i* gene which behaved as i^-.

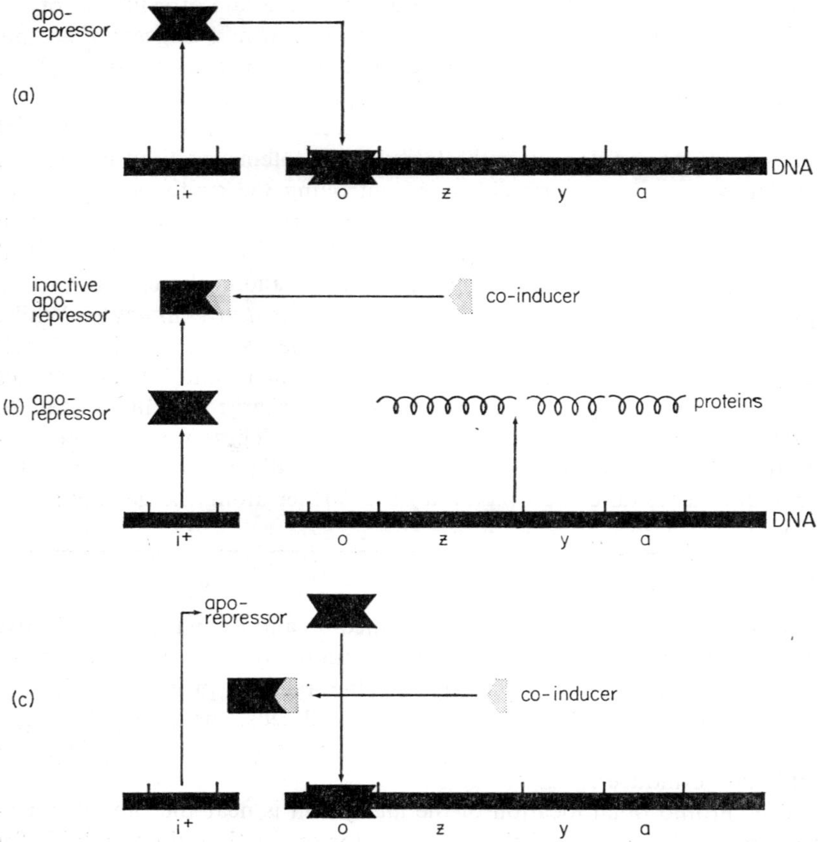

Figure 10.1. Regulation of the *lac* operon. (a) The regulator gene, i^+, synthesizes an apo-repressor which binds to the *lac* operator and prevents the operon from producing enzymes. (b) After the addition of a co-inducer, the apo-repressor suffers a conformational change which results in the loss of its ability to bind to the operator. As a result, the operon is no longer switched off and the structural genes synthesize proteins. (c) A representation of this system. (⊥) indicates a switch-off of the apo-repressor by the co-inducer. Thus in the absence of the co-inducer the apo-repressor switches off the operon, but the addition of the co-inducer inactivates the apo-repressor and prevents it from exercising this function

Jacob and Monod (1961) suggested that the i^+ gene governs the synthesis of an *apo-repressor*. The specificity of action of the i gene in affecting only the *lac* system, and its pleiotropy in controlling the synthesis of all three enzymes, implies that this apo-repressor acts upon some feature of the system common to all the structural genes. This is the *operator* (o) region, which is responsible for controlling the synthetic activity of the *zya* segment. In the absence of a co-inducer, the apo-repressor combines with the operator and this interaction switches off the *zya* structural genes. The presence of a co-inducer inactivates the apo-repressor so that it no longer combines with the operator; in the absence of such interaction the system resumes protein synthesis. This is illustrated diagramatically in Figure 10.1. (The prefixes apo- and co- serve to distinguish the regulating molecules synthesized by the bacterium and supplied externally by the medium, respectively.)

In i^- mutants the apo-repressor is either inactive or absent, so that the operator cannot be switched off and the three structural genes synthesize their enzymes continuously. The normal i^+ allele would be dominant to the i^- because its presence restores normal apo-repressor able to switch off the system, but responsive to inactivation by co-inducer. This is illustrated in Figure 10.2.

Another type of mutant in the i regulator gene was also isolated; i^s(*uninducible*) mutants lack the capacity to synthesize the *lac* enzymes, even in the presence of a co-inducer. This is not due to deletion of the structural genes but rather to an altered regulator. In diploids i^s/i^+, i^s is dominant, and these properties were accounted for by the proposal that the i^s allele produces an altered apo-repressor, which although still active in switching off the operator is no longer subject to inactivation by a co-inducer. This is dominant, because the presence of i^s means that the altered apo-repressor switches off the operon; the presence of a normal i^+ cannot interfere with this action. (See also Willson and coworkers, 1964, for details of i^s mutations.) This situation is illustrated in Figure 10.3.

The operator was identified and mapped genetically by the finding of o^c mutants. These were isolated by selecting for constitutivity in cells diploid for the *lac* system, that is $i^+z^+y^+a^+/\text{F-}i^+z^+y^+a^+$. Because the i^- constitutive mutant is recessive, mutation in both i^+ alleles would be required to produce i gene constitutivity, and such mutation is very rare. Constitutive operator mutants could arise from impairment of the affinity for apo-repressor so that the structural genes cannot be switched off. Such mutants must be dominant since the presence of a second operator sensitive to apo-repressor cannot affect the inability of the mutated operator to respond to control. If mutation from o^+ to o^c results from a loss of

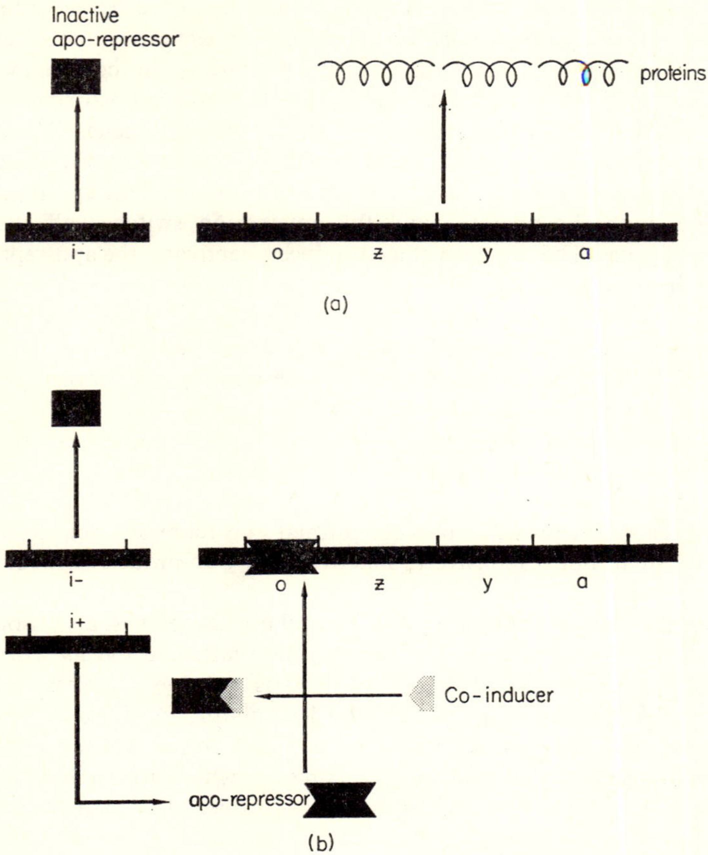

Figure 10.2. Constitutive regulator mutants in the *lac* system. (a) The i^- gene synthesizes an inactive apo-repressor which cannot switch off the operator. (b) However, if an i^+ gene is also present, it synthesizes the normal apo-repressor, thus restoring the response to the co-inducer. Hence i^+ is dominant over i^-

sensitivity of the operator to apo-repressor, the o^c cells must be insensitive to the presence of the altered apo-repressor synthesized by the i^s mutant; it was confirmed experimentally that o^c constitutivity holds over i^s uninducibility.

The o^c mutant is pleiotropic, affecting all the three structural genes of the system. It is *cis* dominant but *trans* recessive, that is it exerts dominant control over the genes in its own segment of DNA but has no effect on

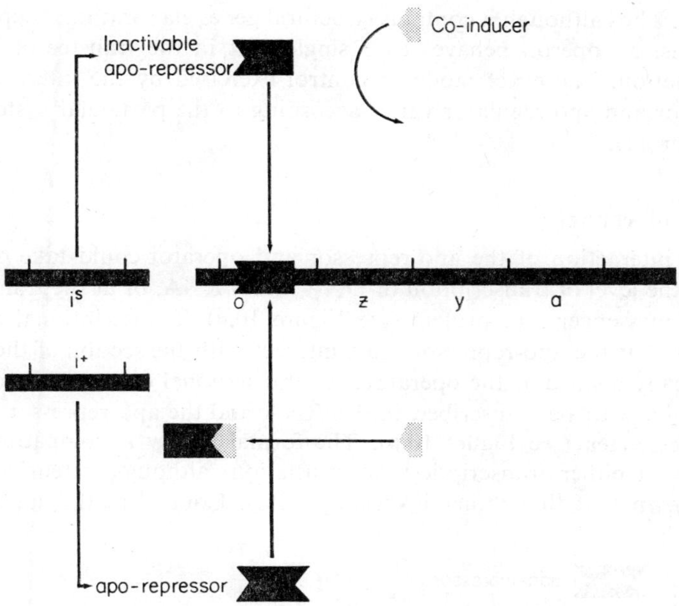

Figure 10.3. Uninducibility in the *lac* system. i^s produces an apo-repressor which cannot be inactivated by the co-inducer; it therefore binds to the operator and switches the operon off. The presence of an i^+ gene cannot affect this; thus i^s is dominant to i^+

those in a different DNA molecule. According to the complementation test, the *o* locus can be regarded as part of the same cistron as *z*, or indeed, similarly, as part of the same cistron as *y*. This excludes the possibility that some product of the operator gene can associate in the cytoplasm with the products of the structural genes to govern their behaviour. Jacob and Monod (1961) proposed that the operator must control some integral property of the adjacent *zya* segment, either on the genome by affecting transcription, or on the corresponding messenger molecule by controlling translation.

The term *operon* was proposed to describe the system as a whole, comprising both structural genes and regulatory functions. The structural genes, which code for a series of metabolically related proteins, comprise a cluster on the genetic map. Their expression is under coordinate genetic control and is governed by the interaction of an operator situated at one end of the gene cluster with an apo-regulator synthesized by a regulator gene—this latter need not necessarily map near the structural gene cluster. Thus although it contains structural genes elaborating independent proteins, an operon behaves as a single unit in the transfer of genetic information. The exact mode of control exercised by the interaction of operator and apo-regulator varies according to the particular system (see Chapter 11).

Levels of control

The interaction of the apo-repressor and operator could take place at either the level of transcription of DNA into *m*RNA, or at the translation of the messenger into protein (see Figure 10.4). Transcriptional control implies that the apo-repressor must interact with the section of the DNA genetically defined as the operator; for translational control this segment would have to be transcribed into *m*RNA and the apo-repressor act on the messenger (see Figure 10.4). The formal theory is compatible with control at either transcription or translation, although extensive work has shown that the former invariably holds. Control at the subsequent

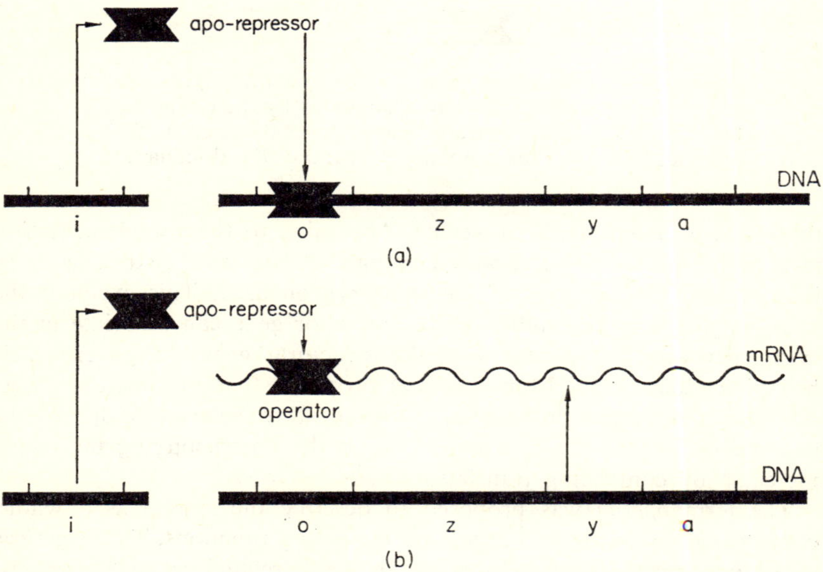

Figure 10.4. Levels of control. (a) The apo-repressor interacts with the operator region of the DNA and prevents transcription of messenger RNA. (b) The apo-repressor interacts with the operator region which has been transcribed into messenger and prevents the production of proteins. In this case, the operon would always make messengers

step of protein activity is termed enzyme inhibition (or activation), and in some instances the same co-effector may influence both production of an enzyme and its activity.

Both indirect experiments in which transcription is blocked, and direct detection of *m*RNA, have shown that all the bacterial operons so far investigated appear to be subject to control at the transcriptional level. Nakada and Magasanik (1964), working on the *E. coli lac* system, added the pyrimidine analogue 5-fluorouracil to cells five minutes before the addition of co-inducer. This resulted in failure of the enzyme activity to appear. The 5-FU is thought to be incorporated into *m*RNA to produce a messenger which codes for an altered, enzymically inactive protein—the presence of such an inactive protein related immunologically to β-galactosidase was found in these cells. However, when the analogue was added after enzyme induction, it had no effect. This indicates that an *m*RNA specific for β-galactosidase is produced during the induction phase, and this directs the subsequent synthesis of enzyme. No additional β-galactosidase *m*RNA was formed after removal of the co-inducer, but that which was already produced continued to synthesize protein even in the co-inducer-free medium.

The induced messenger species are *polycistronic*, each molecule representing transcription of the complete set of structural genes. A useful technique is to reveal the differences between induced and uninduced bacteria by labelling them with the same component—usually the RNA precursor uridine—carrying H^3 in one case and C^{14} in the other. The C^{14}/H^3 ratio is then constant in all RNA fractions except for the induced species. Guttman and Novick (1963) found only one change in the cellular RNA following induction of β-galactosidase, and this was the production of a new RNA sedimenting at 30S. Although it is difficult to decide exactly how many nucleotides this species comprises, it is certainly larger than would be expected if it were to represent only the *z* gene and code for merely the β-galactosidase enzyme. Martin and coworkers (1963) found that *Salmonella typhimurium* producing the ten enzymes responsible for histidine biosynthesis have extra RNA species compared with those in which the production of these enzymes is repressed. An RNA sedimenting at 34S is probably of the order of magnitude of the messenger which would be required to code for all the enzymes of the operon, and is certainly very much larger than could be accounted for by any one of its individual enzymes.

Imamoto and coworkers (1965) obtained phage $\phi 80$ particles carrying the five structural genes of the *E. coli* tryptophan (*trp*) operon. Hybridization experiments between RNA and the denatured transducing phage

DNA were performed both on cells grown under conditions in which the operon is active and on cells grown in the presence of tryptophan when it is repressed. This demonstrated that RNA complementary to these genes is produced only when the operon is active. Further work showed that each *trp* *m*RNA molecule carries all the genes of the operon.

Kiho and Rich (1965) found polysomes to which nascent β-galactosidase enzymes were attached. Genetic modifications to the genes of the *lac* operon were associated with changes in the polysome manufacturing the enzyme; in bacterial strains with deletions in *z* or *a* this polysome was much smaller. Baydasarian and Cièsla (1968) have detected an enzyme of the histidine operon (phosphohistidine phosphatase) in association with polysomes much too large to be responsible for only its synthesis—they are probably polycistronic messengers representing the whole operon. Thus the polycistronic messenger is used as such—it is not broken down into smaller one-cistron units for translation.

The kinetics of induction show that the addition of a co-inducer causes synthesis of protein at the maximum rate within a very short 'induction lag', and its removal causes cessation of synthesis just as rapidly. This rapid cessation of protein synthesis excludes the possibility that the *m*RNA formed is stable; either it must be inherently unstable or it must be destroyed enzymically very rapidly. Jacob and Monod (1961) introduced a chromosome loaded with P^{32} from Hfr z^+ cells into recipient z^- cells by conjugation; radioactive decay subsequently inactivates the introduced z^+ gene. They found that β-galactosidase is only produced in the progeny cells so long as integrity of the z^+ gene is maintained—this is the result that would be expected from a short-lived messenger template.

*E. coli lac m*RNA decays with a half life of about 2–2·5 minutes, and the rate of decay is independent of the presence or absence of a co-inducer, or other factors such as the overall rate of protein synthesis or energy production (Baker and Yanofsky, 1968a; Leive and Kollin, 1967; Nakada and Magasanik, 1964). This short life time is due to degradation of the messenger by nucleases very soon after its transcription has commenced (see pages 277–280). Thus the intracellular level of this messenger—and presumably of others also—depends only upon its rate of synthesis and the activity of the enzymes responsible for its degradation.

The apo-repressor

The role demanded of the apo-repressor would suggest a protein since it is able to exist in active or inactive forms as the result of its interaction with a small molecule co-inducer. This is most readily explained as an allosteric effect, the protein possessing a binding site for the operator which is affected by the binding of the co-inducer at a second, allosteric, site. This can also most easily account for the altered behaviour of *i* gene regulatory mutants—i^s mutants, for instance, are those which have lost this second, co-inducer, binding site.

Indirect evidence (Bourgeios, Cohn and Orgel, 1965; Garen and Garen, 1963; Muller-Hill, 1966) indicates that the regulator codes for a protein. Constitutive regulator mutants of both the alkaline phosphatase and *lac* systems of *E. coli* can be suppressed by an external nonsense codon suppressor; this phenomenon is known to be exerted at the level of translation into protein. Bourgeios, Cohn and Orgel (1965) were also able to demonstrate complementation in diploids between certain i^- mutants and the normal i^+ to give the uninducible i^s phenotype—this interaction suggests that the *i* gene product is polymeric, as would be required for a protein subject to allosteric control. Only the *E. coli lac* repressor and the phage λ repressor (which is responsible for maintenance of the lysogenic state) have been investigated in detail, and the latter also appears to be polymeric; Lieb (1969) has demonstrated intragenic complementation between certain pairs of temperature-sensitive mutants.

Gilbert and Muller-Hill (1966) used its property of binding to the (radioactively labelled) IPTG co-inducer to isolate the *lac* apo-repressor. This proved to be a tetramer of molecular weight about 150,000, and present to the extent of some five to ten copies per cell. Each subunit binds one molecule of IPTG. As expected, proteins isolated from the wild type and from *i* gene mutants showed different affinities for the co-inducer. As well as this bacterial species, the apo-repressors of two bacteriophages have also been isolated. Ptashne (1967a) isolated the apo-repressor of phage λ by comparing the proteins of the wild type with those of *amber* mutants of the C_1 regulator gene (this is genetically defined as the cistron responsible for the production of the lysogenic apo-repressor). The protein sedimented at about 2·8S, which would correspond to a molecular weight of around 30,000. Pirrotta and Ptashne (1969) isolated the corresponding species from phage 434, which is closely related to λ. This comprises a similar protein, with a molecular weight some 2000–3000 lower than that of the λ apo-repressor, and is also slightly more basic.

If the apo-repressor protein interacts with DNA to prevent transcription, rather than with RNA to prevent translation, it should bind *in vitro* to DNA containing its operator, but not to DNA lacking the recognition site. Gilbert and Muller-Hill (1967) reported that the protein isolated by

IPTG binding would bind only to DNA possessing the region defined genetically as the *lac* operator, and not to o^c mutants. As confirmation of the postulated mode of apo-repressor action, addition of the IPTG co-inducer released the protein from the DNA. Ptashne (1967b) compared the binding of apo-repressor to normal phage λ and to a mutant which differs from the wild type only in that it synthesizes the apo-repressor of phage 434 rather than that of λ. DNA from each type of phage was mixed with radioactively labelled apo-repressor protein and was sedimented through a sucrose gradient; with the wild type the radioactivity became associated with the DNA, but not with the mutant.

The promotor

In their original scheme, Jacob and Monod (1961) suggested that the operator might possess the distinct properties of both comprising the recognition site of the apo-repressor and of initiating the transcription (or possibly translation) of *m*RNA. This idea was supported by the finding of o^o mutants, in which the ability of the operon to produce its enzymes is completely lost. These are recessive to o^c and o^+, and map very close to the o^c mutants; thus they were interpreted as representing some form of inactivation of the operator, possibly a loss of the ability to initiate transcription. However, Beckwith (1964) showed that o^o mutants could be restored to normal activity by external suppressors, that is, they represent chain terminating codons which act by preventing translation rather than transcription, and are merely examples of extreme polar mutants (see pages 263–266).

Jacob, Ullman and Monod (1964) observed that treatment of *E. coli* cells with 2-aminopurine or ultraviolet irradiation does not induce constitutive mutation in the *lac* region, and that o^c mutants do not revert to the wild type o^+. This suggests that o^c mutation is the result of deletion in the operator region. However, despite the absence of the operator recognition site for apo-repressor control, these mutants still have enzyme activity, so that transcription can still occur. This implies that the operator is not the site for the initiation of transcription, but that some separate locus must bear the responsibility for this. A site which is indispensable for activity of the operon was isolated by examing revertants to *lac y* gene activity from diploid i^s (uninducible) mutants. Because i^s is dominant over i^+ or i^-, these revertants arise almost entirely from deletion of the operator to yield o^c mutants. When they were genetically mapped, it appeared that none of these deletions extended over the operator into the *z* gene. However, the absence of the operator or part of the *z* gene alone does not prevent expression of the *y* gene, so this indicates that there must be some site between *o* and *z* which is essential for operon function, and its deletion prevents expression of all three structural genes.

This locus could be the site where RNA polymerase is recognized and commences transcription—the *promotor*. Further support for the concept of a distinct promotor has come from studies which have isolated strains with properties which might be expected to result from mutation at this locus (Arditti, Scaife and Beckwith, 1968; Ippen and coworkers, 1968; Scaife and Beckwith, 1966). These are *cis* dominant and *trans* recessive, and reduce the maximum rate of operon expression without affecting the mechanism of regulation. Certain single base transitions can lower the activity of the *lac* operon fifteen-fold, presumably by altering its affinity for RNA polymerase. (These may be analogous to certain o^c mutants recently found which can be reverted by 2-aminopurine). Ippen and coworkers (1968) reported that certain strains used in the original deletion mapping of the promotor had been incorrectly identified, and a re-examination of the data showed that the site reducing maximum operon expression lies between *o* and *i*. It is this site which probably acts as the initiation point for transcription, and the region between *o* and *z* may correspond to the initiation site for translation. The order of the *lac* operon is thus *i-p-o-z-y-a*, and the *p–o* region is roughly one hundred base pairs long (Miller, Beckwith and Muller-Hill, 1968).

The mode of action originally proposed for the apo-repressor envisaged its binding at the operator as inhibiting the binding of RNA polymerase, and thus preventing transcription of the operon. The binding of apo-repressor to the operator at a site located between the promotor and the structural genes suggests that rather it blocks the progress of the polymerase into the operon, and thus prevents it from transcribing the structural genes, as illustrated in Figure 10.5. Experimental data from the tryptophan operon of *E. coli* support this concept (Imamoto, 1968b). When this operon is repressed, transcription currently proceeding can continue to completion of the polycistronic messenger, except for polymerase molecules located at the beginning of the operon. The cessation of transcription in this region is rapid after the addition of co-repressor. This suggests that the apo-repressor binds at the operator and prevents the polymerase molecules from continuing transcription beyond this point. This implies also that the operator is transcribed into messenger after all (although it need not be translated).

E. coli strains have been isolated with deletions which remove some or all of the control elements of the *lac* system, and thus fuse the remaining and intact structural genes of the operon to the *trp* operon adjacent on the chromosome (see Reznikoff and coworkers, 1969). In these,

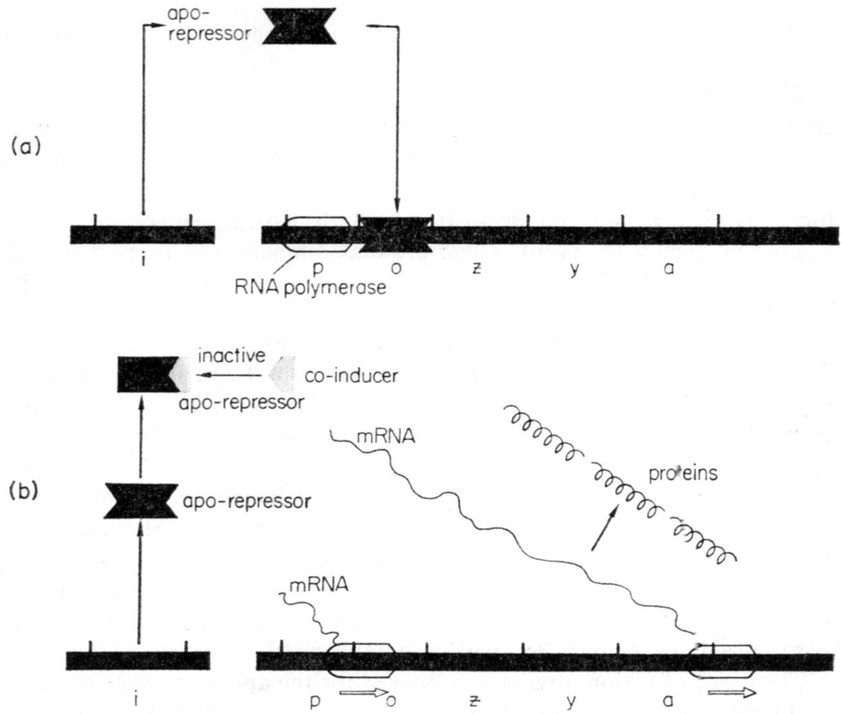

Figure 10.5. The action of the repressor. (a) Under repression conditions, the apo-repressor binds to the operator and prevents progress of the RNA polymerase into the operon. There is no transcription. (b) Upon the addition of co-inducer, the apo-repressor is converted into an inactive form which no longer binds the operator. There is now no impediment to the progress of the polymerase, and transcription produces a messenger RNA for the operon which is translated into proteins

activity of the *lac* operon is controlled by the *trp* system, so that repression of both operons results from the addition of tryptophan. Transcription of the *lac* operon is then initiated at the beginning of the *trp* operon. When the deletion does not remove the *i* gene, read-through is prevented by a barrier to the continuation of *m*RNA synthesis at the end of the *i* gene nearest to the *lac* promotor. Deletion of *p* and the remaining part of *i* removes this barrier; this indicates that there must be a 'stop transcription' signal at the end of the *i* gene. However, the introduction of an $F'i^+$ allele then prevents transcription of the *lac* structural genes. This would be consistent with the apo-repressor binding at the operator to prevent progress of the polymerase.

At this point, we should mention the elegant experiments of Shapiro and coworkers (1969), which have resulted in the purification and isolation *in vitro* of the *E. coli* DNA corresponding to much of the *lac* operon. This work has made use of two specialized transducing bacteriophages—strains of λ and ϕ80—which carry the *lac* system. The particular phage species used were derived from bacterial strains with transpositions of the *lac* operon to sites on the genome near to the attachment sites of the phages. Because the *lac* operon has been inserted into the genomes of these two bacterial strains in opposite polarity, and because both λ and ϕ80 integrate into the bacterial chromosome with the same overall orientation, the resultant transducing phages have *lac* operon DNA inserted into their genomes with opposite orientations. After denaturation of the DNA of these phages, the single strands can be separated by their different densities into a 'heavy' (H) strand, and a 'light' (L) strand (see p. 143). As the *lac* operon is present in the two different phages with reversed polarity, when this treatment is applied to preparations of transducing phage, in one case the sense strand of the *lac* operon is contained in the H strand of the phage, and in the other case it is present in the L strand (see Figure 10.6). If the separated H strands of the two phages are then annealed, the bacterial *lac* sequences will be complementary and can renature to form duplex DNA. However, because the phage sequences represent the same (H) strands, they will be non-complementary and unable to anneal, and are therefore compelled to remain single stranded. This yields the four-tailed structure shown in Figure 10.6. The single strands, that is, the phage regions, can then be removed by treatment with a nuclease specific for single-stranded DNA, leaving a duplex corresponding to only *lac* DNA.

This procedure relies upon absolute non-complementarity between the H strands of the two phages (otherwise duplex regions corresponding to phage genes would be produced), and there is, in fact, no detectable homology between them. Contamination of the duplex DNA produced could still arise, however, from the presence in the phages of other bacterial genes able to behave in the same manner as that of the *lac* operon. Indeed, this is a real difficulty, for the transducing phage preparations almost always contain other bacterial genes as well as those of the *lac* system. To avoid this and ensure that the two phage strains have in common only chromosomal material derived from the *lac* operon, one of the phages used was obtained in the state of carrying no bacterial genes other than those from the *lac* operon. This was achieved by preparing it from bacterial strains with two deletions, one extending from the middle of the *y* gene to past the end of the operon, and the other extending into

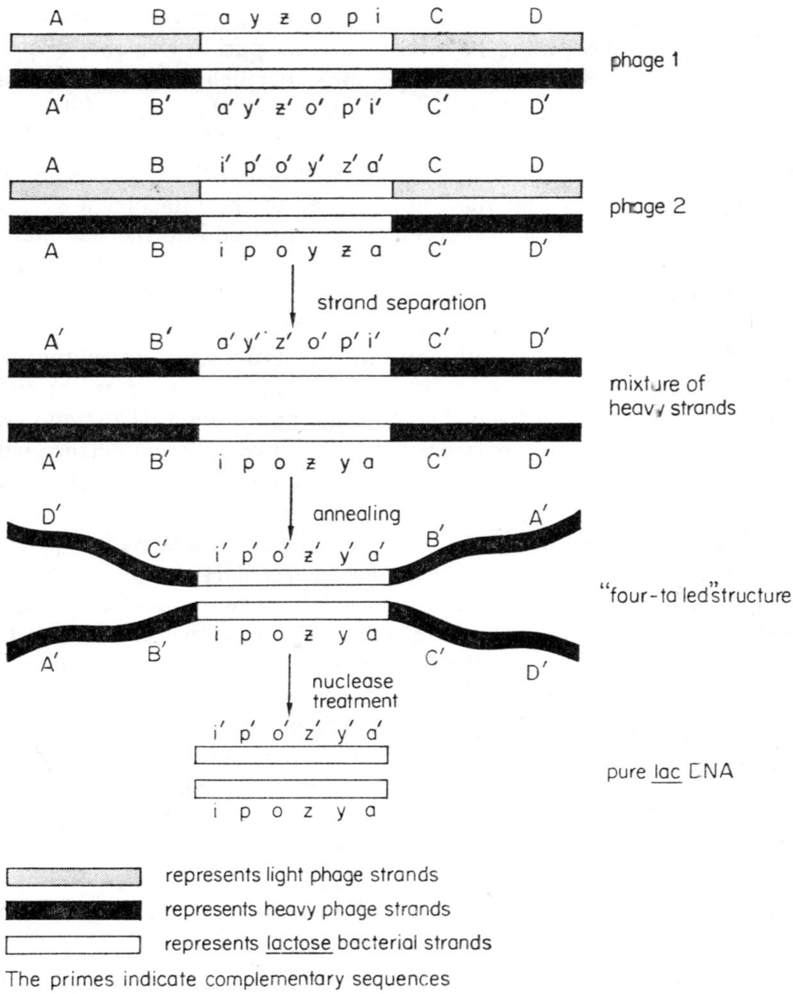

Figure 10.6. The isolation of pure *lac* operon DNA. The two transducing phage preparations have the genes of the *lac* system inserted in opposite orientation. After denaturation and centrifugation, the two separated H strands are allowed to anneal with each other. The only complementary sequences are those of the *lac* genes, which thus form a duplex. The phage DNA remains single stranded. Treatment with a nuclease specific for single-stranded DNA leaves only the duplex region representing the pure *lac* DNA (after Shapiro and coworkers, 1969)

the operon from the other end and terminating in the *i* gene. This means that the isolated duplex segment does not contain the whole operon, but has intact the promotor, operator, and *z* gene. Electron microscopy on the preparation of annealed DNA showed the expected 'four-tailed' structures, and after digestion with nuclease showed the presence of pure duplex regions. These were some 1·4–1·5 μm long, which would correspond to about 4250 base pairs; this is in good agreement with the estimated size of the *lac* operon (about 4700 base pairs for the region from *i* to *y*). Hybridization experiments with *lac* *m*RNA synthesized *in vivo* demonstrated that this duplex region represents *lac* DNA.

The isolation of this segment of DNA has many implications for studies on the control mechanisms of the operon. Not least, it should prove possible to extend the techniques used to make possible the isolation of other bacterial operons. It has proved difficult to investigate *in vivo* the interactions which take place between apo-repressor and operator, and between RNA polymerase and promotor, and this system should assist determination of the precise mode of action of the apo-repressor. Since this segment of DNA contains only a single promotor, it should also make possible further investigations into the action of the sigma factor and the initiation of transcription by RNA polymerase. A use of the pure DNA of the *z* gene will be to measure by hybridization the corresponding *m*RNA—such measurements have proved difficult to date. A particularly pressing problem is that of the mode of recognition of nucleic acids by proteins, that is, the mechanism by which a protein recognizes a base-pair sequence in DNA. This should be very much easier to investigate using the isolated components of the system.

Glucose repression

The rate of synthesis of β-galactosidase (and also of other catabolic enzymes) is lowered when cultures of *E. coli* are grown in the presence of glucose. This occurs under three conditions. When the *lac* system is induced, the rate of synthesis of β-galactosidase is considerably reduced in cultures growing upon glucose, compared with cells for which some other metabolite is provided as the carbon source. This has been termed *catabolite repression* (or *permanent repression*). The absolute extent of the repression varies with the particular bacterial strain, but the rate of β-galactosidase synthesis may be reduced to about one-quarter to one-half of that shown on non-glucose media.

Transient repression is exhibited when glucose is added to cultures of *E. coli* growing upon some other carbon source, for example, glycerol.

In this case, there is a severe repression of β-galactosidase synthesis and no enzyme is produced for roughly the next half generation. After this period, the enzyme is synthesized at the rate characteristic of normal growth on glucose cultures.

Both catabolite and transient repression result from the effect which glucose exerts on lowering the level of 3′–5′ cyclic AMP (*c*AMP) in the cell. The addition of *c*AMP to cultures overcomes either type of glucose repression by stimulating the production of β-galactosidase. (There is a third effect in that glucose lowers the uptake of the co-inducer and thus its intracellular level; this is not related to the level of *c*AMP.)

The effect of *c*AMP in stimulating the synthesis of β-galactosidase is exerted at transcription by increasing the rate of synthesis of its *m*RNA. However, the effect is not mediated through the operator–repressor control of the operon since the addition of *c*AMP is as effective with operator and repressor regulator mutants as it is with wild-type strains. Rather, it acts through the promotor, as shown by the altered response to glucose repression and *c*AMP stimulation exhibited by promotor mutants. In a culture with a promotor mutant which permits the *lac* system to make only 5 per cent. of its normal level of enzymes when induced, glucose is unable to cause transient repression when added to glyceryl-grown cells (Pastan and Perlman, 1968). Although catabolite repression is shown, it can only be overcome by the addition of cyclic AMP at rather higher concentrations than usual (Perlman, De Crombrugghe and Pastan, 1969). However, another promotor mutant, mapping at a distinct site, has no effect upon the response to *c*AMP.

This suggests that the recognition which occurs between promotor and RNA polymerase is subject to a precise quantitative control, and different mutants affect this to different extents (see also pages 274–275). It is not yet clear exactly how *c*AMP influences this interaction, but Jacquet and Kepes (1969) have obtained data which suggest that transient repression may influence the initiation of *m*RNA chains; they found that glucose represses the rate of initiation of *lac* *m*RNA by about 90 per cent., and cyclic AMP restores the rate to a level of about 115 per cent. It is noteworthy also that *c*AMP is able to exert its effect *in vitro*; Chambers and Zubay (1969) found that their DNA-dependent *in vitro* system for synthesizing β-galactosidase was stimulated eight- to thirteen-fold by the addition of *c*AMP. Finally, we should mention that the type of influence which *c*AMP exerts upon the *lac* promotor may not account for its effect upon the synthesis of other catabolic enzymes, for its stimulation of the synthesis of tryptophanase is achieved at the translational rather than the transcriptional level (Pastan and Perlman, 1969).

CHAPTER 11

Systems of regulation

Types of control system

Control systems of induction and repression can be characterized by the type of interaction exerted between the apo-regulator and co-effector. Systems of induction show activity in the presence of the co-inducer whereas systems of repression fail to show activity in the presence of the co-repressor. In systems under positive control, the operon is maintained in a non-functional state unless it is switched on by interaction of the operator with an apo-inducer. The converse is true of negative control systems, which are characterized by functioning unless switched off by interaction of the operator with an apo-repressor. Thus deletion of the operator in positive control systems results in the inability of the operon to function, and deletion under negative control has the consequence that the structural genes cannot be prevented from functioning.

In positive induction, the regulator gene synthesizes an apo-inducer which is inactive until it has interacted with the co-inducer; this gives an active species which causes expression of the structural genes of the operon. Deletion of the operator produces uninducibility. A defective regulator gene resulting in the absence of apo-inducer produces uninducible mutants; these are recessive since introduction of a normal apo-inducer restores the response to the co-inducer. Constitutivity can be achieved by a mutant regulator gene which directly synthesizes an altered apo-inducer activating the operator without need for a co-inducer; this is dominant since its activity is unaffected by the presence or absence of a normal apo-inducer. Positive induction control systems are illustrated in Figure 11.1, and the L-arabinose system of *E. coli* may be under this type of control.

In a negatively controlled induction system, the regulator gene synthesizes an apo-repressor which causes the operator to switch off the operon; expression of the system is achieved through inactivation of the apo-repressor by the co-inducer. Operator deletion gives constitutive synthesis because it is not possible to switch off the operon. An inactive

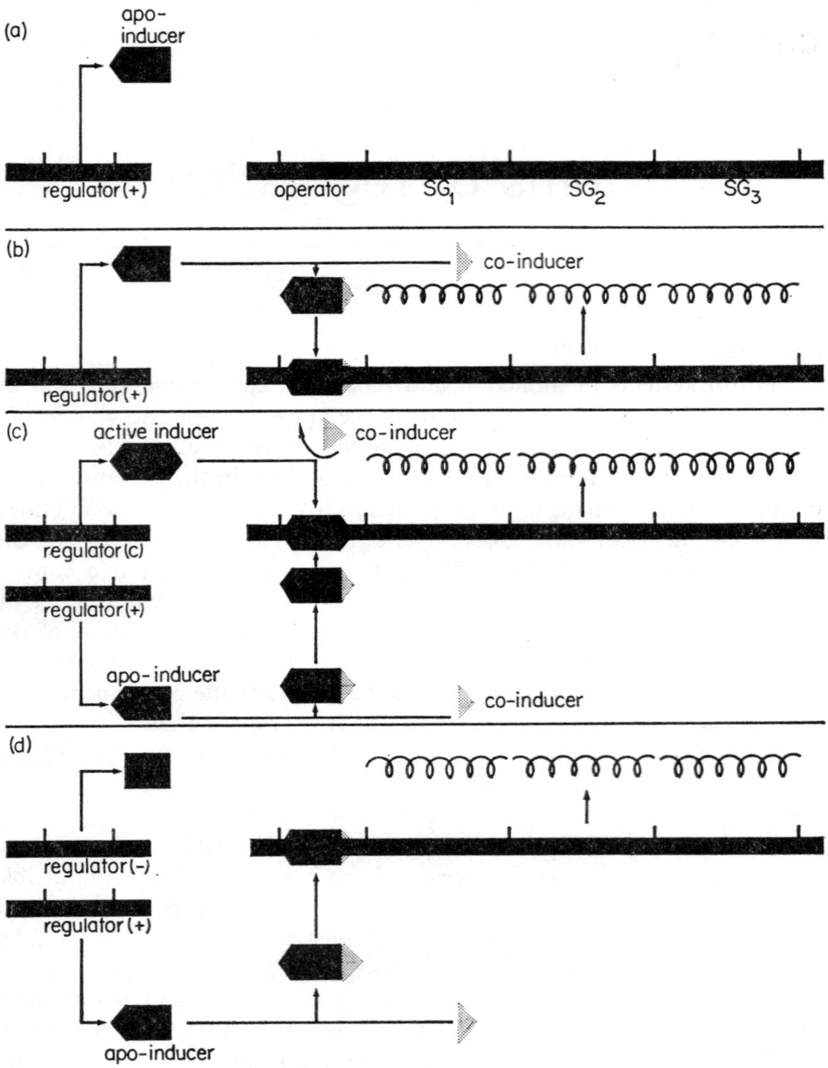

Figure 11.1. The positive control of induction. (a) The apo-inducer is produced, but is inactive in the absence of a co-inducer. (b) The addition of the co-inducer activates the apo-inducer, and the operon is switched on. (c) Constitutive action of the operon results from synthesis by the regulator of an apo-inducer which is active even in the absence of a co-inducer. The presence of a normal regulator does not interfere with this; thus constitutivity is dominant over inducibility. (d) Uninducibility results from synthesis of an active apo-inducer. However, the introduction of a normal regulator causes synthesis of an apo-inducer which can be activated by a co-inducer, and therefore restores normal control. Hence uninducibility is recessive to inducibility

regulator gene results in the absence of the apo-repressor so that there is constitutive production of the proteins specified by the operon. This mutation is recessive since the presence of normal apo-repressor restores the response to the co-inducer. Uninducible mutants result when the regulator produces an apo-repressor locked in its active form, that is, although it still switches off the operator it cannot be inactivated by the co-inducer. This is dominant because the presence or absence of a normal apo-repressor cannot affect it. This is the control system exercised in the *lac* system of *E. coli*, and is shown in Figures 10.1, 10.2 and 10.3.

Regulatory mutations in systems of repression can give the states of *de-repression* and *super-repression*. In de-repression, the system no longer reponds to the co-repressor and thus cannot be switched off; this is analogous to uninducibility in an induction system. Super-repression is analogous to constitutivity; just as a constitutive induction system is permanently switched on, even in the absence of a co-inducer, so a super-repressed system is permanently switched off, irrespective of the presence of a co-repressor.

In positive control of repression, the operator requires the presence of an apo-inducer synthesized by the regulator gene if it is to be active and cause expression of the operon. Thus operator deletion results in super-repression. The co-repressor normally inactivates the apo-inducer, thus repressing the system. An inactive regulator results in absence of the apo-inducer with consequent super-repression; this is recessive to the restoration of control by production of the apo-inducer from a normal regulator. The synthesis of a mutant apo-inducer which still recognizes the operator but is no longer inactivated by the co-repressor results in dominant de-repression. Figure 11.2 illustrates the positive control of repression.

In a negatively controlled repression system, the operon is functional until switched off, so that deletion or inactivation of the operator results in de-repression. The regulator gene sythesizes an apo-repressor which must interact with the co-repressor before it is active in repressing protein synthesis by the operon. An inactive regulator results in the absence of the apo-repressor with consequent de-repression, and this state is recessive to the restoration of control by a normal apo-repressor. The production of a mutant apo-repressor which is altered so as to be directly active in repressing the operon, without the need for a co-repressor, results in dominant super-repression. Figure 11.3 illustrates the negative control of repression.

The four types of straightforward control system outlined above can be characterized by the dominance relationships of their regulatory mutants.

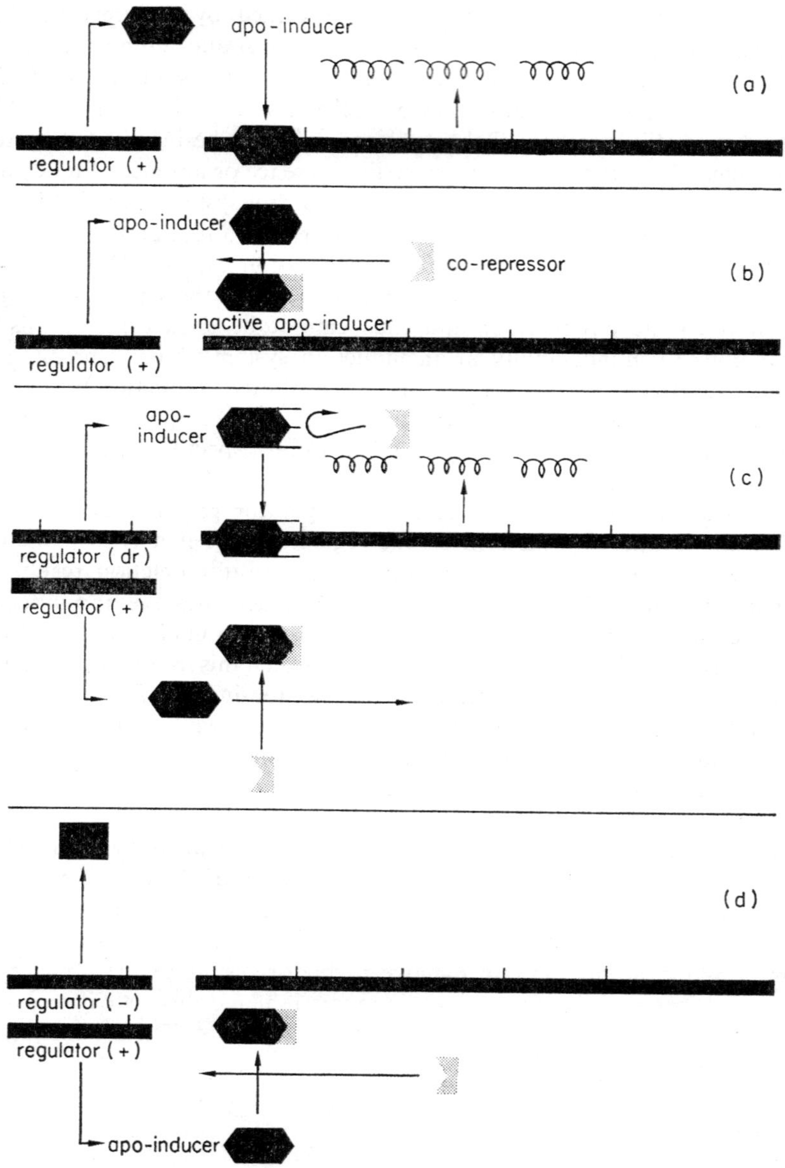

apo-inducer
(a)
regulator (+)
apo-inducer
co-repressor
(b)
inactive apo-inducer
regulator (+)
apo-
inducer
(c)
regulator (dr)
regulator (+)
(d)
regulator (−)
regulator (+)
apo-inducer

Operator loci are defined as *cis* dominant and *trans* recessive; their deletion results in switch-off for positive control systems and switch-on for negative. There are two classes of regulatory gene mutants, one of which results in a switch-on of the operon, and the other of which causes switch-off. Whether control is positive or negative is defined by which of these is recessive and which dominant; regulator deletions show as the recessive variant. The effects of regulatory mutants in positive and negative induction and repression can thus be summarized as shown in Table 11.1.

Table 11.1. Characteristics of control systems

	Positive	Negative
	Operon only functions after operator is switched on by the apo-inducer	Operon functions unless operator is switched off by apo-repressor
Induction	Co-inducer activates apo-inducer Operator deletion: uninducible Regulator gene mutants: constitutive, dominant, uninducible, recessive	Co-inducer inactivates apo-repressor Operator deletion: constitutive Regulator gene mutants: constitutive, recessive uninducible, dominant
Repression	Co-repressor inactivates apo-inducer Operator deletion: super-repressed Regulator gene mutants: super-repressed, recessive de-repressed, dominant	Co-repressor activates apo-repressor Operator deletion: de-repressed Regulator gene mutants: super-repressed, dominant de-repressed, recessive

Figure 11.2. The positive control of repression. (a) The apo-inducer switches the operon on. (b) The co-repressor inactivates the apo-inducer; this leaves the operon in a non-functional state. (c) A de-repressed regulator mutant produces an inducer species which is active in switching the operon on, but can no longer be inactivated by the co-repressor. The presence of a normal regulator gene cannot interfere with this; de-repression is therefore dominant to repression. (d) An inactive regulator mutant causes super-repression. However, the introduction of a normal regulator allele effectively restores control in response to the co-repressor. Super-repression is thus recessive to repression

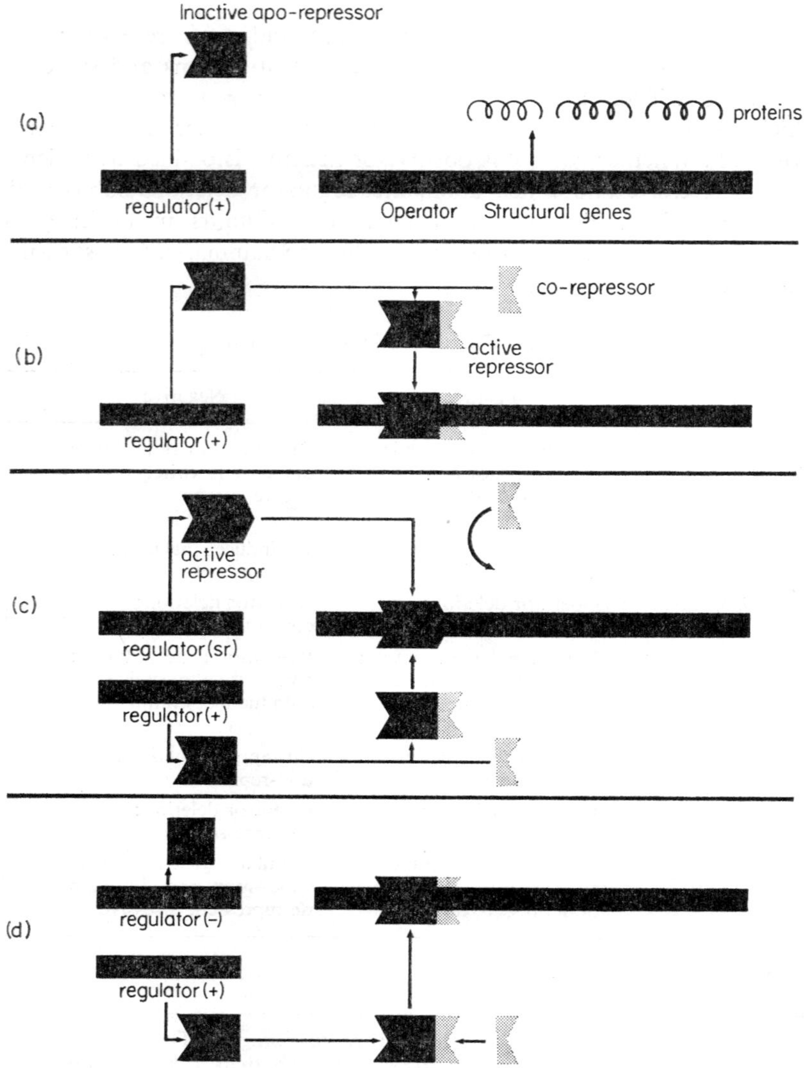

Figure 11.3. The negative control of repression. (a) The regulator gene sythesizes an inactive apo-repressor which cannot switch the operon off. (b) The addition of the co-repressor activates the apo-repressor and switches the operon off. (c) Super-repression is achieved by a regulator mutant which synthesizes a repressor species which is active, even in the absence of a co-repressor. The presence of a normal regulator has no effect on this; super-repression is therefore dominant to repression. (d) An inactive regulator synthesizes an apo-repressor with no activity. However, the presence of a normal regulator restores the response to the co-repressor. Thus de-repression is recessive to repression

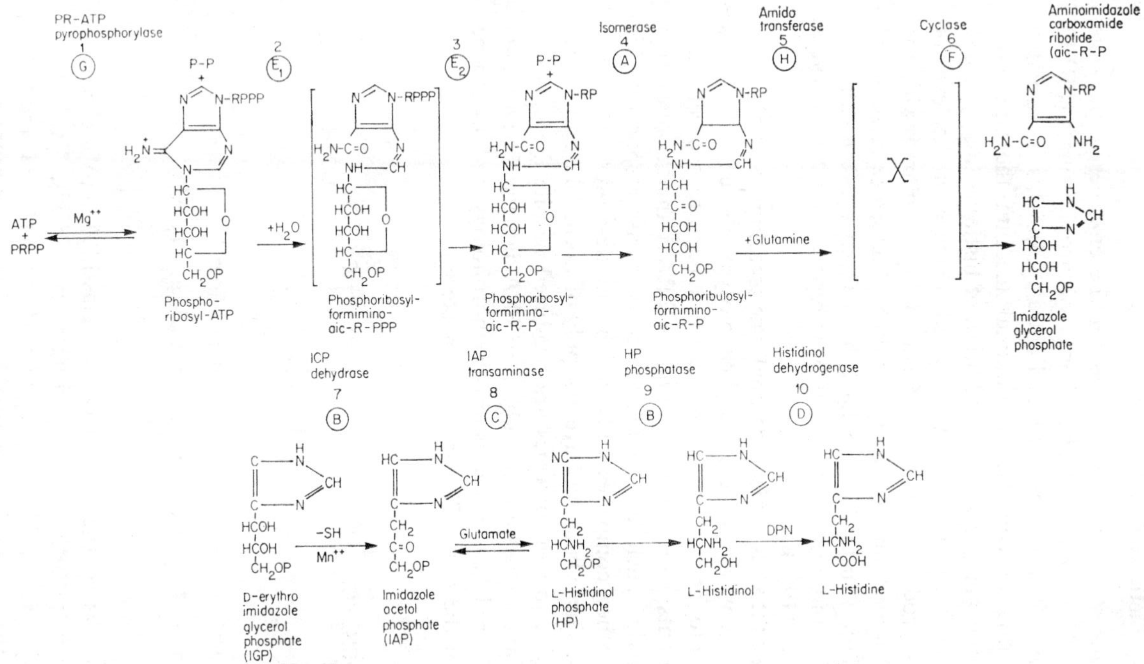

Gene:	E	I	F	A	H	B	C	D	G	O
Step:	2	3	6	4	5	7,9	3	10	1	
Enzyme:	PR-ATP pyrophospho hydrolase	PR-AMP hydrolase	Cyclase	Isomerase	Amido-transferase	Dehydrase phosphatase	Transaminase	Dehydrogenase	PR-ATP pyrophosphorylase	

Figure 11.4. The histidine biosynthetic system. (After Ames and Hartman, 1963; Roth and coworkers, 1966)

The histidine biosynthetic system

A cluster of nine structural genes codes for the enzymes responsible for the synthesis of histidine from phosphoribosylpyrophosphate (PRPP) and ATP in *Salmonella typhimurium* (see Figure 11.4). This is regulated as a coordinate unit of control; on minimal medium the wild type synthesizes all the enzymes, but on the addition of histidine this synthesis is repressed. There is about a twenty-five-fold increase in the level of histidine enzymes produced on histidine starvation compared with the level produced on a medium containing histidine. Detailed references and a review of regulation in this system have been given by Roth and co-workers (1966), and more recent work has been reported by Anton (1968) and Fink and Roth (1968).

Triazolealanine (TRA) is a histidine analogue (see Figure 11.5) which acts as a false co-repressor of the histidine biosynthetic enzymes. It is misincorporated into protein with consequent damage to the cell, and selection for mutants which are resistant to TRA damage gives two classes. One is trivial, such as mutants in the permease responsible for uptake, but the other comprises cells which are de-repressed for the histidine enzymes. These bacteria have a supply of normal histidine since the appropriate biosynthetic enzymes are produced despite the presence of a co-repressor, and so they are not subject to TRA damage. The mutants which are de-repressed have a characteristic morphology and so can be distinguished from the other TRA-resistant bacteria.

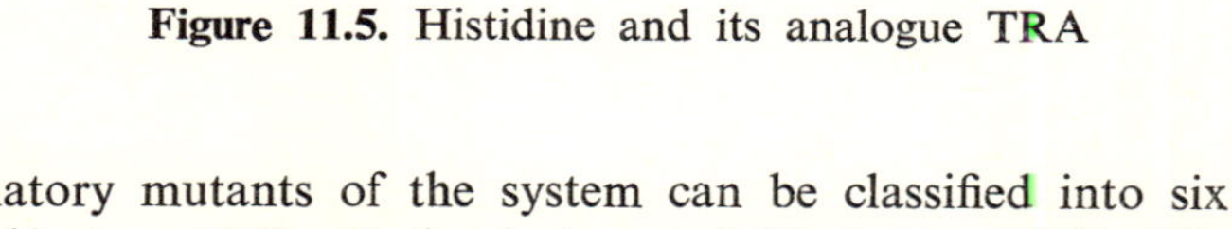

Figure 11.5. Histidine and its analogue TRA

Regulatory mutants of the system can be classified into six genetic groups (Anton, 1968; Roth, Anton and Hartman, 1966). These have been termed *his*-O, -R, -S, -T, -U and -W and are scattered about the genetic map. All these mutants are capable of further de-repression to give even higher levels of the *his* enzymes when grown under histidine starvation; this suggests that they represent alteration but not destruction of regulator genes.

Mutants in *his*-O are located at the right end of the operon, suggesting that this locus represents the operator, with transcription and translation starting at this end of the operon. Data on polarity (see Chapter 12) support this concept, and Fink and Roth (1968) demonstrated that this gene has the expected *cis*-dominant and *trans*-recessive properties. Some de-repressed *his*-O mutants appear to result from small deletions, and this would suggest a negative control system.

The *his*-S gene is the structural gene for histidyl-*t*RNA synthetase, and *his*-S mutants have an altered activating enzyme with a considerably lowered activity for histidine, although the same quantity of synthetase is present in the cell. Roth and Ames (1966) showed that the addition of sufficient histidine to these mutants could overcome the effects of mutation, and this led to the suggestion that both growth of the mutants and repression of the histidine operon are limited by the supply of charged histidyl-*t*RNA.

Silbert, Fink and Ames (1966) found that *his*-R mutants had only about 55 per cent. of the normal histidine accepting *t*RNA activity, and suggested that this gene is involved, directly or indirectly, in the production of a histidine-specific *t*RNA. Although attempts to separate more than one such *t*RNA have failed as yet, it has been proposed that there may be two species, one of which is inactivated by this mutation. There are also indications that *his*-W too may be concerned in the synthesis of a functionally active $t\mathrm{RNA}_{\mathrm{his}}$. No biochemical alteration has been detected in *his*-T or *his*-U mutants, and their role is unknown. All the mutants *his*-R, -T, -U and -W are recessive to their wild-type alleles (Fink and Roth, 1968), excluding the possibility that they represent mutations of minor non-histidine *t*RNA species to accept histidine.

Schlessinger and Magasanik (1964) grew bacterial cells in the presence of α-methylhistidine. These have de-repressed levels of the histidine biosynthetic enzymes, resulting in the accumulation of histidine. They showed that the analogue does not inhibit any of the enzymes concerned; this excludes the possibility that the effect is due to a lowering in the intracellular concentration of histidine, which would result in de-repression. They observed that the analogue inhibits the rate of formation of histidyl-*t*RNA by competition with histidine for its activating enzyme—the α-methylhistidine cannot be transferred to *t*RNA.

This led to the suggestion (see Roth and coworkers, 1966) that the histidine operon is an example of regulation by negative repression, with histidyl-*t*RNA as the co-repressor necessary for activation of the apo-repressor synthesized by one of the *his* regulator genes. This model is illustrated in Figure 11.6. *His*-S and *his*-R are thus not true regulator

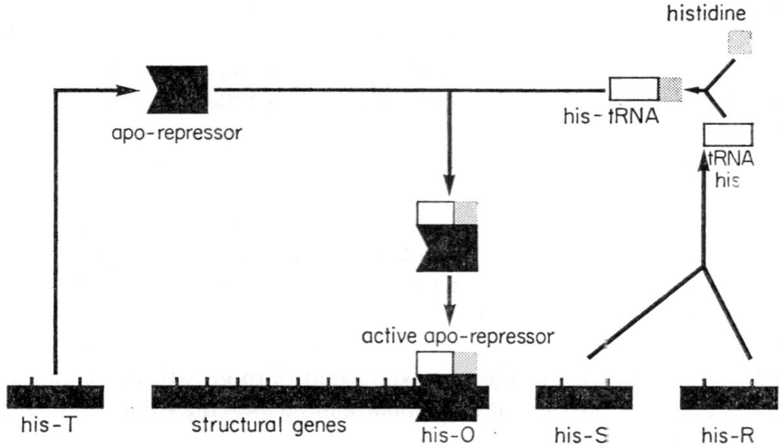

Figure 11.6. The control of the histidine operon in *S. typhimurium*. The *his*-T gene synthesizes an inactive apo-repressor. The *his*-S and *his*-R genes are both concerned in the production of $t\text{RNA}_{\text{his}}$, which after reaction with histidine to form his-*t*RNA interacts with the apo-repressor to activate it and thus switch off the operon. This is a negative control system for repression (see Figure 11.3)

genes but are only concerned in the production of the co-repressor from histidine itself. Different analogues of histidine must have different effects upon the operon, depending on the extent to which they mimic histidine. Those with sufficient resemblance to be incorporated into protein (such as TRA) act as co-repressors, but those which merely compete for the activating enzyme (such as α-methylhistidine) prevent synthesis of the his-*t*RNA co-repressor, thus de-repressing the system. A drawback of this scheme is that his-*t*RNA is a rather larger molecule than would be expected to exert an allosteric effect on the regulator protein. However, Eidlic and Neidhardt (1965) have observed that valyl-*t*RNA acts as a co-repressor for the valine-repressible enzymes; a mutant of *E. coli* with a temperature-sensitive valyl-*t*RNA synthetase was found to synthesize increased quantities of these enzymes at high temperature. It is possible, therefore, that co-repression by aminoacyl-*t*RNA may be more general control phenomenon.

Control of L-arabinose metabolism

L-arabinose is metabolized in *E. coli* by three enzymes which convert it to D-xylulose-5-phosphate; these are regulated as a unit of coordinate control by induction. Mutants which cannot utilize the pentose as carbon energy source (*ara*⁻ phenotype) can be grouped by genetic and physiological criteria into five genes. Genes *ara*A–D are closely linked in the sequence DABC, the first three representing the structural genes for the enzymes of the metabolic pathway, epimerase, isomerase and kinase, respectively. Gene *ara*E controls the L-arabinose permease responsible for its uptake into the cell, and is unlinked to the other genes as determined by transduction. The system is illustrated in Figure 11.7. Detailed references and a general review of the system have been given by Engelsberg and coworkers (1965), and more recent work has been reported by Sheppard and Engelsberg (1966, 1967), Engelsberg, Squires and Meronk (1969) and Engelsberg and coworkers (1969).

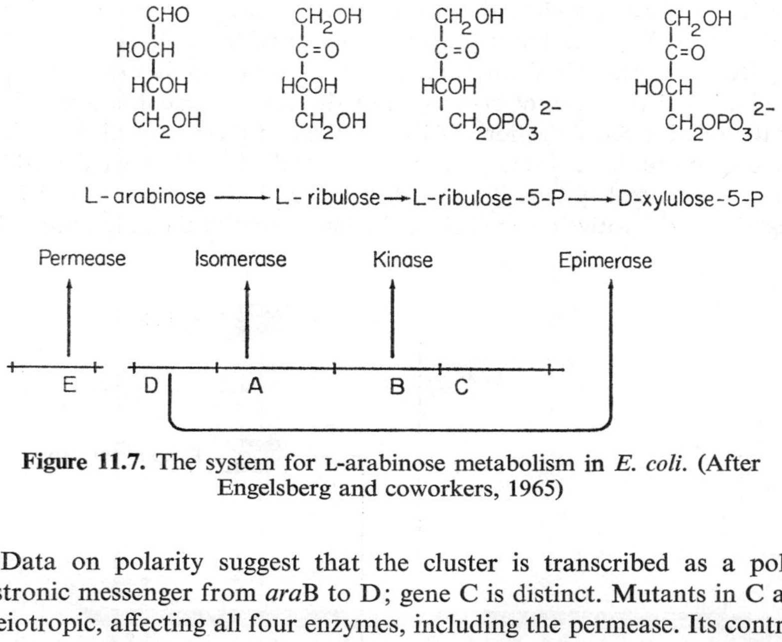

Figure 11.7. The system for L-arabinose metabolism in *E. coli*. (After Engelsberg and coworkers, 1965)

Data on polarity suggest that the cluster is transcribed as a polycistronic messenger from *ara*B to D; gene C is distinct. Mutants in C are pleiotropic, affecting all four enzymes, including the permease. Its control of the unlinked *ara*E gene excludes the possibility that the C cistron is an operator region; *ara*C⁻ mutants are unable to synthesize any of the enzymes, whilst *ara*C^c show constitutive synthesis. Deletions in the gene result in the *ara*C⁻ phenotype, as would be demanded of a positive control system. The construction of heterogenotes for this region showed that the uninducible *ara*C⁻ gene is recessive to the normal *ara*C⁺ allele, suggesting a positive induction system (see Figure 11.1), in which *ara*C produces an apo-inducer (P_1) which is inactive until converted by L-arabinose into the active species (P_2). Diploid *ara*C^c/*ara*C$^-$ strains demonstrate that constitutivity is dominant over uninducibility; this is accounted for if *ara*C^c produces a species P_3 which activates the operon directly without need for interation with the co-inducer.

However, for normal positive control *ara*C^c should be dominant over *ara*C$^+$, but in fact, the reverse occurs. This unexpected relationship can be reconciled with the other results if P_1 prevents the action of P_3 in some manner. The dominance of inducible over constitutive regulator suggests negative control through the production of an apo-repressor which is inactivated by L-arabinose (see Figure 10.1) Thus Engelsberg has proposed that the *ara*C gene product, P_1, switches the operator off. Its allosteric interaction with L-arabinose gives P_2, which is an inducing species required to switch the operon on. P_1 does not interact with the unlinked *ara*E gene, but P_2 does so to provide the positive control required to release it from whatever control system it otherwise falls within. The basal level of expression of the BAD segment is determined by the relative concentrations of P_1 and P_2, and their respective affinities for the control sites (*ara*O and *ara*I) at which they bind on the DNA. Thus this control system, illustrated in Figure 11.8, has features of both negative and positive control, since the interaction of the co-inducer with

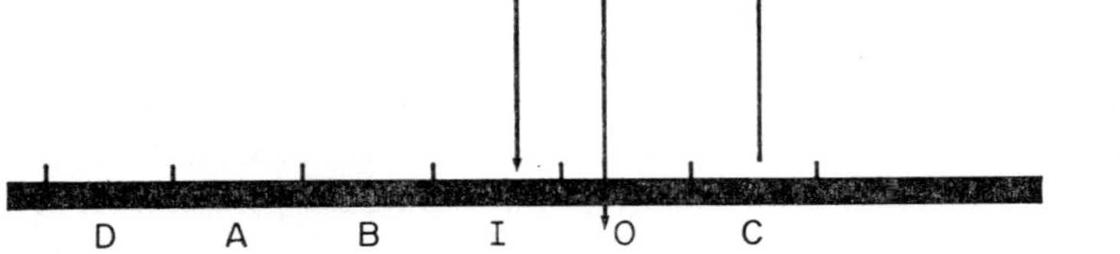

Figure 11.8. A repressor–inducer model for control of the L-arabinose system. The *ara*C gene synthesizes an apo-inducer (P_1) which acts at *ara*O to prevent functioning of the operon. The addition of L-arabinose converts P_1 to an activating species (P_2) which must act at *ara*I if the operon is to function. P_2 can also interact with gene *ara*E to cause synthesis of its product. (After Engelsberg Squires and Meronk, 1969)

the apo-repressor does not merely inactivate it, but actually converts it to an inducing species necessary for expression of the operon.

Two sites between the *ara*C and B genes have been identified as the loci where the P_1 repressing species and the P_2 activator may exert their effect. The operator-like locus where P_1 may act appears to lie immediately adjacent to the C gene. Deletions which extend over *ara*C into *ara*B result in the *ara*⁻ phenotype, with pleiotropic loss of expression for all three of the BAD structural genes. This is not caused by a malfunction of the C gene because the same phenotype is shown even in mutants in which *ara*C is functional. The mutation is *cis* dominant and *trans* recessive, as demanded of an operator. A second control site, at which P_2 may bind, has been identified by the isolation of revertants from *ara*C^- mutants. These contain a second mutation at a site adjacent to the B gene (termed *ara*I), and exhibit a *cis* dominant–*trans* recessive low level of constitutivity.

Further support for this control system is provided by the behaviour of two different deletion mutants. Δ719 represents a deletion covering all known mutant sites in C. Its behaviour can be interpreted as demonstrating that the deletion extends over *ara*O but not to *ara*I. Δ766 is a deletion with its left end terminating within *ara*C, so that both the control sites between C and B remain intact. Both Δ719 and Δ766 exhibit the *ara*⁻ phenotype because their C gene is inactive. However, their responses differ when an *ara*C^+ gene is introduced as an exogenote (in the absence of coinducer L-arabinose), as shown by the data:

	I^+ Δ719	I^+ Δ766	I^c Δ719	I^c Δ766	
araC^-	1	1	4	4	(relative
C^+ (trans)	35	2	8	0·8	expression)

Whereas Δ719 shows a thirty-five-fold increase in operon expression (as measured by the activity of the isomerase coded by *ara*A), Δ766 shows only a two-fold increase. Because there is no operator in Δ719, the introduction of *ara*C^+ cannot result in any repressive effect, but only in stimulation of activity. With Δ766, both the P_1-*ara*O and P_2-*ara*I interactions can take place, so that normal control is restored. If the two mutants have the *ara*I^c constitutive allele in place of the normal *ara*I^+, their basal level of activity is increased. The introduction of *ara*C^+ then causes a two-fold increase in activity with Δ719. However, with Δ766, a five-fold decrease in activity is observed. This is explained if the introduction of an active *ara*C^+ gene allows formation of P_1, which can bind to the operator site and prevent transcription of the operon.

The alkaline phosphatase system

In wild-type *E. coli* the rate of biosynthesis of alkaline phosphatase can vary over a thousand-fold range in response to the concentration of phosphate in the medium; the maximum rate is obtained when it is low, and as its concentration is increased, production of the enzyme is repressed. Three genes are involved in synthesis of the enzyme; the *P* gene is the structural cistron which specifies the phosphatase, and genes R_1 and R_2b are concerned with regulation of its rate of formation. A fourth gene, R_2a, is concerned with uptake of the coeffector phosphate. R_1 is unlinked to the other three genes, which map together. This work has been reported by Echols (1961); Garen and Echols (1962a, 1962b); Garen and Otsuji (1964).

De-repressed mutants in which the enzyme is synthesized even in the presence of phosphate can be isolated in both the R_1 and R_2b genes. The study of heterozygotes showed that R_1^- and R_2b^- mutants are both recessive to their normal alleles, suggesting a negatively controlled repression system (see Figure 11.3) in which both genes are somehow involved.

Super-repressed mutants causing a reduction in the rate of enzyme formation in low phosphate are found to occur only in the R_1 gene. However, these are also recessive to the normal allele, suggesting that R_1 is involved in positive control (see Figure 11.2).

The dual effects of the R_1 locus can be explained by a model which incorporates features of both positive and negative control. R_1 synthesizes an apoinducer which is necessary for the cell to achieve a high level of alkaline phosphatase. The phosphate co-effector does not merely inactivate this species, but converts it to a repressor necessary for complete repression of the operon. However, the product of the R_2b gene is required at some stage to mediate this conversion. This scheme is illustrated in Figure 11.9.

De-repressed R_1^- mutants fall into two classes. In one, although repression in high phosphate is prevented, the capacity to synthesize enzyme under conditions of low phosphate is decreased. This is explained if the mutation has resulted in the production of an apo-inducer which can no longer be converted into a repressor, and so has an impaired inducing capacity. Both these changes would be recessive to a normal apo-inducer. The second class show full enzymic activity in low phosphate, but some (although less than normal) repression is possible at high concentrations. This could arise as the result of mutation to produce an apo-inducer with unimpaired inducing activity, but with partial block in its ability to transform to a repressor. De-repressed R_2b mutants result from an absence of the enzyme activity which is involved in the conversion of an apo-inducer to a repressor.

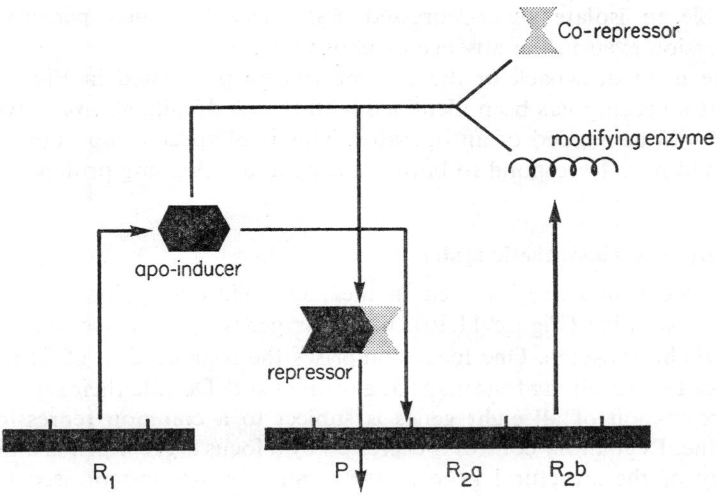

Figure 11.9. The control of alkaline phosphatase. Gene R_1 synthesizes an apo-inducer which is required to switch gene P on. This species interacts with the co-repressor and a modifying enzyme specified by gene R_2b to yield a repressing species which switches gene P off

Mutants in R_2a also show the same de-repressed phenotype as R_2b mutants, and at one time it was thought that this locus might also be a regulator. However, a comparison of the proteins of wild-type and R_2a^- strains by ammonium sulphate precipitation followed by disc electrophoresis identified the species coded by this gene. It has a high affinity for phosphate ions, and probably represents a phosphate permease responsible for its cellular uptake. Its formation appears to be under the same control as that of the adjacent phosphatase gene, being repressed under the same genetic and environmental conditions. Thus it is not a regulator gene, but appears to be necessary for repression of alkaline phosphatase because it controls uptake of the co-repressor.

The exact role of the R_2b protein is not specified. This leaves the possibility that it may fulfil some function similar to that of R_2a, in the sense that its action on the phosphate is required if the latter is to be able to act as a corepressor. In this case, it would not be a true regulator gene, but might comprise part of an operon with P_1 and R_2a. On the other hand, it could specify an enzyme which converts an apo-inducer to a repressor in the presence of phosphate. If this were so, it should be possible to isolate super-repressed R_2b mutants which perform this conversion even in the absence of phosphate.

The main drawback of the control scheme postulated in Figure 11.9 is that no region has been identified with the *cis* dominant–*trans* recessive properties demanded of an operator. This is of special importance since it would have to respond to both inducing and repressing protein species.

The arginine biosynthetic system

Eight enzymes are involved in the metabolic conversion of glutamic acid to arginine (Figure 11.10), but their genes map at five loci on the *E. coli* chromosome. One locus comprises the four genes *arg*ECBH, and each of the remaining four maps at a distinct site. Despite their separation, the expression of all eight genes is subject to a common repression by arginine. Regulatory control is exercized by a locus *arg*R which is unlinked to any of the structural genes; *arg*R$^-$ mutants are de-repressed for all eight enzymes.

The enzyme ornithine transcarbamylase (OTC), specified by gene *arg*F, is the easiest to assay because its specific activity varies most widely with changes in arginine concentration. Maas and coworkers (1964) and Maas and Clark (1964) studied its repression by arginine in both the temporarily diploid zygotes resulting from a mating between Hfr *arg*R$^-$ and F$^-$ *arg*R$^+$, and in strains which are stable diploids for some one-third of the chromosome, including the *arg*R locus. This showed that the presence of an *arg*R$^-$ allele does not affect the behaviour of *arg*R$^+$, that is, de-repression is recessive to repression. Subsequent work has confirmed that the *arg*R$^-$ allele is inactive by demonstrating that *amber* mutants in the regulator produce the de-repressed phenotype. These results suggest a negative control system (see Figure 11.3) in which *arg*R synthesizes an aporepressor which is activated by the arginine co-repressor.

Although most strains of *E. coli*, including K, C and W, show repression of all eight enzymes in the presence of arginine, strain B is an exception. This has a slightly repressed level of these enzymes in the absence of arginine, but an induction of the system results from its addition. Conjugation between K donors and B recipients shows that the insertion by recombination of *arg*R$^+$ into the B genome introduces repression in response to arginine. This suggests that the unusual control behaviour of B cells is due to the presence of a different allele (*arg*R^B) at the regulatory locus. Jacoby and Gorini (1967) confirmed this by transduction experiments in which the *arg*R locus was exchanged between K and B strains. This has the advantage over conjugation that only a

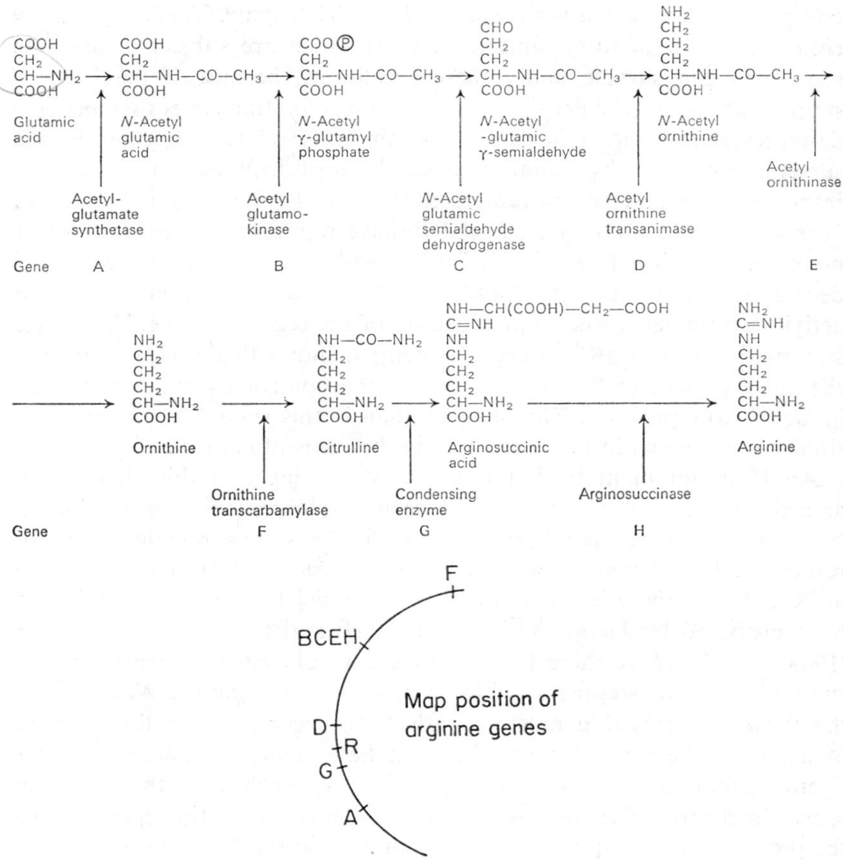

Figure 11.10. The arginine biosynthetic system of *E. coli*

small amount of the donor genome is transferred; thus replacement of the host regulator by recombination with the allele received from the donor inserts only the latter into the genetic background of the host cell. The mode of control found in recombinants is that of the donor allele, that is, the pattern of regulation of the structural genes is determined by this locus alone.

Regulator mutants in strain B were isolated by growing bacteria carrying a streptomycin-suppressible defect in *arg*F on medium containing streptomycin and lacking arginine. Because streptomycin exerts only a low degree of suppression, cells which are able to grow under these conditions must have a high level of the OTC enzyme. This can arise by either of two regulatory mutations. Cells which are subject to arginine repression can survive under these conditions, because if the level of suppression is low the end-product arginine is limiting; thus they become de-repressed and manufacture large amounts of the enzyme. Hence survival can result by mutation from the *arg*R^{B} absence of repression response to arginine to the repressibility control shown by other strains. Several survivor strains were found to have repression control, although none was identical to strain K. The second type of survivor represents de-repressed mutants, and these were found to have a uniform high activity, presumably due to inactivation of the regulator gene. The single step mutation of *arg*R^{B} to arginine control shows that a change in just the allele present at this locus can vary the control characteristic from induction to repression. This suggests that, in this system, these are very similar phenomena in terms of their mechanisms of control.

*Arg*R^{B} is dominant to de-repression, suggesting that this allele plays an active role in regulation. In *arg*R^{B}/*arg*R^{+} diploids, control is dominated by *arg*R^{B} at low levels of arginine, that is the system remains partially repressed. Under conditions of arginine excess, control is exercised by *arg*R^{+}, that is there is full repression. A model to account for this has been proposed by Jacoby and Gorini (1969) and Karlstrom and Gorini (1969). In *E. coli* K there is a negative control system in which *arg*R^{+} synthesizes an apo-repressor which is activated by arginine. With *arg*R^{B}, the form of interaction must be rather more complex. In the absence of arginine, the regulator protein must have a greater affinity for the operator than the apo-repressor sythesized by *arg*R^{+}, so that there is a certain degree of repression. Arginine interacts with this species, and the presence of large quantities of the amino acid must then reduce rather than enhance its affinity for the operator.

This system shows that it is possible to exercise control of the synthesis of the enzymes comprising a metabolic pathway, even when the genes which code for them are not gathered in a cluster. This implies that each of the five loci representing the *arg* structural genes must possess an operator which recognizes the same repressor protein synthesized by the *arg*R regulator. The system may thus be considered as under *parallel* control, in that all its genes respond to the same regulatory mechanism; however, this control is *non-coordinate* since, when synthesized, the enzymes are not produced in equimolar proportions (as they would be by the polycistronic messengers representing clustered gene systems). This may be accounted for in terms of properties of their various operator–promoter control regions. In fact, there are two possible explanations, both of which require that the interaction between nucleic acid and protein is such that changes in the nucleotide base-pair sequence can quantitatively change its affinity for the protein. The first possibility is that the various operators, although all responding to the same repressor, are subtly different, so that they have different affinities for it. The other is that it is the promoters of the various gene loci which may function with different degrees of efficiency in recognizing RNA polymerase and in achieving transcription. (Of course, these are not mutually exclusive.)

Some of the postulated operator loci have now been identified. Jacoby and Gorini (1969) reported that a site *arg*O_F closely linked to *arg*F can be mutated to give *cis* dominant–*trans* recessive synthesis of OTC without affecting the levels of the other enzymes of the system. The de-repression is only partial, and there is still some response to arginine. However, the mutation alters the recognition properties of the operator for different repressor proteins in a different manner. Whilst it still shows de-repression with the regulatory locus *arg*R^+, the combination of the mutant operator (*arg*O_F^c) and the regulator *arg*R^B introduces repression of OTC in response to arginine. This implies that a mutation in the structure of this operator causes a specific change in its recognition of repressor protein such that it has a decreased affinity for the *arg*R^+ product, but an increased affinity for the *arg*R^B protein when combined with arginine. Indeed, this is precisely the same type of change in the characteristics of operator recognition postulated above, and implies that the events observed at this one locus may be representative of a more general phenomenon.

The four-gene (ECBH) cluster appears to behave not as one operon, but as two, with the CBH genes expressed coordinately and the *arg*E gene under parallel but non-coordinate control (Baumberg, Bacon and Vogel, 1965; Elseviers, Cunin and Glansdorf, 1969). Although both the *arg*E and *arg*CBH respond to the same repression control, whereas the activity of the single E gene can be repressed only twenty-fold, synthesis of the enzymes specified by the CBH genes is coordinately repressible up to fifty to seventy-fold. The CBH segment is read from *arg*C toward *arg*H, as shown by the finding that *amber* mutants in *arg*B and *arg*C can influence the expression of *arg*H. However, such mutants in *arg*E do not interfere with the expression of the other three loci, suggesting that *arg*E is responsible for synthesizing an *m*RNA for its protein and that *arg*CBH synthesizes a tricistronic messenger, which directs synthesis of all three of their proteins. When the boundary between the *arg*E and *arg*C genes is removed by a deletion, the repressibility of *arg*H is reduced to the extent typical of *arg*E. This suggests that there is an operator–promoter control unit for *arg*CBH at this boundary, and its removal places these three genes under the control of a separate operator and promoter which normally control only *arg*E. In a mutant strain of *E. coli* in which the *arg*E gene has been translocated away from the others of the cluster, its behaviour in response to repression is unaltered. This too suggests that it possesses its own operator and promoter. In a sense, the situation within this gene cluster mirrors the general behaviour of the genes coding for the enzymes of the *arg* pathway; that is, control is parallel but non-coordinate, with more than one operator responding to the same repressor.

Control systems in fungi

Generally, higher organisms do not appear to possess the extent of functional grouping present in bacteria. However, gene clusters for some systems have been identified in some fungi, and it appears that in other systems a coordinate control may be exerted over the synthesis of related enzymes despite the dispersion of their genes. Although a conventional operon (the galactose system) has been identified in yeast, the gene clusters found in *Neurospora crassa* appear to represent proteins which are organized as a functional aggregate, that is a multienzyme complex. The control mechanisms postulated for the fungal systems so far studied are rather more complex than those which operate in bacterial regulation.

In *E. coli* galactose induces a fifteen-fold rise in the levels of the three enzymes responsible for its conversion to UDP-galactose and the permease responsible for its uptake. The metabolic enzymes—kinase, transferase and epimerase—constitute an operon and appear to be transcribed and translated as a polycistronic messenger from one end. An unlinked regulator gene shows behaviour similar to that of the *lac* system *i* gene, and this suggests the same type of negative-induction control system. In yeast three closely linked genes synthesize the three metabolic enzymes, and Douglas and Hawthorne (1966) demonstrated that these are induced by galactose and, under control from an unlinked regulator (*i*), are identified by recessive constitutive mutants. This suggests a similar control to the bacterial system.

However, mutants at another locus (gal_4), which segregates independently of the structural genes, result in a failure to synthesize all three enzymes. This region must be translated since its mutants display interallelic complementation and are susceptible to correction by external suppressors. Further, dominant constitutive mutation occurred at a site (the C locus) immediately adjacent to gal_4. These mutants showed the behaviour characteristic of operators, since their expression required the presence of a normal gal_4 allele in *cis* configuration.

The model illustrated in Figure 11.11 was proposed to account for these properties. The *i* regulator produces an apo-repressor which acts at C but may be inactivated by the galactose co-inducer. C is an operator controlling the expression of the contiguous *gal4* gene, which synthesizes an inducer exerting positive control of the expression of the structural genes. Inactivation of the apo-repressor causes constitutive mutation because it is not possible to prevent synthesis of the *gal4* inducer. Inactivation of *gal4* means that the inducer cannot be synthesized, so the system is uninducible. Modification of C so that it no longer recognizes the *i* gene apo-repressor results in an inability to prevent inducer synthesis and thus dominant constitutivity.

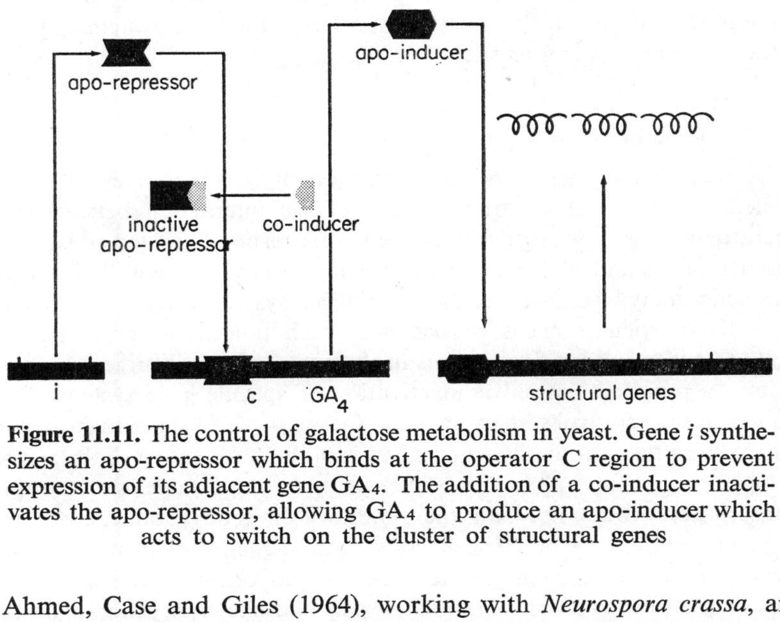

Figure 11.11. The control of galactose metabolism in yeast. Gene *i* synthesizes an apo-repressor which binds at the operator C region to prevent expression of its adjacent gene GA4. The addition of a co-inducer inactivates the apo-repressor, allowing GA4 to produce an apo-inducer which acts to switch on the cluster of structural genes

Ahmed, Case and Giles (1964), working with *Neurospora crassa*, and Fink (1966), working with yeast, showed that the genes of the histidine system are scattered over the different chromosomes. However, in each case there is a locus which is functionally more diverse than a single gene; this comprises the three genes which synthesize enzymes 3, 2 and 10 of the pathway (see Figure 11.4). In each species there are defective mutants which show a polarized loss of all three of the enzymes cyclohydrolase, pyrophosphohydrolase and dehydrogenase. The results can be interpreted as due to the production of a polycistronic messenger which represents the whole cluster and is translated as such. Ahmed (1968) reported that the proteins which this messenger synthesizes in *Neurospora* are organized as a multifunctional complex.

Giles and coworkers (1967) found that a cluster of five structural genes in *Neurospora* codes for the enzymes of the polyaromatic pathway, which converts dehydroquinolinic acid to chorismic acid. Mutants resulting in the simultaneous loss of all five enzymes show polarity behaviour consistent with transcription and translation of the gene cluster as single polycistronic *m*RNA. Case and Giles (1968) reported that the product of these genes forms an aggregate of all five enzymes. Although both the *his* and *arom* gene clusters produce the polycistronic messenger characteristic of an operon, no regulator or operator loci have been identified for these systems. This raises the possibility that the clustering is necessary for some function concerned with formation of the multiprotein aggregate rather than for control as such.

The evolution of control systems

In essence, the induction and repression of bacterial operons is controlled by the effect on transcription of the interaction between their operator and an apo-regulator protein; this responds to the presence or absence of a small molecule co-effector in the environment. This type of interaction may be considered as a specialized system which has evolved to meet the particular needs of bacteria; such unicellular organisms are likely to suffer sudden fluctuations in their external environment and thus require a rapid activation or inactivation of specific gene systems. Thus while bacterial control systems demand the synthesis of short-lived messengers, so that they can be switched off as rapidly as on, much of the *m*RNA in higher organisms appears to be considerably more stable; thus stopping transcription would not halt the synthesis of gene products. Another important difference between bacteria and higher organisms resides in the organization of their genetic material. No other species appears to possess the functional clustering shown by the bacterial genome, so that only bacteria (and fungi in one or two instances) have their genes arranged for coordinate control. The much increased quantity, and the more complex state, of the DNA in the nucleoprotein structures of higher organisms also imply that the mechanisms concerned with gene control may perhaps be of a different nature. Thus bacterial control systems may well have evolved to meet their specific environmental conditions. Although it is difficult to exclude the possibility, there is as yet no evidence to support speculation that similar systems may be involved in the differentiation and development of higher organisms.

Experiments in selection have shown that when a bacterial strain loses its proper control mechanism it is at a selective disadvantage. Zamenhof and Eichoran (1967) showed that in the *Bacillus subtilis* tryptophan system (the operon is repressed by tryptophan), de-repressed mutants are at a selective disadvantage, presumably because of their wasteful production of tryptophan. Baich and Johnson (1968) compared the growth of wild-type *E. coli* with a strain lacking the ability to control proline synthesis (by end-product) inhibition. In a mixed culture, there was a steady increase in the number of wild type at the expense of the mutant.

It is clear that a bacterial strain could gain a selective advantage by acquiring a mechanism to repress the production of an enzyme when it is not required. Protein synthesis is energetically expensive (a total of three high energy bonds per amino acid), and so repression would lower the demands made on the energy-producing system of the cell. Thus the evolution of negative control systems, in which the regulator gene synthesizes an apo-repressor, is readily explained. It is less clear, however, how positive control systems, where the regulator synthesizes an apo-inducer, would evolve, since in this situation the cell must originally have had the capacity to synthesize the enzymes concerned, irrespective of external conditions. One explanation might be a change in the function of the apo-regulator protein with evolution; systems in which the co-effector converts an apo-repressor into an inducing species might represent an intermediate stage in this process. However, if the apo-repressor protein is to act by blocking the progress of RNA polymerase into the operon, it is difficult to see just what role such an inducing species might play.

In previous discussion, negative and positive control systems have been treated in formal terms as merely different aspects of the same fundamental type of control mechanism, without considering the molecular basis for the action of positive control. If negative control is exerted by the binding to DNA of a protein which prevents the initiation of transcription, the analogous action for positive control would be the binding of a protein whose presence is necessary if transcription is to be initiated. The recent discovery of the role played by the sigma factor in assisting RNA polymerase to initiate transcription (see pages 141–142) suggests that it could fulfil this demand.

If a sigma factor is to act as an apo-inducer, it must have two properties. First, each operon to be controlled must have a factor which recognizes only its promoter. Thus the action of each such factor will be to cause the initiation of transcription of its operon. There is evidence that this type of control is exercised during bacteriophage infection of *E. coli.*

During infection of the bacterial cell, phage T4 directs synthesis of a sigma factor with a different affinity for sites on the phage genome compared with that of the *E. coli* factor; that is, the *E. coli* and T4-coded sigma factors each cause the transcription of different phage genes by the *E. coli* minimal enzyme (Travers, 1969).

Further, however, a sigma apo-inducer would have to possess the ability to combine with a co-effector, either for activation or inactivation. It is interesting that the implication of such a control mechanism is that the promoter would fulfil the role exerted in negative systems by the operator and promoter together. Essentially, the ability of the RNA polymerase both to recognize this site and to start transcription would be controlled by the availability of the appropriate sigma factor in active form. The direct evolution of this type of system is easy to account for. RNA polymerase must have the ability to recognise the promoter, and if this resides in a distinct polypeptide chain, it would be comparatively straightforward for the cell to develop the ability to provide further such chains, modify their recognition specificities, and allow their activities to become subject to an interaction with small molecule co-effector. (We should observe, however, that there is at present no evidence for more than one bacterial sigma factor, or that such interaction can occur.)

CHAPTER 12

Translational control mechanisms

The expression of an operon is controlled at the level of transcription by the interaction between its operator and an apo-regulator protein responsive to a co-effector. Although this interaction primarily operates a qualitative 'on/off' switch, it may have quantitative aspects, as shown by the different extents to which the genes of the non-clustered arginine system may be repressed (see pages 254–258). When switched on, the numbers of messengers which an operon synthesizes may be accounted for in terms of the specific interaction between its promotor and RNA polymerase. That this sets messenger production at a precise level is shown by the way in which it is influenced by the concentration of cyclic AMP (see pages 237–238), and by the existence of point mutations in the *lac* promotor which change the maximum level of expression of the operon (page 233). In addition to these mechanisms, discussed in Chapter 10, further systems must operate upon the synthesis of proteins from polycistronic messengers to ensure that each message is translated the correct number of times. This control may depend to some extent upon the relative timing of events in transcription and translation, which appears to be very precise. The phenomena described in this chapter support the concept that each stage in the expression of a gene is under specific control, from the qualitative decision as to whether it is to be switched on or off, to the rate at which its messengers are sythesized, translated and degraded.

Polarity

Ames and Hartman (1963) reported that some half of the point mutations isolated in cistrons of the histidine biosynthetic operon of *Salmonella typhimurium* have dual effects. As well as lacking the enzyme specified by the mutated gene, the *quantities* of all the enzymes coded by genes on the operator–distal side of the mutated cistron are reduced (see

Figure 12.1). The amounts of enzymes synthesized by genes on the operator–proximal side are unaffected. It has now been demonstrated by numerous workers that all mutational events which produce polar effects do so through the introduction of a nonsense codon, and *amber*, *ochre* and UGA triplets are all equally effective (for review see Zipser, 1969). As might be expected, suppressor genes which act on nonsense codons partially relieve polarity; the extent of the relief is about equivalent to the extent of suppression characteristic of the suppressor used (Imamoto, Ito and Yanofsky, 1966). Since the suppression of nonsense codons takes place at translation, this implies that it is at this level that polarity exerts its effect.

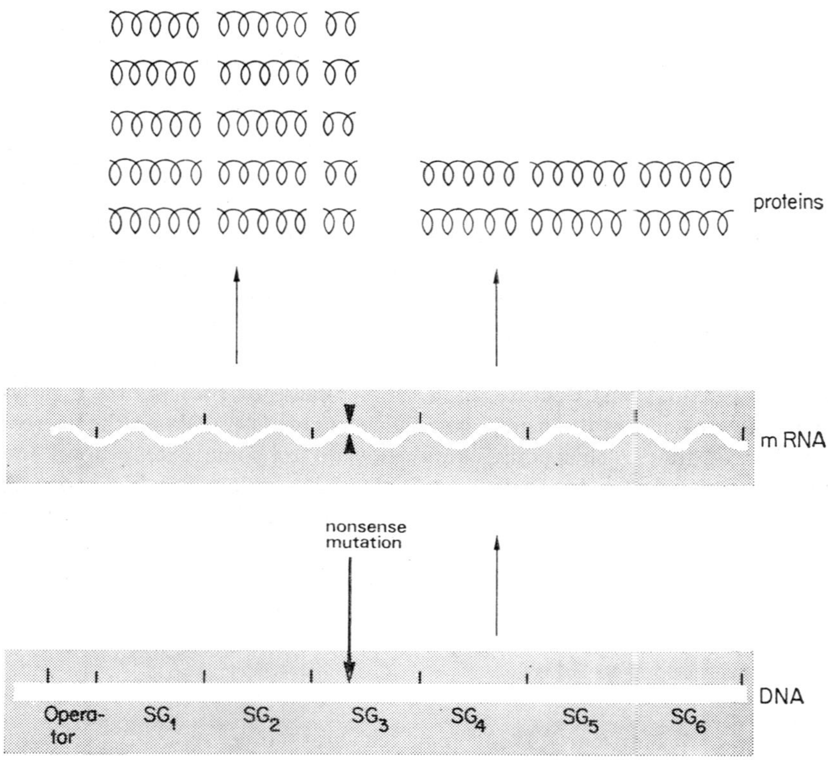

Figure 12.1. Polarity. A nonsense mutation in SG_3 causes premature termination of protein synthesis in that gene. However, it also reduces the *numbers* of proteins synthesized by each of the subsequent genes. Each of the genes SG_{4-6} suffers the same extent of reduction

Newton and coworkers (1965) investigated the effect of nonsense mutations in the *z* gene upon translation of the *E. coli lac* operon. The positions of various z^- mutant sites within the gene were located by deletion mapping, and the mutant strains assayed for permease and acetylase activities. The extent of the reduction in synthesis of these enzymes was found to depend upon the location of the nonsense codon within the first gene; mutants near the operator were completely polar, allowing no production of permease and transacetylase, whereas mutants at the operator–distal end of the gene allowed the production of the second two enzymes at levels approaching that of the wild-type strains. There is 'conservation of polarity' in that nonsense mutants in *z* create the same degree of polarity for the expression of both *y* and *a*. The gradient of polarity with map position is illustrated in Figure 12.2. The effect of nonsense codons in creating polarity is not limited to the first gene; *amber* mutants in *y* reduce the level of transacetylase synthesized by the *a* gene.

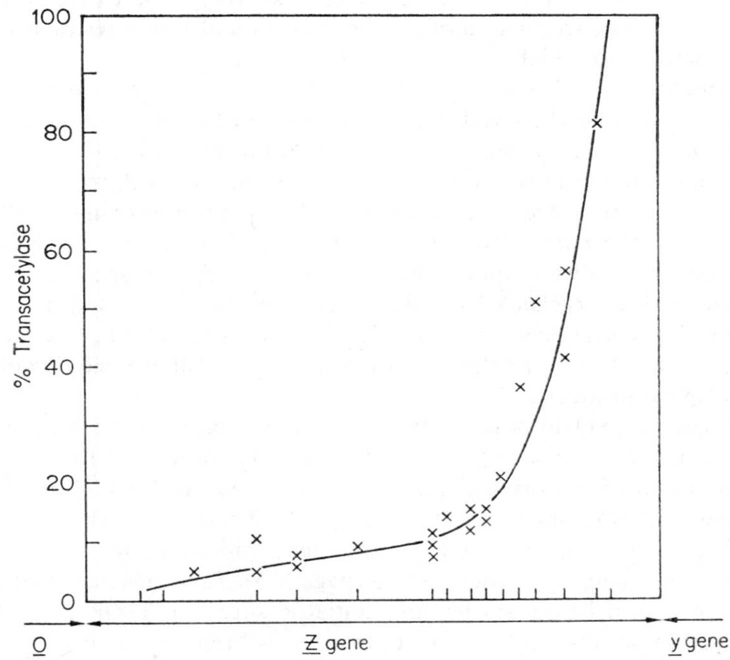

Figure 12.2. The gradient of polarity with map position of the nonsense mutant in the *E. coli lac* operon. (After Newton and coworkers, 1965)

The deletion of a large segment of *z* lying between the site of the mutation and the end of the gene was found to restore permease to strongly polar nonsense mutants in *z*. This suggests that the critical distance in the creation of polarity is that from the mutation to the next cistron, rather than that from the mutant site to the beginning of the gene. Zipser and Newton (1967) confirmed this by constructing double mutants in *z* which possessed a deletion either operator–proximal or –distal to a nonsense codon. Whereas the former did not influence the degree of polarity, the polar effect was partially suppressed by the insertion of a sufficiently large operator–distal deletion. In effect, such deletions 'move' the polar mutant to a new point on the gradient. Grodzicker and Zipser (1968) observed that, if the degree of polarity depends upon the distance from the nonsense mutation to the initiation site beginning at the next cistron, reversion of polarity should result from a further mutation in which a new polypeptide chain initiation site has arisen between the polar mutant and the end of the cistron. Such revertants of a polar mutant in *z* were found; a new initiation site very close to the original mutation allowed translation of the distal portion of *z*, and restored *y* and *a* gene activity to a high level.

Imamoto, Ito and Yanofsky (1966) showed that nonsense codons can also cause an *antipolar* effect, that is, the level of the enzyme specified by the gene immediately *preceding* the one with the nonsense codon is lowered. Antipolarity is also oriented in the same way with respect to the mutant position in the gene; the nearer the operator–proximal end, the more severe the effect. However, Balbinder and coworkers (1968) have observed that, whilst antipolar effects are found in the *trp* operons of both *E. coli* and *S. typhimurium*, they have not been detected in other systems. They suggested that this effect may thus represent some special feature of this system related to its formation of multienzyme complexes by two gene products.

A theory to explain polarity on a translational basis has been proposed by Martin and coworkers (1966a, 1966b) and Yanofsky and Ito (1966) as an expansion of the original theory of Ames and Hartman (1963). This suggests that ribosomes can attach to a polycistronic messenger only at the first cistron, and not at the subsequent initiation sites which must exist at the beginning of each cistron (page 125). The ribosome must then commence translation at the first initiator site, but, even after chain termination at the end of the cistron, must remain attached to the messenger and continue to travel along and initiate and synthesize each of the subsequent proteins coded by the messenger. If attachment is possible only at the beginning of the message, chain termination by polar mutants should have the same effect on the syntheses of every protein coded by a gene beyond the mutant cistron; that this is so is demonstrated by the conservation of polarity. The idea that each ribosome which attaches to the messenger translates every cistron is supported by the finding that the *his*, *trp* and *gal* operons all produce equimolar quantities of each gene product (for review see Epstein and Beckwith, 1968). (In the *lac* system there is a twenty-five-fold difference between the numbers of β-galactosidase and transacetylase enzymes synthesized—Zabin, 1963—but this may be a special occurrence related to the apparent lack of a function for the latter enzyme.)

The theory suggests that when a ribosome encounters a nonsense codon and chain termination occurs, it slips down the subsequent portion of the messenger in a disorientated manner with no defined phase. Since detectable amounts of the enzymes coded by genes beyond the block are produced, the reading frame of the ribosome must be restored at some point. This implies that all cistrons must start with an initiation signal (although this may be more complex than just an initiation codon) capable of providing the realignment. The gradient of polarity is accounted for by the supposition that there is a high probability that the ribosome will dissociate from the messenger when it is in this unstable state—somewhat proportional to the distance it must travel. If a ribosome is not realigned by the next initiator signal which it meets, it must traverse such a large segment of messenger to reach another that it has only an extremely low probability of ever arriving there.

The theory proposed by Martin and coworkers (1966a, 1966b) differed from that of Yanofsky and Ito (1966) in ascribing a role in polarity to the chain initiation region at the beginning of the next cistron. The resumption of polypeptide synthesis would depend upon the efficiency of the particular initiation signal, so that the extent of polarity would be determined not only by the distance which the ribosome must travel in unstable association with the message, but also on the realignment region—the more efficient the latter, the lower the degree of polarity. This might depend upon the particular initiation codon (AUG or GUG), or upon factors such as the number of nucleotides between the termination codon at the end of one cistron and the initiation codon at the beginning of the next.

Mutants in the C region of the histidine operon provide evidence for the multiple role of chain initiation regions. This region comprises two cistrons which specify different polypeptide chains of about the same molecular weight; however, mutants in the first gene are strongly polar, but those in the second are only weakly so. Since in this case differing lengths cannot determine the different extents of polarity, this might be due to reinitiation with differing efficiencies. Martin and Talal (1968) reported further evidence in support of this conclusion. Several mutants in the more operator–distal C gene were isolated, all with the same polarity. However, whilst most mapped in the middle of the gene, one mapped at the operator–distal end; this makes it unlikely that the distance to the next gene can be the only parameter determining the extent of polarity.

If ribosomes are to provide the free 30S subunits necessary for initiation at internal cistrons, they must separate into subunits upon chain termination at the end of the previous cistron. If the 30S subunit remains attached to the messenger, it can undertake reinitiation, associate with another 50S subunit, and continue translation. One possible mechanism to account for the retention of the small subunit by the messenger suggests that, although nonsense codons themselves cause termination of polypeptide chain synthesis, a more complex signal is required to cause dissociation of the ribosome from the messenger. In this case, the ribosome would not encounter such a signal at a polar mutant, since it would be present only at the inter-cistronic boundary. The ability of ribosomes to undertake reinitiation of synthesis at sites within a gene (discussed below) suggests that chain termination as such must involve separation of the subunits, even although the 30S does not dissociate from the messenger. However, upon *in vitro* translation of RNA bacteriophages bearing nonsense mutants, the ribosomes dissociate from the messenger at the mutant site (page 287). This suggests that a termination codon should cause loss of the 30S subunit from the messenger. A scheme to account for this would suggest that this does not happen at the end of cistrons in a polycistronic messenger because the close proximity of the initiation codon for the next cistron prevents loss of the 30S subunit and holds it to the messenger. However, this would predict that polar mutants in bacterial *m*RNA cause loss of the attached 30S subunits. The basis for their retention (as required by the conservation of polarity) is not clear, and it is worth noting that, even assuming they are retained, it is not known whether they would be capable of the postulated phase-less movement.

Recent theories to explain polarity have placed increasing emphasis upon the role of reinitiation. Fink and Martin (1967) proposed a theory based upon the isolation of polar mutants in the C aminotransferase gene of the histidine operon which possess low but detectable levels of this gene product. They suggested that there might be two situations in which nonsense mutants could show activity of the enzyme in whose cistron the mutation occurs. If the nonsense triplet is near the end of the cistron, the length of the polypeptide chain already synthesized might be sufficient to provide enzyme activity. Alternatively, its location at the beginning of a cistron, if followed by a phasing signal within a short distance to cause rapid reinitiation, might allow sufficient of the protein to be synthesized for activity. However, both these classes should show little, if any, polarity, as opposed to the mutants actually isolated. They therefore suggested that reinitiation of the terminated polypeptide chain can occur at various sites after the nonsense mutation but before the start of the next cistron. Because the ribosome is in a phase-less condition, this might take place in any reading frame. If the frame is wrong, it is likely that there would soon be a termination. Thus the extent of polarity would be determined by the ratio of translation in-frame/out-frame (and by the varying efficiencies of different initiators).

Although the in-frame/out-frame model does not predict the gradient of polarity, the behaviour of double mutants which possess two nonsense codons in the same gene supports the suggestion that reinitiation events play a role in polarity. When a double mutant comprises nonsense codons located in two successive cistrons of an operon, each nonsense triplet independently reduces by its usual proportion the synthesis of the enzymes coded by genes distal to it; thus the double mutant is more polar than either of the single mutants (Newton, 1969; Yanofsky and Ito, 1966). This implies that a reinitiation event takes place at the inter-cistronic boundary, thus permitting expression of the second mutant.

Different results are obtained when both nonsense codons are located in the same gene. Yanofsky and Ito (1966) found that, in this case, the polarity exhibited is that typical of strains possessing only the more operator–proximal nonsense triplet. This is understandable if the ribosome passes out of phase after passing the first codon; the second triplet must then be unable to exert any influence on it. This implies that the first efficient reinitiation event after passing the first mutant site must lie beyond the second site, most probably at the next inter-cistronic boundary (although it does not exclude the possibility that an *inefficient* low level of reinitiation occurs between the sites).

Newton (1969) found that the polar behaviour of double nonsense mutants in the *z* gene of the *lac* operon depends upon their precise location. When their sites lie in the operator–distal portion of the gene, the double mutant shows the same polarity as a strain containing only the more operator–proximal mutant. However, if the operator–proximal mutant lies in the early or middle part of *z*, the introduction of a more distal mutation causes an increase in polarity. Indeed, the nearer the first mutant is to the operator end of the gene, the greater the influence of the second mutant. This indicates that the more operator–proximal the first mutant, the greater is the chance that a reinitiation of polypeptide synthesis can occur before the ribosome reaches the next mutant site. If such a reinitiation does occur, the second mutant will be read as chain-terminating and will thus exert a polar effect. The actual appearance of such effects suggests that the *z* gene may contain several sites at which protein synthesis can be reinitiated if it has been prematurely terminated, and the probability that one such site will be included between any two mutant sites depends upon their distance apart.

Newton suggested that the shape of the gradient of polarity (see Figure 12.2) can be accounted for in terms of the presence of reinitiation sites. The steep gradient in the last quarter of *z* probably reflects the distance of mutant nonsense triplets from the efficient reinitiation signal at the beginning of *y*. The flat portion in the middle of the gene may be accounted for by inefficient reinitiation signals which occur within this region of the gene. The extreme polarity exerted by mutants in the beginning of *z* may possibly result if an efficient reinitiation signal is required for continued translation (or transcription) after termination very early in the message.

Michels and Zipser (1969) have mapped reinitiation sites (π sites) in the *z* gene by examining double nonsense mutants which showed an increase in polarity compared with the single more operator–proximal mutant. These must have a reinitiation signal between the mutant sites. Only two efficient reinitiators were found, although the β-galactosidase enzyme has twenty-four methionine residues, so that it must have twenty-four AUG codons in phase, as well as others out of phase. This suggests that the signal for initiation is likely to be more complex than just an AUG triplet, possibly involving a more complex nucleotide sequence, or the location of the initiation triplet in a region with some specific secondary structure. Preliminary work has suggested that the gradient of polarity in the *z* gene may be accounted for by the presence of these two reinitiators; if they play the postulated important role in polarity, their presence should result in a three-part gradient with two discontinuities, and it appears that this may be so.

Just as mutation to a nonsense codon can cause a termination of protein synthesis, so can mutation create new initiation signals. Of course, it is not possible to initiate synthesis at a new point along the messenger unless there has been a previous termination signal. Thus new initiator signals have been identified as revertants of nonsense mutants. Sarabhai and Brenner (1967) found that an *ochre* mutant created by a frameshift in the *r*IIB cistron was reverted by a point mutation which caused reinitiation and synthesis of a polypeptide chain corresponding to the rest of the gene. The effect is exerted at the translation level, and the new signal caused only an inefficient restart to synthesis.

Although many studies have concentrated on providing an explanation for polarity at the translation level, transcription studies have shown that, in some cases, polar mutations also affect the messenger templates. Attardi and coworkers (1963) found that two extreme polar mutants in the *lac* operon did not contain *lac* specific *m*RNA after induction, and Kiho and Rich (1965) observed that modification of the permease and transacetylase genes was associated with changes in the size of the polysome manufacturing β-galactosidase. *Amber* point mutations reduced its size, and reversion to the genetic wild type was associated with reversion to the larger polysome.

Imamoto (1965) demonstrated hybrid formation between *trp m*RNA and the DNA of phage ϕ80 carrying the tryptophan operon of *E. coli*. (The tryptophan synthetic system is illustrated later in Figure 12.7). This hybridization is quite specific for *trp m*RNA since the hybrid forms only under conditions of de-repression of the operon and not after its expression has been repressed by tryptophan. Imamoto, Ito and Yanofsky (1966) and Imamoto and Yanofsky (1967a, 1967b) examined polar nonsense mutants for *trp m*RNA production during de-repression. Total *trp m*RNA and *m*RNA corresponding to particular regions of the operon were measured by hybridization with ϕ80 transducing phage carrying only certain parts of its DNA.

Pulse label experiments showed that polar mutants produce normal numbers of *trp m*RNA molecules relative to the total RNA. However, the overall amount of hybridizable *trp m*RNA is reduced in strong polar mutants. Sucrose density-gradient centrifugation profiles demonstrated that the majority of *trp m*RNA molecules in such mutants is smaller than the messenger for the whole operon produced by wild-type cells. The size of the predominant *trp m*RNA species was found to decrease in relation to the nearness of the location of the nonsense mutant to the operator–proximal side of the mutated cistron. Thus in strong polar mutants, a major consequence of the introduction of the mutant nonsense codon is the appearance of short *m*RNA chains representing the operator–proximal portion of the operon, as shown in Figure 12.3. Although the precise point where the messenger terminates cannot be demonstrated experimentally, presumably it corresponds to the site of the mutation. Such mutants produce some normal full length *trp m*RNA species, and the relative proportions of normal and short molecules seem to correlate reasonably well with the extent of polarity (although, again, the experimental results are not sufficiently precise to decide whether this effect

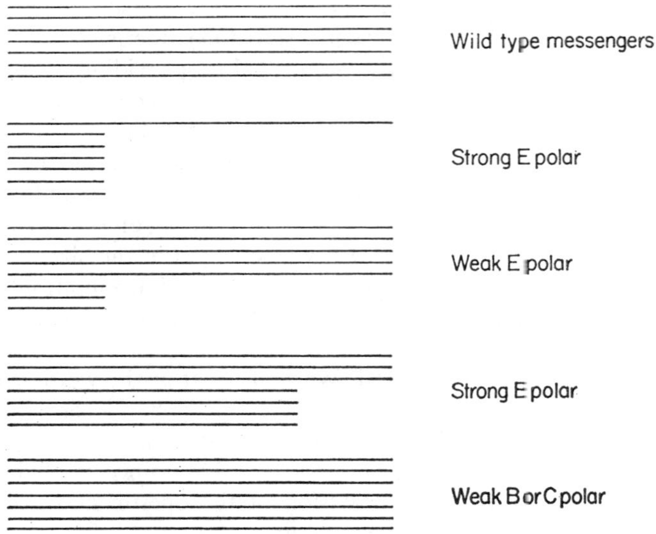

Figure 12.3. Polarity at the transcription level for the *E. coli trp* operon. (After Imamoto, Ito and Yanofsky, 1966)

alone is responsible for polarity). In the relatively weak polar mutants, *trp* *m*RNA size is normal, and it seems likely that in these the polar effect must be explained solely in terms of translation.

One possible mechanism to account for the size reduction would be for messengers to be degraded operator–distal to the site of the nonsense mutant. This might occur as a result of loss of their ribosomes, exposing them to nuclease attack. However, Carter and Newton (1969) argued that, in this case, the functional stability of *m*RNA corresponding to distal genes should be reduced compared to the segments coding for genes operator–proximal to the mutation. They measured the half-lives of the *z* and *a* segments of *m*RNA representing the *E. coli lac* system in both a wild-type strain and in bacteria with a polar mutant in the *y* gene. Experimentally, this was achieved by measuring the rate at which a cistron loses the ability to synthesize β-galactosidase and transacetylase after a short period of induction; after induction has ceased, the capacity of the cell to synthesize these enzymes must reflect the activity of the *m*RNA previously synthesized. This showed that the *m*RNA segments representing the operator–proximal and the operator–distal genes have similar half-lives, both in wild-type and in mutant strains, although the latter produced much less of the transacetylase enzyme. A possible explanation is that the nonsense mutation in some way reduces the rate of transcription of the distal portions of the operon.

However, Moore and Yanofsky (1969) have observed that degradation of *m*RNA distal to a polar mutant does appear to occur in the *E. coli trp* system. They note that the *overall* kinetics of synthesis and degradation of *trp m*RNA appears normal in these mutants, even although the amount of messenger distal to the nonsense codon is greatly reduced. Their results suggest a model in which the RNA polymerase continues normally beyond a nonsense codon, but because ribosomes are discharged at the mutant site, the distal messenger is unprotected against nuclease attack. This implies that degradation of the messenger is initiated by an endonucleolytic attack. After this, an exonuclease 'chases' the RNA polymerase along the messenger, degrading the polynucleotide chain as it is synthesized. However, when the polymerase reaches the initiation region at the beginning of the next cistron, the messenger synthesized at this point may be able to 'reload' with ribosomes, and thus be protected against further nuclease degradation. The gradient of polarity is accounted for if the probability that the nuclease will 'overtake' the polymerase—and as a consequence prevent 'reloading' by degrading the message almost immediately upon its synthesis—is proportional to the length of messenger exposed between the mutant and the reinitiation site. When 'reloading' does take place, the normal relationship of synthesis and degradation becomes re-established, so that the intermediate region of degradation can only be detected by use of a very short labelling period. This was found experimentally.

Fink, Klopotowski and Ames (1967) observed that regulatory mutants in the histidine operon of *S. typhimurium* which produce high constitutive levels of the histidine biosynthetic enzymes cause an inability to grow at 43°C. Reversion of the temperature sensitivity can be achieved either by true back mutation or by a secondary polar mutation in the operon which decreases the levels of its enzymes synthesized. This system has the advantage that polarity can be studied with a high level of the affected enzymes for assay.

Using this approach, Fink and Martin (1967) reported that there is a difference between polarity effects in the first and in the subsequent cistrons of an operon. Whilst the first gene shows a steep gradient of polarity, the internal cistrons show a much more shallow gradient. Data for the histidine operon are shown in Figure 12.4. They proposed that these two types of polarity may be caused by different effects. Polarity in the first gene would be due to a transcriptional effect. The extent of transcription beyond a given nonsense codon in this gene would be

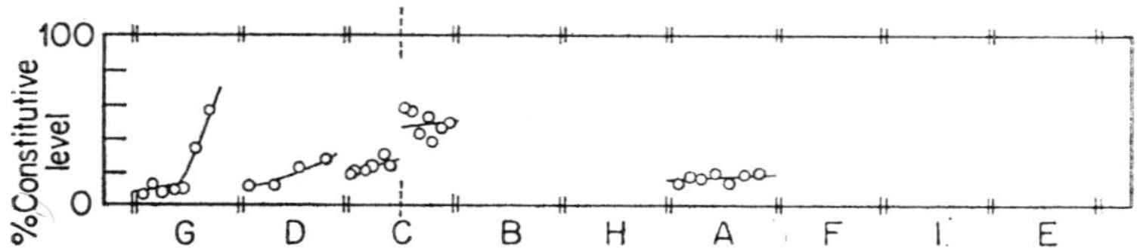

Figure 12.4. Gradients of polarity in the *his* operon of *S. typhimurium*. (After Fink and Martin, 1967)

statistically distributed about some mean; mutants near the operator–proximal end would be strongly polar, with at first a slow and then a more rapid increase in enzyme production proceeding to the distal end, with a lowering-off at this end. The final levelling-off might be lowered by the role of re-initiation in determining polarity; there appears to be some evidence for this. Their predicted curve is in agreement with the data for both the *lac* and *trp* operons as well as the *his* system.

The shallow gradient in internal cistrons suggests that once transcription has passed some critical point, translation and transcription become uncoupled—the basis for such uncoupling could be the initiation of a second round of transcription from the DNA before completion of the preceding polycistronic messenger. This would commence when transcription has proceeded past some critical distance, and would continue even although synthesis of the preceding *m*RNA has ceased. The transcription of this second species could in turn present an impetus for completion of synthesis and release of the first *m*RNA. Polarity in the internal genes would thus be explained largely on the basis of translational effects.

Periodicity of transcription

However, recent work on the intervals between successive rounds of transcription has suggested that this type of coupling does not take place. Morse, Baker and Yanofsky (1968) achieved synchronous transcription of the *trp* operon by addition of the tryptophan analogue indole-3-propionic acid to growing bacteria to de-repress the system. If a high concentration of tryptophan is added subsequently, the operon is repressed, and further initiation of transcription is prevented, although chains of *m*RNA already under synthesis are completed. Baker and Yanofsky (1968b) found that addition of tryptophan before 2·5 minutes have elapsed after de-repression permits only one round of transcription. When tryptophan is added at 4 minutes, an additional *trp* *m*RNA is synthesized, that is, a second round of transcription is initiated between 3 and 4 minutes after de-repression. The periodicity of initiation depends on the rate of transcription (which can be varied by alteration of the growth conditions); at rates of 1200 and 1000 nucleotides per minute, the periodicity is 2–2·5 and 2·5–3·0 minutes, respectively. This corresponds to a spacing between the polymerase enzymes on the operon DNA of about 3000 nucleotides. Surprisingly, however, the periodicity of initiation remains unaltered in a highly polar mutant in which little *m*RNA is synthesized corresponding to the last four genes of the operon.

Imamoto (1968a), using a different method of synchronization, found that transcription proceeds sequentially from the E gene to the operator-distal end of the system (the five genes are in the order EDCBA—see later, Figure 12.7). A second round of transcription begins when the first polymerase has only reached the middle of the operon (around the D gene). The rate of transcription of the operon remains constant, implying that when the third polymerase is about to commence transcription at the initiation site, the first is just completing its polycistronic messenger. Thus at any given moment, two or three polymerase enzymes may be transcribing the operon. Again, however, the same time span separates consecutive initiations in mutants in which most of the terminal portion of the operon has been deleted.

The mechanism responsible for maintaining this periodicity is not known. However, the data obtained from polarity and deletion mutants would seem to contradict the idea that a certain length of messenger must be synthesized, or that a certain number of ribosomes must attach, before the next round of transcription is initiated. Possibly, some (presently unknown) facet of the interaction between promotor and polymerase may provide the explanation.

Transcription and translation feedback

Stent (1964) proposed a model for coupling transcription with a need for simultaneous translation; a ribosome might assist the release of nascent messenger from DNA by translating it during its transcription—this would 'pull' it off the genome, as illustrated in Figure 12.5. In his original model, Stent suggested that if the nonsense triplet were near enough to the initiation triplet at the beginning of the next gene, it might be possible for another ribosome to attach there and continue pulling off the messenger—this would account for the gradient of polarity, since the shorter this distance, the greater the probability of such an event. However, this would demand internal attachment of ribosomes to the

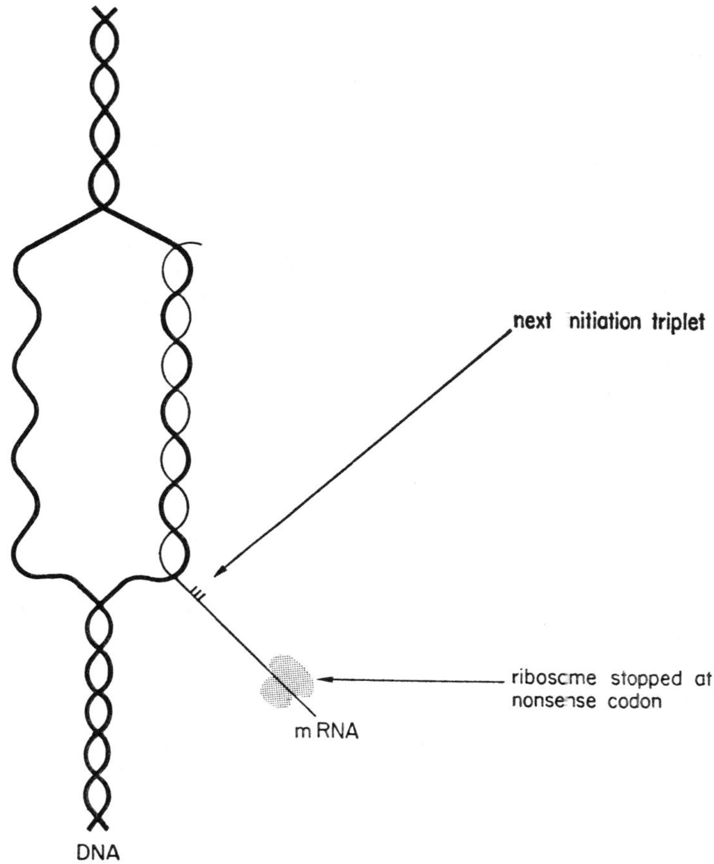

Figure 12.5. A model for coupling transcription and translation. The movement of the ribosome as it translates the messenger helps 'pull off' the *m*RNA from the DNA. If the ribosome is stopped at a nonsense codon, the shorter the distance to the next initiation triplet, the greater the chance that it will be exposed and another ribosome attached to thus overcome the blockage

messenger, which does not appear to happen. Another proposal has been that translation might be 'forcing' on transcription—ribosomes would attach to the messenger immediately behind the RNA polymerase transcribing it, and their movement would 'push' the polymerase along in front of them. Again, the presence of a nonsense codon would abolish this effect.

However, whilst translation of *m*RNA, in *E. coli* at least, may commence before the completion of transcription, there does not appear to be any direct formal coupling between them. Forchhammer and Kjelgaard (1968) argued that dependence of transcription on simultaneous translation would predict that a decrease in the overall rate of translation must result in a decrease in the overall rate of transcription. Further, no matter what the growth conditions, cells would have to maintain a constant ratio of *m*RNA/ribosome involved in protein synthesis. These two predictions were tested experimentally by measuring the rate of *m*RNA synthesis under conditions of both active protein synthesis and amino acid starvation, and by determining the ratio of *m*RNA/*r*RNA in cells growing at different rates (when the number of ribosomes per cell varies).

The concentration of *m*RNA in bacterial cells is determined by an equilibrium between its synthesis and breakdown, so that measurement of the true rate of synthesis requires the use of conditions under which there is no breakdown. This was achieved through the use of an *E. coli* strain auxotrophic for uracil. Starvation for this pyrimidine causes a loss of RNA because it breaks down rapidly due to its low stability; messenger synthesis is resumed upon the restitution of uracil, and the initial rate of synthesis can then be measured independently of *m*RNA degradation. This rate proved to be the same under conditions of both protein synthesis and amino acid starvation, so that *m*RNA transcription must be independent of concomitant protein synthesis.

The messenger activity relative to *r*RNA proved to remain constant in cells varying in growth rate between 23 and 40 minutes per doubling time. However, in a mutant which has normal 30S subunits but in which the 50S is unable to mature, the proportion of *m*RNA per *r*RNA remains the same as in the wild type. This suggests that the level of *m*RNA is not governed by the 70S ribosome, but may be influenced by an interaction with the smaller subunit, possibly through formation of an initiation complex. It is interesting that *in vitro* experiments have shown that 30S subunits, but not 50S subunits or 70S ribosomes, assist transcription under conditions when formation of the initiation complex is possible (see pages 150–151).

Kinetic studies have demonstrated that transcription and translation of the *lac* and *trp* operons in *E. coli* occur simultaneously, even if not dependent upon each other. Leive and Kollin (1967) found that 2·5 minutes are required for transcription of the β-galactosidase gene after induction. If transcription proceeds continuously, another 1·5 minutes would be required to transcribe the remaining genes. The time required for translation and activation of the enzymes from the messenger is some 1·5 to 2·0 minutes for each enzyme. However, β-galactosidase is available in the cell before the 4 minutes that these processes would take if consecutive; thus translation and activation of the first enzyme, β-galactosidase, must occur before completion of transcription of the transacetylase end of the message (see later, Figure 12.6, for an illustration of the state of an operon shortly after induction).

Ito and Imamoto (1968) reported that the enzymes of the *trp* operon appear sequentially on de-repression after tryptophan starvation in the same order as that in which the structural genes of the operon are transcribed. The end of the E (first) gene is transcribed within 2 minutes, and the end of the A (last) gene at about 8 minutes. The enzymes corresponding to each of the five structural genes appear about half a minute after completion of its transcription. Since the E gene thus appears within 2·5 minutes, and the messenger is not completed until 8 minutes, this indicates that ribosomes must attach to nascent *m*RNA molecules before their synthesis upon the DNA has been completed.

Morse, Baker and Yanofsky (1968) investigated the events resulting from each of three successive rounds of transcription of the tryptophan operon. The rate of ribosomal movement along the message appears to remain constant at about 1200 nucleotides per minute (this is estimated by taking the difference between the times of appearance of two enzymes as the time required for the first ribosome to travel from the end of one gene to the end of the other). The RNA polymerase has been estimated to travel at about 1000 nucleotides per minute, so that transcription and translation take place at much the same rate. The time at which the last ribosome passes the end of a gene is given by the moment when the increase in activity of the enzyme which it specifies comes to a halt. This shows that it proceeds at the same rate as the first ribosome.

Measurement of the production of tryptophan synthetase (the product of the last two genes of the operon—see later, Figure 12.7) showed that the number of enzyme molecules synthesized in each round is approximately constant. This suggests that a fixed number of ribosomes (about ninety) translate each *trp* messenger molecule. Since the last ribosome passes the same point as the first ribosome only 1 minute later, these ribosomes must travel along the messenger as a tightly clustered group with virtually no spacing between adjacent ribosomes. The small time between the completion of transcription and translation of a gene implies that the first ribosome of the group must travel along the message almost immediately behind the polymerase as it synthesizes the polynucleotide chain. These results could be accounted for if attachment of ribosomes to *m*RNA can only take place during a limited period, possibly immediately after the initiation of its synthesis, although the basis for such a restriction is not clear in view of the successful translation of isolated messengers *in vitro*.

The low stability of bacterial *m*RNA (the half-life of *lac* operon *m*RNA is only 2 minutes) has led to suggestions that the beginning of the messenger may be subject to enzymic attack before the synthesis and release from the template of its distal end. In other words, the succession of events at the operator–proximal (5′) terminus would be: initiation of transcription, attachment of ribosomes, and degradation by nucleases. This is illustrated in Figure 12.6. Recent work on the tryptophan system of *E. coli* (Morikawa and Imamoto, 1969; Morse and coworkers, 1969) has confirmed this model, and excluded an earlier suggestion (Baker and Yanofsky, 1968a) that degradation proceeds from the distal (3′) end

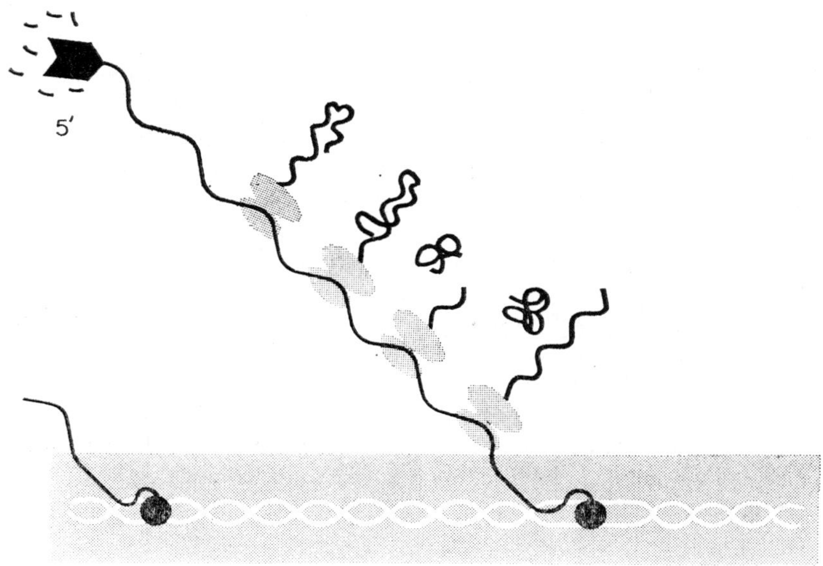

Figure 12.6. The operon shortly after induction. Ribosomes have attached to the messenger produced by the first polymerase before completion of its synthesis and are engaged in translation. The first ribosomes to attach have already completed synthesis and have released the protein specified by the first gene of the operon. A nuclease has attached to the 5′ end of the messenger and commenced degradation. Another polymerase has attached at the beginning of the operon and has begun to synthesize another messenger

of the messenger, in the opposite direction to transcription and translation.

Eight minutes after its de-repression (the time required for completion of the first round of transcription), the rate of transcription of the *trp* operon reaches a steady state. The administration of a short pulse of H^3-uridine during steady state transcription labels all the segments of *trp* *m*RNA, and the subsequent addition of unlabelled uridine then rapidly dilutes the radioactivity in the RNA precursor pools so that little further label is transferred to *m*RNA. The rate of degradation of the *m*RNA can be measured by following the disappearance of the radioactive label from different regions of the messenger; experimentally this is achieved by measuring the level of radioactivity in the *m*RNA which hybridizes with phage ϕ80 DNA carrying different (specified) regions of the operon.

The label in the *m*RNA region corresponding to the E genes begins to disappear rapidly ($\sim$1 minute) after the dilution with unlabelled uridine, and the D and C–A regions of the messenger begin to lose their label at 2–3 and 5 minutes, respectively. Thus degradation must begin at the 5′ end before transcription has reached the 3′ end. Almost all the label is lost from E *m*RNA within 10 minutes, from D within 15 minutes and from C within 25 minutes. This indicates that degradation proceeds in a sequential manner from E to A (5′ to 3′), taking some 15 minutes—about twice the time required for synthesis of the messenger. This must have the consequence that the operator–distal regions of the *m*RNA remain intact for longer than the operator–proximal segments.

The most likely mode of degradation to account for these results would be attack by an exonucleolytic enzyme, and the sedimentation properties of the species remaining after various time intervals are consistent with this. However, as yet, no bacterial ribonuclease with the required 5′$\longrightarrow$3′ activity has been found. One possibility is that an endonucleolytic enzyme might attack at various points along the messenger in a sequential manner, followed by a 3′–5′ exonucleolytic degradation of the fragments released.

Internal initiation of transcription and translation

The tryptophan operon comprises five genes which are regulated as a single unit by an unlinked locus, and show coordinate repression of synthetic activity in response to tryptophan. An operator locus has been identified at one end of the gene cluster by the lack of response of the system to tryptophan after its deletion. This system is the same in both *E. coli* and *S. typhimurium*, and is illustrated in Figure 12.7.

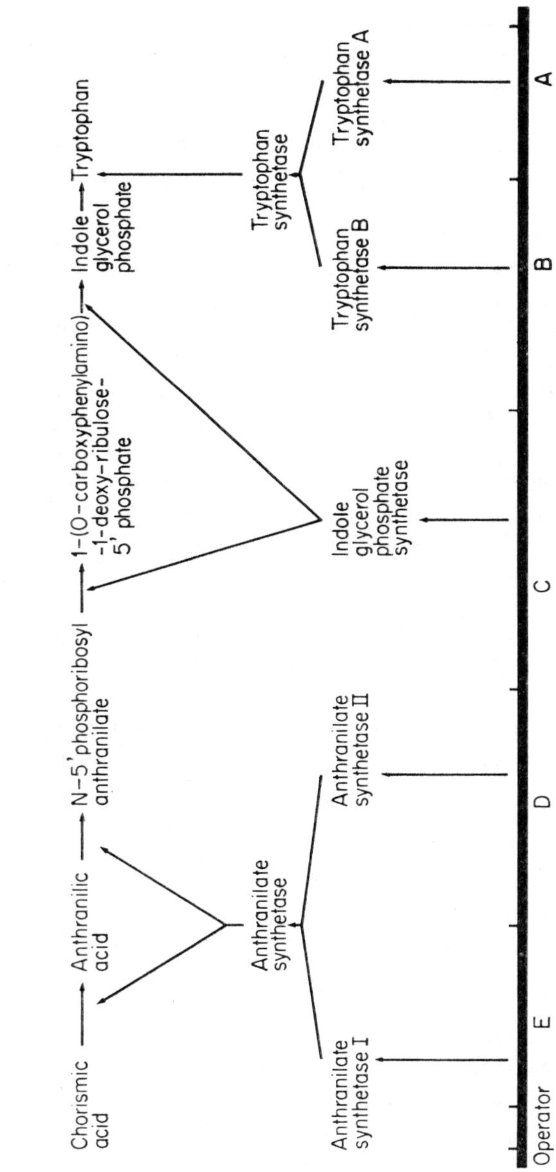

Figure 12.7. The *trp* system of *E. coli*

Despite the coordinate control over expression of its cistrons, in both bacteria transcription and translation can be initiated at internal sites. Bauerle and Margolin (1966) examined enzyme synthesis by the *S. typhimurium* operon in a group of deletion mutants which extend into the operon from the operator end and terminate at varying points within it. In contrast to the effects demonstrated by other operons, they found that the pleiotropy exerted by deletion of the promotor adjacent to the operator does not extend to all five cistrons. All strains with deletion ending in either of the first two genes lack expression of both these cistrons, but retain a low level of functioning of the remaining three genes. Expression of the last three genes is lost whenever the deletion extends beyond the boundary between the second and third cistrons. Polar mutants in the first gene are pleiotropic and affect all five genes, but the pleiotropy is not coordinate for the remaining four genes; expression of the second gene is more severely limited than are the last three. This can be accounted for if there is a second initiator element, located internally between the second and third genes, where either transcription or translation (or both) can be initiated. This results in a higher, repressor-resistant basal level of expression of the last three genes, creating the apparent non-coordinate expression of the first two and last three genes. Deletion of this internal site was found to abolish the non-coordinate effect.

Morse and Yanofsky (1968) reported that the *trp* operon of *E. coli* also shows non-coordinate expression of its last three genes. The basal level of synthesis of these enzymes is about five times that of the E and D gene products, so that the CBA segment shows low-level enzyme synthesis, even under conditions of repression, at about 2 per cent. of the de-repressed rate. After de-repression, there is coordinate synthesis of all five enzymes, due to production of a polycistronic *m*RNA representing the whole operon, which is translated to give equimolar synthesis of all its enzymes. This suggests that the non-coordinate synthesis is unlikely to result from translation of just the last three cistrons of such a messenger, but more probably results from the synthesis of a *m*RNA molecule corresponding only to the CBA segment of the operon. The level of synthesis required would be about one transcription event in every generation. The difference in levels of *m*RNA synthesis from each of the two promotors implies that the precise structure of this DNA segment can control the efficiency of its interaction with the polymerase enzyme.

Imamoto and Ito (1968) and Imamoto (1969) found that, under certain conditions, both transcription and translation can be initiated at (probably several) internal sites within the *trp* operon. When de-repressed bacteria are treated with DNP (2,4-dinitrophenol), transcription is inhibited, but regains its normal rate within one minute of the removal of the inhibitor. Usually, after de-repression, the *trp m*RNA synthesized during the next two minutes corresponds to only the E gene. However, the *trp m*RNA synthesized after DNP treatment may contain molecules corresponding to other regions of the operon. Hybridization studies showed that after a DNP treatment of 1·5 minutes, the *m*RNA synthesized upon its removal corresponded to only genes E and D; after 4–9 minutes of treatment, the messenger synthesized during the first minute after removal of the inhibitor corresponded to EDC and part of B; and after 11–29 minutes, this *m*RNA corresponded to the entire operon. If this is due to simultaneous initiations of transcription as well as at the operator–proximal promotor, the *trp m*RNA made under these conditions should be smaller than usual; this was confirmed experimentally by sedimentation studies. This demonstrates that although transcription is not usually initiated internally, under these abnormal conditions this becomes possible. When normal transcription is inhibited by DNP treatment, the sites available for initiation spread from E toward A with increasing times of treatment. One possible scheme which may account for this suggests that DNP allows RNA polymerase to move along the operon, but without catalysing the synthesis of *m*RNA. Upon removal of the inhibitor, the polymerase molecules can resume synthesis.

If a de-repressed culture is incubated with DNP, and tryptophan is added upon removal of the inhibitor, the amino acid prevents initiation of transcription at the operator–proximal promotor. The synthesis of the proteins specified by the E, C and A genes was measured after one round of transcription. After normal de-repression, synthesis of these enzymes is sequential, but after this treatment there is a simultaneous appearance of both the C and A proteins (the same result may hold for the D protein, which was not measured). There is no synthesis of the E gene product. The simultaneous production of the C and A proteins indicates that there are (at least) two internal sites where ribosomes can attach to *trp m*RNA molecules lacking the operator–proximal segment, although these sites are not used under normal conditions when ribosomes attach only at the end of the messenger during or soon after transcription.

One explanation for the usual inability of ribosomes to attach internally to messengers would be that, although each cistron commences with an initiation signal (which may itself be more complex than just an AUG triplet), there is only one site for ribosomal attachment, located at the 5′ end of the messenger. However, the internal attachment of ribosomes to *trp m*RNA (albeit under abnormal conditions), their ability to enter a circular messenger directly (see page 127), and the internal initiation of protein synthesis which occurs in the *his* system (discussed below), indicate that sites at which ribosomes can attach *do* exist within the messenger. Thus an explanation is required as to how ribosomes are excluded from using these sites under normal conditions.

An answer to this may be provided by the finding that, for the *trp* system, ribosomes attach to the 5′ end of the messenger only during a short period of time immediately succeeding the initiation of transcription. At this time, any sites promoting internal attachment have probably not yet been transcribed into messenger. The attached ribosomes translate the message by following as a group very close behind the polymerase synthesizing the messenger. As the message is transcribed and internal attachment sites appear, the ribosomes already on the messenger may be so near that they inhibit the attachment of any new 30S particles. In any case, as they move along they will shortly pass over the site(s) and preclude access by other ribosomes. Once the last ribosome has passed the site, the messenger may already be under enzymic attack, and thus unable to promote any further attachments.

Although some such control may apply for the *lac* and *trp* systems, experiments with the *his* system cannot be explained on this basis. Under certain conditions, the operon which specifies the enzymes responsible for histidine biosynthesis in *S. typhimurium* suffers internal initiations of translation of its messenger (Berberich, Kovach and Goldberger, 1967; Berberich, Venetianer and Goldberger, 1966; Venetianer, Berberich and Goldberger, 1968). In wild-type strains, the kinetics of de-repression upon the withdrawal of histidine are in accord with the usual sequential of an operon, the temporal sequence of appearance of the enzymes corresponding to the positional sequence of their genes. However, several mutants in this system exhibit an alternative mode of de-repression in which all the enzymes are synthesized simultaneously after the removal of histidine.

An examination of the *m*RNA produced by both normal and mutant strains showed that this effect is mediated at translation rather than transcription. The *m*RNA corresponding to the operon was identified by comparing the species present in strains possessing the operon with the RNA molecules produced by a strain which has a deletion covering most of this region; a large polycistronic *m*RNA was present in the former but not in the latter case. The same species proved to be present under both the conditions of sequential and simultaneous de-repression. Similarly, hybridization to DNA derived from either normal or deletion strains showed that the messenger corresponding to the operon is unaffected by the mode of de-repression.

A feature common to all the mutants which suffer simultaneous de-repression is the presence of a defect in any one of the enzymes which catalyse the first six steps of the biosynthetic pathway. Mutants in the last four enzymes exhibit the usual sequential de-repression. This suggests that it is the absence of the product(s) of reaction 6 which causes de-repression to be simultaneous. This was confirmed by the addition of phosphoribosyl-AIC (PR-AIC), which is produced by this reaction, to a strain showing simultaneous de-repression; the result was to convert it to the more usual sequential expression. This compound is converted through a series of steps in a pathway which finally yields AMP. The first step in the pathway is the formylation of PR-AIC to formyl-PR-AIC, and the donor for this reaction is N^{10}-tetrahydrofolate, also responsible for providing the formyl groups necessary for the synthesis of the fmet-tRNA$_f$ translation initiator. In a mutant in which this step is blocked and the PR-AIC cannot be formylated, simultaneous de-repression is exhibited under conditions which would otherwise result in the usual sequential expression. This suggests that the PR-AIC exerts its effect through the use of the formyl donor, and experiments in which compounds were added to the culture to either enhance or reduce formylation are consistent with this conclusion. It seems, therefore, that when the formylating capacity of the cell is high, initiation of protein synthesis takes place at each cistron on the messenger, but when its capacity is reduced in order to meet other demands, translation is initiated only at the operator-proximal end of the polycistronic messenger.

It is difficult to understand this finding in terms of the type of control postulated for other systems, that is, in terms of mechanisms for the control of ribosomal attachment. Rather, in this case, the mode of attachment appears to depend only on the quantity of fmet-tRNA$_f$ available for initiation. However, it does support the conclusion that the signal for initiation is more complex than just an AUG triplet—otherwise 30S subunits should be able to attach to any AUG (whether in phase or out of phase) in the messenger. Presumably, there must be a specific ribosomal attachment site at the beginning of each cistron on the messenger.

Translation of RNA bacteriophages

As indicated by their name, RNA bacteriophages have RNA as genetic material in place of the usual DNA. This means that an infecting parental phage genome must play two roles; first, it acts as a direct template for protein synthesis, without the benefit of a messenger intermediate, and then it is replicated through a double-stranded form by one of the enzymes which it codes. A group of related phages with *E. coli* as host (f_2, R17, MS2) all carry the information for the synthesis of three proteins: coat protein, phage-specific RNA replicase and 'maturation' protein. The positions of the three genes cannot be mapped by conventional means because there is no recombination between RNA phages. However, the latter two proteins of phage f_2 RNA contain histidine, which is absent from the coat protein, so that its incorporation can be considered a measure of their translation. The use of this technique has shown that when the RNA is translated the synthesis of coat protein predominates; this is the major product and is synthesized in great excess of the other two proteins (see below).

These phages comprise a system which is controlled at the level of translation. Nonsense mutants in the coat protein gene may exert a polar effect upon the translation of further genes, and Lodish (1968a) examined the effect of a highly polar mutant (*sus*-3) on each stage of the synthesis of further phage RNA in infected bacterial cells. There was a marked depression in the synthesis of RNA, but no effect of lowered transcription specifically from the mutation point (such a lowering would be expected if there were some obligatory coupling between transcription and translation). All the effects of the mutation in reducing the synthesis of progeny RNA molecules could be ascribed to a general reduction in the level of the phage-specific RNA replicase coded by the gene immediately subsequent to the mutated coat protein gene. This implies that, in this system at any rate, polarity must be accounted for on a purely translational basis.

Zinder and coworkers (1966) and Engelhardt, Webster and Zinder (1967) reported that, when translated *in vitro* by a bacterial extract, phage f_2 RNA exhibits many of the characteristics of polarity found *in vivo*. When an *amber* codon is located at the site coding for the sixth amino acid from the *N*-terminus of the coat protein (mutant *sus*-3), there is extensive polarity, with little translation beyond this point. By contrast, when the *amber* triplet is located at site 70 (mutant *sus*-11), there is little or no polarity, and synthesis of the replicase gene product is achieved at its normal level, even though the nonsense triplet is read as chain-terminating and releases an *N*-terminal coat protein fragment. Capecchi (1967b) was able to confirm that the chain termination is responsible for the polar effect by finding that, when the Mg^{2+} concentration is raised, the loss of translational fidelity results in the incorporation of an amino acid at the *amber* site, with consequent loss of polarity.

Webster and Zinder (1969) have used the *in vitro* translation of this phage to develop a system in which the messenger can be translated by only a single ribosome. This allows the fate of the ribosome to be followed upon chain termination at a nonsense codon, either naturally at the end of a cistron, or prematurely at a mutant site. In principle, this is achieved by allowing only one initiation of protein synthesis, after which any further ribosomal attachment is prevented. Experimentally, the extract for translation *in vitro* is deprived of asparagine (the third amino acid of the coat protein). This permits formation of an f_2–ribosome–peptidyl-*t*RNA complex; the binding of further ribosomes can then be prevented either by adding an excess of some other *m*RNA to remove them, or by the addition of aurintricarboxylic acid (a specific inhibitor of polypeptide chain initiation). Upon restoration of asparagine to the mixture, the complex already formed is permitted to resume protein synthesis.

Wild-type phage f_2 RNA is not released from the ribosome until after 6 minutes, when protein synthesis is completed. However, if the RNA messenger is derived from the *sus*-3 mutant phage, the messenger is released from its ribosome within 0·75 minutes of the resumption of protein synthesis. Similarly, the mutant *sus*-11 is also released from its ribosome at the time predicted for the reading of codon 70. However, despite the fact that both *sus*-3 and *sus*-11 dissociate from their ribosome upon termination, the former is strongly polar both *in vivo* and *in vitro*, but the latter is not. The ability of *sus*-11 to synthesize the replicase coded by the gene subsequent to the coat protein gene indicates that it must be capable of initiating polypeptide chain synthesis internally. Since the difference between the two mutants is only a matter of how far the ribosome is able to progress before it must dissociate from the message, we are led to the conclusion that it is necessary to read a certain length of the first gene of the message (somewhere between six and seventy codons), in order to permit ribosomes to attach to any subsequent signals for internal initiation. The *sus*-3 mutant is thus polar because it does not permit this reading. (It is worth emphasizing here that this explanation for polarity must be peculiar to this system; as discussed earlier, the internal attachment of ribosomes has been excluded as an explanation of the polarity found in the *lac*, *trp* and *his* bacterial systems.)

Upon infection of its host, the phage RNA shows differential expression of its three genes. Coat protein is synthesized at a rapid rate through most of the infectious cycle, but RNA replicase and the maturation protein are produced at rapidly diminishing rates. Nathans and coworkers (1969) and Eggen and Nathans (1969), working with the similar phage MS2, reported that the production of these enzymes is regulated through a repression exerted by the phage coat protein, that is the major gene product. *In vitro*, coat protein can combine with the phage RNA to selectively repress the synthesis of non-coat proteins; the interaction of coat protein and RNA yields a complex which acts as a template with full activity for the synthesis of coat protein, but with greatly reduced activity for the synthesis of the other two proteins. In fact, one mole of coat protein binds specifically and tightly to 1 mole of R17 RNA in the environment required for *in vitro* protein synthesis (Spahr, Farber and Gesteland, 1969). That this control is used *in vivo* is shown by the finding that the synthesis of RNA replicase is increased in cells infected with (non-polar) mutants in the coat protein gene, compared with infection by wild-type.

It is difficult to follow the translation of RNA phages because, although the messenger directs synthesis of three proteins, it is not usually possible to identify the maturation protein. Lodish (1968c) used a fingerprinting technique to detect the *N*-terminal sequences of the proteins synthesized *in vitro* by f_2 RNA. S^{35}-fmet-*t*RNA$_f$ was used to label the *N*-termini, and these were then identified by two-dimensional chromatography of the fragments resulting from trypsin and chymotrypsin digestion. This showed that the wild-type phage synthesizes proteins in the ratio 100 coat protein : 30 RNA-replicase : 5·5 maturation protein. The highly polar *sus*-3 mutant was found to reduce replicase synthesis to 25 per cent. of its previous level, although there was a slight rise in the synthesis of maturation protein. This suggests that whilst synthesis of the replicase protein depends upon the translation of (at least the early part of) the coat protein gene, the initiation of synthesis of maturation protein may be independent. This is supported by the kinetics of appearance of these products; whilst the appearance of replicase occurs subsequent to that of coat protein, maturation protein is produced at the same time as the latter.

However, Spahr and Gesteland (1968) found that enzymic cleavage of R17 phage RNA can yield two major fragments, each of which can direct protein synthesis. The fragment with the original 5′ terminus directs the synthesis of coat-like material, although less efficiently than wild type, and the species lacking the original 5′ proximal end synthesizes RNA replicase *in vitro*. Thus the fragment with a new 5′ terminus can direct ribosome attachment, although it does not do so when present in the intact three-cistron species. Lodish (1968b) isolated a mutant of phage f_2 which has lost the 5′ end of its RNA; this directs the synthesis of very little coat protein (less than 3 per cent. of wild type) as judged by tryptic fingerprints, but produces the same quantities of peptides as wild type for the other two proteins. That is, despite the loss of at least part of this gene, the remaining phage genes can be translated efficiently.

Removal of the coat protein gene abolishes the lag in synthesis of the subsequent proteins; this suggests that whilst under normal conditions ribosomes are unable to attach at the replicase gene, they can do so after removal of the coat protein gene. This can be accounted for if changes in the secondary structure of *m*RNA regulate the initiation of translation by controlling the availability of internal attachment sites to the ribosome. The intact messenger would have a structure such that attachment takes place to only the coat and maturation genes; the site for initiation of replicase synthesis would not usually be exposed until translation of the coat protein gene changes the secondary structure of the RNA so as to make it available to the ribosome. The physical removal of the coat protein gene would cause a similar change in secondary structure.

Support for this idea is lent by the recent finding that R17 RNA does indeed appear to have an ordered secondary structure. Adams and coworkers (1969) have sequenced fifty-seven nucleotides of its first gene which code for amino acids 81–99 of the coat protein. This sequence can be written in the form of a simple loop in which nineteen of the twenty-four base pairs are complementary. According to this structure, all its guanylic acid residues are involved in pairing, explaining why their bonds are not attacked by T ribonuclease (which specifically splits single polynucleotide chains on the 3′ side of guanylic acid residues). That the phage exists *in vivo* in this form is suggested by the isolation of this oligonucleotide unfragmented after a partial enzyme digest of the RNA. Indeed, the highly specific fragmentation which results from the wide variation in rates with which the enzyme splits the chain at different guanylic acid residues, suggests that the phage secondary structure is highly organized. This may serve two purposes; it may be important for the packaging of the genome within the phage coat, and it may also provide a control of translation. If ribosomes are to progress along the RNA, it must unfold to a single-stranded form, and this change may influence its ability to undertake the translation of subsequent genes.

The existence of such a secondary structure implies that the phage must suffer two different selective forces. Selection for an effective coat protein must act directly on its amino acid sequence, and thus only indirectly on the nucleotide sequence. But at the same time, there must also be a direct selection for nucleotide sequences which permit the formation of secondary structure. This may perhaps be an example of the use of the degeneracy of the code in evolution.

Stringent and relaxed control

In *stringent* (wild-type) *E. coli*, the synthesis of RNA depends upon an adequate intracellular supply of amino acids; starvation of an auxotroph for its amino acid not only blocks protein synthesis, but also causes a large reduction in RNA synthesis. In certain *relaxed* mutant strains, however, RNA synthesis continues normally during amino acid starvation. This results from mutation at a single locus, which replaces the stringent (RC^{str}) gene with its relaxed (RC^{rel}) allele. Most explanations for this effect have postulated some form of feedback from the protein synthetic system to the transcription process, which is present in stringent cells, but has been abolished in relaxed strains (for review see Edlin and Broda, 1968).

One such proposal was that uncharged *t*RNA might repress RNA synthesis by exerting an inhibitory effect on RNA polymerase (Kurland and Maaløe, 1962; Stent and Brenner, 1961). Amino acid starvation would cause the accumulation of uncharged species, but this would have no effect in relaxed cells, since in these the polymerase would not respond to the *t*RNA. However, experimental evidence has demonstrated that *t*RNA does not inhibit RNA polymerase molecules engaged in transcription, and although there is an inhibition of free enzyme, the difference in effect between *t*RNA and aminoacyl-*t*RNA is only slight. Further, experiments in which the intracellular concentration of charged *t*RNA was increased four- to five-fold by incubating amino acid-starved bacteria with chloramphenicol (Ezekial and Valulis, 1966) showed that this has no effect upon RNA synthesis.

Certainly, however, it does not appear to be the lack of amino acids as such which inhibits RNA synthesis. Cells with a temperature-sensitive mutation in valyl-*t*RNA synthetase charge the *t*RNA with valine at 30°C, but not at 42°C. Stringent strains are prevented from synthesizing RNA by an increase in temperature (which blocks protein synthesis), but this has no effect upon relaxed cells. This suggests that the critical step in stringent control is concerned with some reaction at or after the charging of *t*RNA (Neidhardt, 1966). Shih, Eisenstadt and Lengyel (1966) blocked protein synthesis with trimethoprin; this inhibits the enzyme dihydrofolate reductase and thus prevents synthesis of fmet-*t*RNA_f (although all other *t*RNA species remain fully charged). This inhibitor prevents RNA synthesis in stringent but not in relaxed strains, that is, the critical reaction occurs at some stage of protein synthesis subsequent to chain initiation.

Another proposal has depended upon the observation that amino acid starvation causes conversion of polysomes to monosomes in stringent but not in relaxed cells (Morris and De Moss, 1966). This could be due to either the breakdown of polysomes or to prevention of their formation. It has been suggested that this raises the level of free ribosomes, and these repress RNA synthesis in some manner. However, Friesen (1968) demonstrated that the precise nature of this effect appears to depend upon the bacterial strain used and the particular amino acid witheld, so that the significance of this connection is not entirely clear.

Gallant and Cashel (1967) and Cashel and Gallant (1968) have argued that the effect may be exercized by substrate regulation. They found that the incorporation of uracil or UMP into RNA was strongly dependent upon the presence of amino acids in stringent but not in relaxed strains. This occurs because amino acid starvation reduces the rate of conversion of UMP to UTP and CTP (by about half) in RC^{str} cells but not in RC^{rel} cells. The use of actinomycin to inhibit transcription does not prevent the formation of UTP and CTP; thus nucleoside triphosphate synthesis is not lowered as a result of the inhibition of RNA synthesis. They proposed that the RC control system acts directly on triphosphate synthesis, and this in turn causes the dependence of RNA synthesis (although this interpretation has been contested by Edlin and Broda, 1968).

However, these explanations would imply a coordinate control over all RNA synthesis. Although there is an *overall* reduction in RNA synthesis upon amino acid starvation of stringent strains, an examination of the RNA which *is* synthesized shows that it comprises messenger species (Sarkar and Moldave, 1968). This synthesis probably remained undetected previously because the entry of radioactively labelled nucleotides to intracellular pools is restricted under conditions of amino acid starvation (Nierlich, 1968). Measurement of the production of specific messengers in both stringent and relaxed strains has confirmed these results. Morris and Kjelgaard (1968) examined the rates of synthesis and breakdown of β-galactosidase *m*RNA after induction of amino acid starved bacteria; although the net rate of RNA synthesis in a stringent strain was reduced to 10 per cent. of the relaxed level, the messenger was not affected. Edlin and coworkers (1968) and Lavallé and de Hauwer (1968) followed the rate of transcription and translation of the *E. coli trp* operon during amino acid starvation. The results showed that it continues to be synthesized and translated at the same rate in both instances.

Stubbs and Hall (1968a, 1968b) developed a more sensitive hybridization technique for the detection of this messenger. *Trp* ϕ80 DNA was hybridized with radioactively labelled *m*RNA, and the level of *trp m*RNA in other preparations was then determined by measuring its ability to compete with the label for binding sites on the DNA. This determines the *trp m*RNA content in relation to the total RNA. The same results were obtained with both stringent and relaxed bacteria. In repressed bacteria, there are two or three messengers in every cell, and de-repression by tryptophan starvation raises the level to about eight *trp m*RNA molecules.

Genetically de-repressed mutants show an even higher level at twenty-five *trp* messengers per cell. However, starvation for arginine causes a five- to seven-fold reduction in *trp* *m*RNA synthesis in both RC^{rel} and RC^{str} cells, although the total *m*RNA synthesis is not affected. They suggested that this effect might be due to some control system distinct from the RC locus; amino acid starvation and the consequent halt of protein synthesis probably cause the accumulation of precursors, some of which might act as effectors for a generalized repression of amino acid synthesis by switching off several groups of operons. Certainly, many intermediates must accumulate under these conditions, and some must act to repress the expression of operons, although it is not known whether more than one operon would be affected.

There is thus no coordinate control over RNA synthesis, and the RC locus affects only the stable species (*r*RNA and *t*RNA). This suggests that the function of this control is to adjust the rate of *r*RNA and *t*RNA synthesis in response to the rate at which amino acids enter protein. That is, the level of the protein synthetic machinery is adjusted to a level corresponding to the availability of amino acids. The observation that starvation for any of several amino acids results in essentially the same overall reduction in the rate of RNA synthesis suggests that this deprivation activates some general mechanism. As yet, the basis for the action of such a control system is not clear.

Control of polysome activity

Ames and Hartman (1963) proposed a model (termed *modulation*) in which the degeneracy of the genetic code forms the basis for controlling the rate of translation. They suggested that the particular nucleotide sequence at any place could affect the reading of the message; out of all the sixty-one coding triplets a number—modulating codons—would code for minor *t*RNAs present in only small amounts, so that they would become rate limiting in reading the message. These would thus have the functions both of serving as adaptors for amino acids and of acting as modulators for controlling the rate of translation at their respective codons. Modulating triplets might also affect the probability of a ribosome leaving the messenger—for example, dissociation of *m*RNA and the ribosome might result from a long wait for a necessary *t*RNA. Roth and coworkers (1966) proposed that histidyl-*t*RNA might exert its regulatory effect on histidine biosynthesis if a codon requiring it were present at the *his*-O region. Stent (1964) extended the theory by observing that nucleases which remove the terminal *p*C*p*C*p*A sequence of a *t*RNA could control its level of availability for aminoacylation. However, the only enzymes with the properties required for such *t*RNA control are active generally, and not specifically directed against any particular *t*RNA. It is worth noting, however, that Hunt, Hunter and Monro (1969) have observed that α-chains of hemoglobin are translated faster than β-chains in the same cells, that is, the two *m*RNA species are read at different rates. This implies that the rate of translation may depend upon the messenger, possibly as a function of the precise nucleotide sequence. In this case, translation might be predicted to proceed at different rates along different parts of the messenger. (But this does *not* happen in bacteria.)

Cline and Bock (1966) have proposed a mechanism for controlling protein synthesis at the translation level in which repression is visualized as related to feedback inhibition rather than as a separate phenomenon—it has been shown in some systems that a single mutation can affect both levels of control. The primary assumption of this model is that the conformation of a partially completed polypeptide chain can influence its further synthesis. It is thought that the folding of the peptide chain takes place as the primary structure is sequentially assembled, and this could cover a considerable surface area—in a system involving subunit aggregation this might even approach the size of the ribosome itself. Whilst still under assembly, the protein might be able to exist in alternative conformations, and one of these might block protein synthesis, possibly by preventing access of aminoacyl-*t*RNA to the ribosome, or by interfering in some manner with the enzymes catalysing peptide bond synthesis. The protein might be interconverted between these different conformations by the same mechanism involved in allosteric feedback controls.

There is evidence that control mechanisms may operate at the translation level in the synthesis of hemoglobin and immunoglobulin G, although these may be specialized control mechanisms rather than of general application. Immunoglobulin G comprises two pairs of peptide chains; the light has a molecular weight of 20,000, and the heavy of 55,000. Williamson and Askonas (1967), using a mouse tumor system to study their synthesis, found two classes of polysomes, one for the synthesis of each chain type. They proposed a system in which completed light chains are autonomously released from their polysomes into a small intracellular pool. Assembly of the complete molecule could then be achieved through the free light chains assisting and controlling the release of heavy chains from their polysomes.

The α- and β-chains of hemoglobin are synthesized on different messenger templates (Kazazian and Friedman, 1968). Baglioni and

Campana (1967) observed that completed α-chains, no longer bound to *t*RNA, were present both free and associated with polysomes from lysed reticulocytes. No completed β-chains were found free, but were present in the globin form, that is the tetramer protein lacking the heme group, which interacts at some subsequent stage. Chase experiments showed that α-chains accumulate on polysomes and disappear from them at the same rate. Considerable numbers of α-chains are found free before the time at which the maximum number is present on polysomes, suggesting an equilibrium state in which they are released from their polysomes upon completion of synthesis, pass through a small pool of free α-chains and then associate with nearly complete nascent β-chains still under synthesis upon their polysomes. The lack of free β-chains suggests that these can only be released in the form of the αβ aggregate, and control of this release could be exercised at the level of the α-β interaction.

IV

The Reproduction of DNA

CHAPTER 13

Replication

The mode of replication

The model which is commonly accepted for replication supposes that the two strands comprising the parental duplex unwind and each acts as a template for the synthesis of its complementary strand by utilizing Watson–Crick hydrogen bonding to assemble the appropriate nucleotides (see Figure 2.9). Each of the two daughter duplexes is identical in its content of genetic information, that is in its sequence of base pairs, to the parental molecule. Physically, each consists of one of the original parental strands and one strand which has been newly synthesized. Thus replication is said to be *semi-conservative* in that the physical unit conserved between parent cell and progeny is one of the two single strands comprising the parental duplex.

Meselson and Stahl (1958) tested the prediction that each daughter duplex should consist of one 'old' and one 'new' strand. Bacteria were grown in a medium containing the heavy isotope N^{15}; this was incorporated into their DNA, altering its buoyant density and hence its equilibrium position on a CsCl density gradient (see Chapter 2). Culture conditions were then changed by substitution of the normal isotope and the DNA extracted at intervals during the next few generations. After one generation, all the DNA of the progeny bacteria showed a buoyant density in between that of the heavy (N^{15}) DNA and normal (N^{14}) DNA. This intermediate value can be accounted for by the semi-conservative mode of replication illustrated in Figure 13.1, in which each of the progeny DNA duplexes in the first daughter generation has one heavy strand (parental) and one normal strand (newly synthesized). As expected from this pattern of replication, after the next generation, half the DNA duplexes remained hybrid in buoyant density and the other half sedimented at the light density corresponding to duplexes with both strands of normal isotope. Replication was followed through several generations, and the experimental results agree with the predictions of semi-conservative replication.

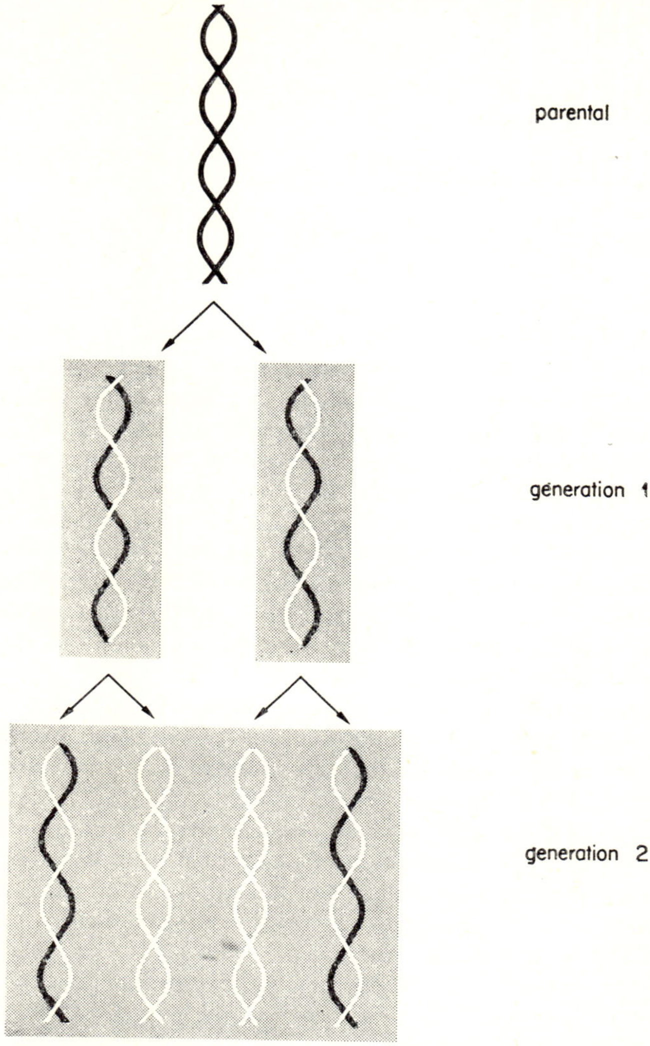

Figure 13.1. Semi-conservative replication. Parental DNA is heavy due to growth in medium containing N^{15}. The DNA present in the daughter cells one generation after the switch to light (N^{14}) conditions is all hybrid in density, that is it contains one heavy and one light strand. One generation later, half the DNA remains hybrid but the other half sediments at a light buoyant density. This can be accounted for by the semi-conservative mode of replication shown

Denaturation of the hybrid species confirmed that the conserved unit comprises a single polynucleotide strand. Further, entirely light chains (duplexes consisting of two newly synthesized strands) appeared only after all the parental molecules had replicated to the hybrid form. Thus, under these growth conditions, a second round of replication does not commence until the first has been completed. This suggests that replication is sequential; synthesis commences at a fixed point on the genome (the *origin*) and proceeds from there until completed at the *terminus*. If the chromosome is circular, the origin and terminus will lie adjacent to each other, rather than at opposite ends as with a linear genome.

Cairns (1963) radioactively labelled the DNA of log phase *E. coli* cells for $1\frac{1}{2}$ generations, lysed the cells and examined their genomes by autoradiography. Some 1 per cent. of the chromosomes were displayed as tangled circles engaged in replication. (This is found only comparatively rarely because the extraction procedures are likely to break the DNA). Thus the chromosome of *E. coli* can not only be represented by a circular genetic map, but also has a circular physical form. The stage of replication observed by the autoradiography was that of the second round of replication after addition of the radioactive label; the results show a single level of radioactivity along the entire length of the chromosome (resulting from the first round of replication) and a double dose along the part the replication so far during the second round. This is what would be expected to result from sequential replication, as shown in Figure 13.2.

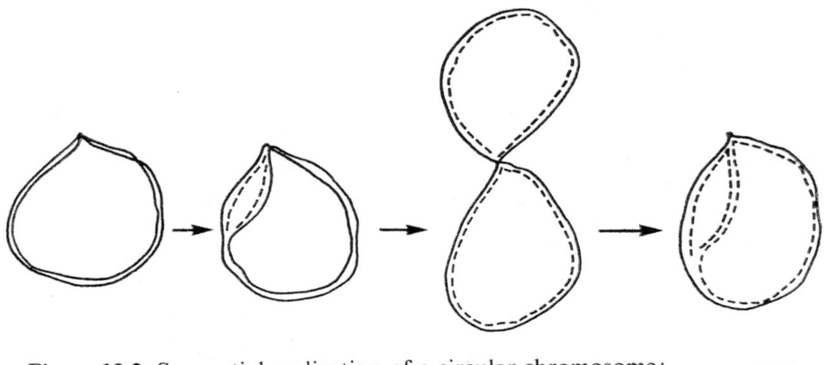

Figure 13.2. Sequential replication of a circular chromosome; ———— indicates parental strands (unlabelled), - - - - - - - indicates newly replicated radioactive strands. The last stage, which represents the beginning of the second replication cycle, was identified by the autoradiography of Cairns (1963)

Hirt (1969) has obtained a series of electron micrographs of polyoma DNA under replication, which are shown in Figure 13.3 (Plate IV). These clearly show the replication of one circular virus genome into two, with the growing point moving sequentially round the circle. Each strand visible in the photograph represents a duplex of DNA.

Lark, Repko and Hoffman (1963) confirmed this conclusion by a direct experiment which combined the techniques of radioactive and density labelling. They used an *E. coli* strain which is unable to synthesize thymidine and must therefore obtain it from the medium. First, an exponentially growing culture was labelled with H^3-thymidine for one-tenth of a generation, and then the cells were transferred to a culture containing the thymine analogue 5-bromouracil (5BU). This is incorporated into DNA in place of thymine, but an essential difference is that 5BU is much heavier, so that DNA containing it can be identified by density gradient centrifugation in CsCl. Samples were removed at various times after transfer of the culture into the 5BU medium, DNA was extracted as small fragments, and the distribution of radioactivity was measured in both the normal and the 5BU-substituted DNA.

At first, the radioactivity is located in a stretch of DNA which has normal weight on both strands, but when this region is replicated again the strand complementary to the radioactive one will be labelled with 5BU and this segment will have hybrid density (Figure 13.4). Thus transfer of the radioactivity from normal to hybrid density demonstrates that the small region labelled by the radioactive pulse has replicated. The results showed that replication of the chromosome has to continue for a whole generation before this occurs, that is replication is sequential. If parts of the chromosome were replicated in a radom order, the transfer would occur much sooner.

Nagata and Meselson (1968) performed a similar experiment in which cells were labelled with a pulse dose of H^3-thymidine, grown on normal medium for several generations and then pulse-labelled with 5BU. The radioactive label will only be found at hybrid density if replication of this segment is proceeding at the time of addition of the density label. They found that the H^3 label is associated with the 5BU only when the interval between the addition of the radioactive and density pulse labels equals a multiple of the generation time. An interval of one generation between successive replications of any given site on the chromosome demonstrates that the synthesis of DNA must occur sequentially from the same origin in each round of replication.

If protein synthesis is prevented in a bacterium, for example, by starvation of an auxotroph for its required amino acid, the current round

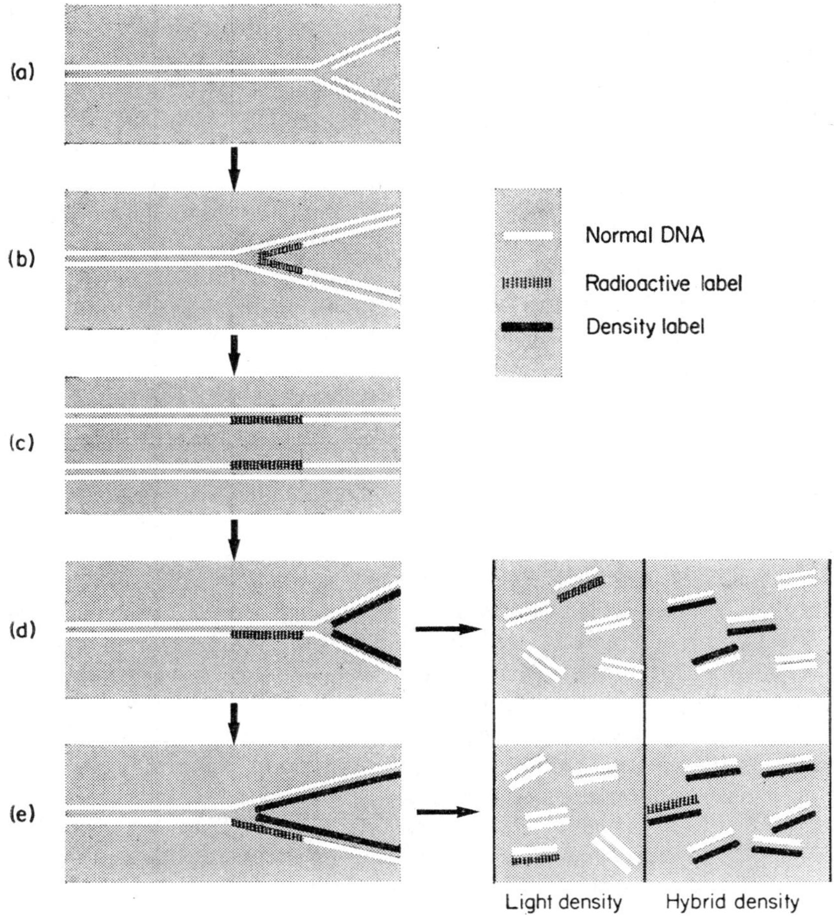

Figure 13.4. Sequential replication of DNA: (a) chromosome engaged in replication; (b) incorporation of radioactive label; (c) daughter chromosomes bearing radioactive label after completion of synthesis; (d) next round of replication commences in the presence of a density label; this is not incorporated into the radioactive region because this has not replicated. After fragmentation, the radioactive label is found only in light DNA; (e) replication has proceeded as far as the radioactive region; after fragmentation radioactivity is found to be in a hybrid density fraction. This does not happen until one generation time after addition of the radioactive label, implying that synthesis is sequential as shown

of replication is completed, but the cell is incapable of initiating another round (this is discussed in detail later). Upon the restoration of amino acids, protein synthesis is resumed and a new round of replication can be initiated. Lark, Repko and Hoffman (1963) used this procedure to obtain synchronous initiation of replication in all the cells of a culture, and the segment of the chromosome first replicated, that is the portion adjacent to the origin, was labelled by administration of an H^3-thymidine pulse dose at the time when amino acids were restored. The cells were then grown in an unlabelled medium for several generations, after which the starvation procedure was repeated, but with the difference that when amino acids were restored 5BU was provided in place of thymine. The segment of the chromosome previously labelled by the H^3 isotope appeared in the DNA sedimenting at hybrid density, demonstrating that the origin for replication remains the same over (at least) several generations.

Sueoka and Yoshikawa (1963) argued that if chromosome replication occurs from a fixed point which is the same in all cells, genes near the origin should be present in a greater frequency than markers at the terminus end; in a growing culture the former will be replicated early in the division cycle and there will be two copies present in the cell until the next cell division, which cannot occur until the remainder of the chromosome has been replicated. The frequencies of genes located at various sites on the *Bacillus subtilis* genome were determined experimentally by transformation. It is not possible to find marker frequencies directly from a transforming DNA preparation because the extent of transformation varies from marker to marker, depending upon differences in their efficiencies of integration. However, if a stationary phase culture is used as control, all its markers are present with equal frequency, and the differences in frequency observed between this and a growing culture can be attributed to differences in the gene frequencies of the latter—the integration efficiency of any particular marker will be the same in each preparation. When log phase *B. subtilis* were compared with those of the stationary phase, the relative frequencies varied from almost two for some markers to one for others. Using this ratio as a criterion, a linkage map was constructed to give the nearness of each marker to the origin. This corresponded with the genetic map already established, thus confirming that replication is sequential from a fixed origin.

There has been some confusion as to the site of the origin for replication in *E. coli*. Nagata (1963) suggested that in *Hfr* strains the site of integration of the *F* factor acts as the origin and the polarity of its insertion determines the direction of replication. In other strains there would be no unique origin, and replication could be initiated at any of several sites. Vielmetter, Messer and Schutte (1968) also obtained results in agreement with this conclusion. However, other workers have found that the sex factor is not concerned with determining the origin for replication, and a variety of evidence has accumulated to support the conclusion that replication proceeds clockwise from a unique origin which remains the same in each round of replication and is located in the 7–8 o'clock region of the chromosome (for review see Lark, 1969)—although there do, however, appear to be one or two exceptions to this.

Three experimental approaches have been used to resolve this problem (see Wolf and coworkers, 1968). Berg and Caro (1967) analysed the *E. coli* system by a method similar to that used by Sueoka and Yoshikawa (1963) for *B. subtilis*, except that gene frequencies were determined by generalized transduction with phage P1 instead of by transformation. Pato and Glaser (1968) subjected cells to a brief induction period for various enzymes at different times during the cell division cycle. The rate of enzyme synthesis upon induction appears to be directly related to the number of gene copies present in the cell; when a gene is replicated the level of enzyme produced is doubled. The time at which this occurs measures the map position of the gene relative to the origin. Cerda-Olmeda, Hanawalt and Guerola (1968) treated cells with the powerful mutagen nitrosoguanidine; this mutates the replication point of a chromosome with a much higher efficiency than any other region. Thus a large increase in the mutation rate for any particular gene should occur at the cell age at which it is being replicated. Using a synchronized culture they tested various bacterial functions for a change in activity after mutagenic treatment at different times during the growth cycle. All these studies produced the same result; replication is initiated at a unique origin which is the same in all cells, irrespective of sex type, and proceeds sequentially clockwise from this origin to the terminus.

Although replication is initiated at the origin described above in most *Hfr* strains, exceptions have been found in which a different origin is used. Caro and Berg (1968) labelled a synchronized culture of cells with 5BU at various times during the division cycle. The cells were then infected with P1 and the resulting lysate centrifuged to equilibrium on a CsCl density gradient. This isolates the transducing particles which contain hybrid density DNA, and their relative transducing activity can then be measured for various markers. They found that replication falls either into the pattern described above, or, with certain *Hfr* strains, replication was initiated at a different origin from which it proceeded anticlockwise; this origin, however, was not the site of integration of the

F factor. It seems, therefore, that the integration of the sex factor may affect the origin used for replication in some, although not all, strains.

The results obtained in which replication is initiated at the site of the *F* factor do not agree with these conclusions. However, these may be reconciled if the chromosome has more than one origin at which it is possible for initiation to occur, although the site at 7–8 o'clock predominates in most strains. The other sites may also serve as locations for integration of the *F* factor, and the physological conditions used in some experiments may have the result of shifting initiation from the normal vegetative origin to one of these other sites. Whilst this may be the one at which the *F* factor is inserted, others may also be used.

Much less is known about the mode of replication in higher organisms, although there is evidence that it too is semi-conservative (Taylor, 1964). A notable difference from bacterial systems is that whereas the latter have only one origin at which replication of their chromosome can commence in any given division cycle, in higher organisms there may be many such origins for each chromosome. Huberman and Riggs (1968) demonstrated by autoradiography of pulse-labelled Chinese hamster cells that their DNA comprises many replication sections, with adjacent sections arranged in opposite orientation. This is illustrated in Figure 13.5. A replication section has its origin immediately adjacent to the origin of its neighbour on one side, and its terminus is located next to the terminus of its neighbour on the other side. Replication occurs at fork-like growing points which proceed from the origin of a section to its terminus, but these progress in opposite directions at the adjacent origins. Since adjacent origins commence replication together, regulation may be at the level of pairs of sections rather than of individual units. Different pairs of sections commence replication at different times and must therefore come under independent control.

The accuracy of replication

Enzymes which catalyse DNA synthesis *in vitro* (DNA polymerases) have been isolated from a variety of organisms ranging from bacteria to mammalian systems (for review see Bollum, 1967; Georgiev, 1967; Kornberg, 1966). The enzymes catalyse the reaction

$$n\ d\text{-NTP} \longrightarrow (\text{NMP})_n + n\ \text{PP}_i$$

and require as substrate all four deoxynucleoside triphosphates. With a duplex DNA template the product of the reaction is two duplexes of DNA, each identical in base pair sequence to the parental molecule.

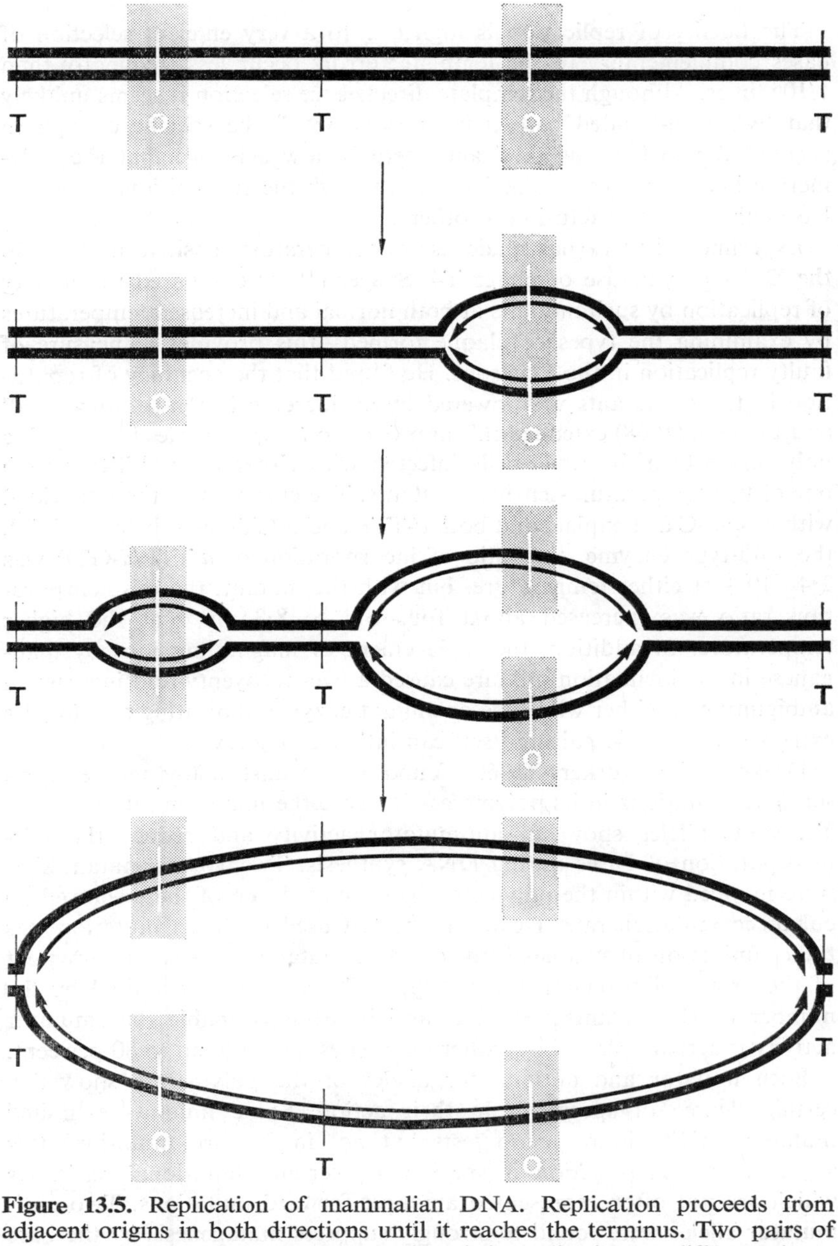

Figure 13.5. Replication of mammalian DNA. Replication proceeds from adjacent origins in both directions until it reaches the terminus. Two pairs of sections are shown here as commencing their replication at different times. Each replication section shares its origin with one neighbour and its terminus with its other neighbour. (After Huberman and Riggs, 1968)

The fidelity of replication is high due to a very efficient selection of bases complementary to the template; errors occur less frequently than $1/10^6$ times. Although the template directs base selection it seems unlikely that hydrogen-bonded base pairing alone would be specific enough to account for so few mistakes, and there is now evidence that the polymerase itself may play an active role through the recognition of correct base pairs and the exclusion of others.

Experimental work has made use of temperature-sensitive mutants in the DNA polymerase of phage T4. Speyer (1965) compared the fidelity of replication by such mutants at both normal and increased temperatures by examining the types of plaque formed—this provides a measure of faulty replication in the *r* cistrons. He found that the accuracy of replication in the *ts* mutants was lowered by an increase in temperature. Hall and Lehman (1968) extended this approach to *in vitro* studies by purifying polymerase from bacterial cells infected with either normal T4 or with one of two temperature-sensitive mutants. The enzyme was then provided with poly-*d*-C as template and both *d*-TTP and *d*-GTP as substrates. With the wild-type enzyme, the ratio of incorporation of *d*-TTP/*d*-GTP was $2{\cdot}4 \times 10^{-6}$ at either temperature, but with the mutant, the misincorporation ratio was increased about four-fold to $8{\cdot}3 \times 10^{-6}$ at the higher temperature. In addition, the replacement of magnesium ions by manganese in the incubation mixture caused a five- to twenty-fold increase in ambiguity with either wild-type or mutant enzyme, indicating that factors extrinsic to the base pairing itself can influence fidelity.

Drake and coworkers (1969) found that whilst many temperature-sensitive mutations in its polymerase increase the mutation rate of phage T4, some alleles show an antimutator activity and reduce the misincorporation of bases during DNA synthesis. Twenty-one mutant sites were mapped within the polymerase gene, and eleven of these showed an enhanced mutation rate. However, the test used to detect mutation was the examination of plaques formed by *r*II mutants, that is measurement of the extent of reversion to wild type. The test is thus limited by the number of *r*II mutants available, and it seems probable that mutator activity is actually shown by rather more sites, possibly up to 80 per cent.

Both mutator and antimutator alleles of the polymerase showed a certain degree of specificity in their action. Some mutants exhibited mutator activity in reversion tests but not in forward mutation; this suggests that their action is specific for certain mutational pathways which are not often represented amongst forward mutations. Two antimutator alleles were found to strongly suppress mutagenesis by the base analogues 2AP (2-aminopurine) and 5BU, to moderately suppress the mutagenic effect of the alkylating agent EMS (ethane methylsulphate), and to exert no effect upon the action of hydroxylamine. Another test of specificity was provided by comparing the reversions caused by chemical mutagens or by an antimutator allele for some particular phage mutant. The results suggested that these two alleles only exert an effect upon mutants which can be induced to revert by a transition from an AT to GC base pair, that is they suppress only this transition.

Thus it appears that the antimutagenic DNA polymerase has an increased ability to discriminate against incorrect base pairing in certain instances. Indeed, the specificity of these effects implies that the differences observed in response to mutagenic treatment of various organisms may represent the different characteristics of their polymerases. In any case, it is clear that mutation rates cannot be considered in isolation as absolutes since several organisms possess mutator genes which can cause appreciable increases in their mutation rates. However, the discovery of antimutator alleles of DNA polymerase does raise the problem of why such improved enzymes should not have evolved during the process of natural selection; presumably, an increase in the accuracy of replication must carry a selective disadvantage of some kind.

Two contrasting models have been proposed to account for the active role which the polymerase must play in base selection. One suggests that the polymerase is responsible for base selection; the template would influence the enzyme in an allosteric manner to direct this selection, but would not interact directly with the incoming base until a later stage. However, Kornberg (1969) has reported that *E. coli* polymerase has only one site for accepting triphosphates, and a role for the enzyme of indirect mediation would demand that this one site could exist in any of four specific conformations—this seems rather unlikely. An alternative model proposes that the template binds a base which the polymerase then accepts or rejects. The basis for this decision would probably lie in a demand for only certain base pairs; all the Watson–Crick base pairs contain regions of identical dimensions and geometry which the enzyme could recognize. When an acceptable base pair is in the active site the enzyme might respond, perhaps by a change in conformation, so that the subsequent catalytic steps can proceed.

Catalytic functions of DNA polymerase

In addition to its role in promoting polynucleotide chain elongation when the template is read from 3′ to 5′, that is in synthesizing a new strand from 5′ to 3′, *E. coli* DNA polymerase has several other catalytic properties (an extensive review has been given by Kornberg, 1969). It can also progressively degrade the single strands of a duplex in either the 3′ to 5′ direction (this has been termed the exonuclease II activity in *E. coli*), or in the 5′→3′ direction, to produce nucleoside monophosphates. The exonuclease activity cannot be separated from the polymerase and is an integral property of the same enzyme molecule. Instead of using hydroxyl ions to achieve hydrolysis, pyrophosphate ions may be used to effect a pyrophosphorylytic degradation from 3′ to 5′ to release triphosphates. Related to the pyrophosphorolysis is the promotion of exchange of free pyrophosphate with the terminal pyrophosphate of free deoxynucleoside triphosphates.

These multiple functions, and the recognition abilities demanded by its active role in base selection, would suggest a subunit structure for the enzyme. However, its molecular weight (109,000) remains unaltered under conditions of denaturation with guanidine hydrochloride and mercaptoethanol, and the enzyme migrates as a single band on polyacrylamide gels. There is a single terminal amino group, one free sulphydryl group and one disulphide bridge per 109,000 daltons; thus the molecule must consist of only a single chain. If the polymerase were roughly spherical, its diameter would be around 65Å (the DNA helix is 20Å across). Further details have been reported by Englund and coworkers (1968) and by Jovin, Englund and Bertsch (1969).

Kornberg (1969) reported evidence that, as might be expected from its multiple functions, there are (at least) five major sites within the active center of the molecule, as shown in Figure 13.6. These may be summarized as follows:

(i) The *template site* binds the chain where a base pair is formed and for some nucleotides on either side.

(ii) The growing chain is located at the *primer site*, and this chain is orientated with a polarity opposed to that of the template.

(iii) The *primer terminus* provides a recognition site for the 3′-hydroxyl group of the terminal nucleotide of the primer; this site is concerned in the 3′⟶5′ hydrolytic and phosphorolytic cleavage as well as in chain elongation.

(iv) There is one *triphosphate site* where all the incoming nucleotide substrates must bind.

(v) A further distinct site provides for the hydrolytic cleavage of 5′-monophosphates from this end of the chain, that is, the 5′⟶3′ nuclease activity.

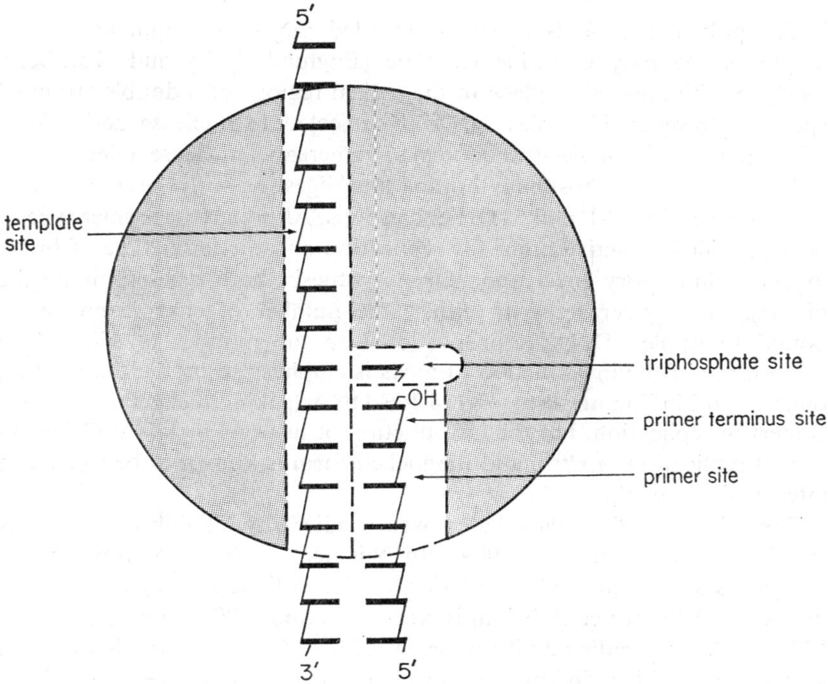

Figure 13.6. Catalytic sites in the active center of DNA polymerase. (After Kornberg, 1969)

The single sulphydryl group is not part of the active center of the molecule since it can be modified without loss of enzyme function. (Jovin, Englund and Kornberg, 1969). The –SH moiety reacts with mercury ions, and the radioactive isotope Hg^{203} was used to provide a fixed radioactive label per unit of enzyme without affecting its activity. This was made use of to measure the binding of polymerase to DNA; mixtures containing the labelled enzyme and DNA bearing an H^3 or P^{32} isotope were layered on top of sucrose density gradients and the resultant complexes were isolated after centrifugation. With an excess of alternating *d*-AT oligonucleotide duplex of either twelve or twenty base pairs in length, all the enzyme sedimented as a complex at 7·7S instead of its usual free 6·1S. An equimolar quantity of the oligomer was incorporated into the complex, implying that the polymerase contains a single binding site for nucleic acid. In the presence of excess polymerase, the oligonucleotide sedimented almost quantitatively with the enzyme, indicating a very high binding affinity.

The polymerase binds to single-stranded DNA in proportion to the length of the polynucleotide template (Englund, Kelly and Kornberg, 1969). No binding takes place to the helical regions of a double-stranded species. However, if duplex DNA is subjected to nuclease action it is 'nicked' to yield single-strand breaks; pancreatic nuclease releases free 3′-hydroxyl and 5′-phosphate termini, and digestion with the micrococcal enzyme gives free 3′-P and 5′-OH end groups. Although the former type of nick provides an active point for replication and the latter type of break exerts an inhibitory effect upon DNA synthesis, both types promote the binding of polymerase. With either, the number of enzyme molecules bound to duplex DNA is proportional to the number of nicks. The importance of the introduction of nicks is emphasized by the finding that there is no binding at all to ϕX174 RF DNA lacking nicks and that this is inert in replication, but the introduction of just one nick is sufficient to permit replication *in vitro*, and the nicked form is known to be the active intermediate *in vivo*.

The binding of triphosphates was studied by equilibrium dialysis (Englund and coworkers, 1969). In this technique a solution of the enzyme is separated from a solution of the radioactive nucleotide by a dialysis membrane; equilibrium is reached in about 90 minutes, and after 120 minutes the radioactivity on each side of the membrane is counted. Experiments with individual triphosphates showed that one molecule of the nucleotide species binds to one molecule of enzyme. The use of each of the six possible pairwise combinations of the four common triphosphates (each member of the pair bearing a distinctive radioactive label) showed that all four species must compete for the same binding site. The specificity of binding at this site depends mostly upon the triphosphate moiety and only to a lesser extent upon the sugar and base. Although triphosphate substrates can bind to this site in the absence of DNA, under this condition there is no polymerization and no pyrophosphate release or exchange. This indicates that the triphosphate is not activated in the absence of a template; however, when polymerization occurs in the presence of DNA, pyrophosphate exchange can take place. This suggests that after a triphosphate has become attached to the enzyme it is activated in some manner which enables it to exchange with a molecule of inorganic pyrophosphate, perhaps through covalent attachment to the primer.

The primer terminus was identified by the use of nucleotide analogues lacking a 3′-hydroxyl group, such as dideoxy-TPP (this is *d*-TTP in which the 3′-OH has been replaced by hydrogen). The chain is then inert to both further extension and to the 3′⟶5′ exonuclease action normally shown by the polymerase, and also to pyrophosphorolysis, although 5′⟶3′

degradation is not affected. This suggests that a chain lacking a 3′-hydroxyl terminus cannot bind properly in the primer site. Further support for the existence of this site was obtained through the demonstration that each of the four deoxynucleoside monophosphates binds to the enzyme and competes for a specific site which is distinct from the triphosphate binding site. The monophosphates inhibit the 3′⟶5′ exonuclease activity, that is they must be binding at the primer terminus site. Replacement of their 3′-hydroxyl group by inert moieties such as hydrogen, or *o*-methylation, prevents their binding to this site. It seems clear, therefore, that there is only one site for both polymerization and the 3′⟶5′ exonuclease action. (The exonucleolytic actions of the enzyme have been discussed by Deutscher and Kornberg, 1969b).

If a linear duplex is partially degraded from a 3′-OH terminus by exonuclease action, the damaged sections are readily repaired by the polymerase as illustrated in Figure 13.7. For this activity the enzyme is bound adjacent to the 3′-OH group on the terminal nucleotide of the primer and is orientated so that it can form a base pair with the template. When the correct triphosphate has been recognized, a covalent bond is formed to the primer. Movement of the polynucleotide chain relative to the enzyme is concurrent with phosphodiester bond formation. The old

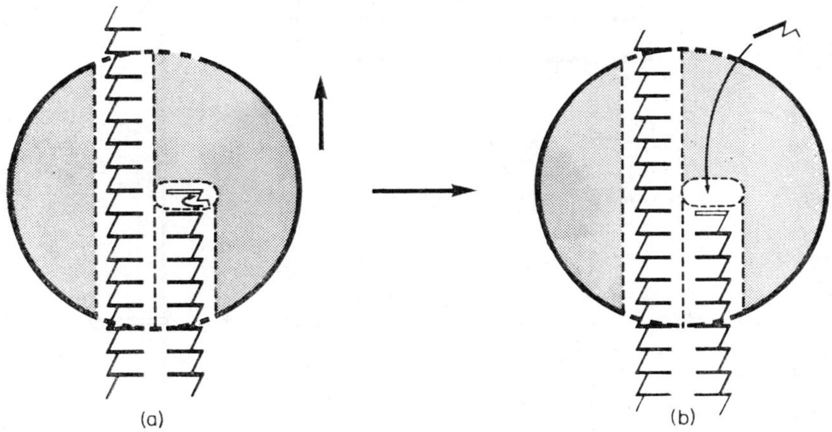

Figure 13.7. Single-strand synthesis from a primer segment. (a) Triphosphate in the triphosphate site is attacked by the 3′-hydroxyl group in the primer terminus to form a phosphodiester bond with the release of pyrophosphate. (b) The enzyme has moved one nucleotide up the chain so that the primer terminus site now contains the newly added nucleotide and the triphosphate site is free to accept the next incoming species

primer terminus is therefore no longer in this site but has been replaced by the newly added nucleotide, which is now ready for addition of the next residue to the chain. Inorganic pyrophosphate is released with the completion of bond formation and chain movement.

Pyrophosphorolysis is the capacity of the polymerase to degrade DNA chains with inorganic pyrophosphate to release triphosphates. Further, an exchange is catalysed between the terminal pyrophosphate of deoxynucleoside triphosphates and free pyrophosphate, although this also requires the primer and template conditions necessary for replication. Both these processes represent a reversal of the mechanisms involved in replication (Deutscher and Kornberg, 1969a). Pyrophosphorolysis is a reversal of the polymerization step, including the concerted chain movement. Inorganic pyrophosphate exchange represents the result of a sequence of a polymerization step and a phosphorolytic step in turn, as shown in Figure 13.8. This model is supported by the behaviour of

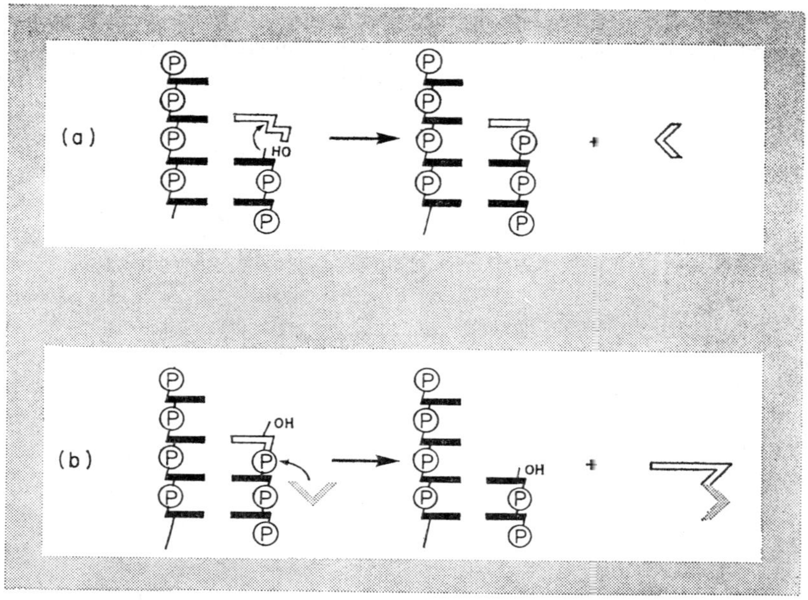

Figure 13.8. Pyrophosphorolysis and pyrophosphate exchange: (a) elongation of the 3′-hydroxyl terminus by one nucleotide; and (b) pyrophosphorolysis after attack by inorganic pyrophosphate on the final phosphodiester bond. The net result of the sequence of both reactions is an exchange of the terminal pyrophosphate on the nucleotide species involved. (After Kornberg, 1969)

chains terminated with dideoxynucleoside triphosphates; these are relatively unsusceptible to the attack of pyrophosphate or hydroxyl ions and support little, if any, pyrophosphate exchange. The lack of a free 3′-OH group prevents binding of the transition state which is formed by attack of the primer terminus on a triphosphate, and the terminal nucleotide in this state cannot be displaced effectively by either pyrophosphate or hydroxyl ions.

It is not entirely clear how DNA polymerase commences replication of duplex DNA, although a nick appears to be the signal which instructs the enzyme to bind to the template. Kornberg (1969) suggests that before replication commences the triphosphate site may be blocked by the preexisting chain, as shown in Figure 13.9. If replication is to proceed, the chain terminating in 5′-P must be displaced from the template strand. There is evidence that during the first phase of replication there is a burst of hydrolysis roughly matching the extent of replication. This is predominantly 5′⟶3′, and the site responsible appears to be distinct from the 3′⟶5′ hydrolytic site. The polymerase is unable to degrade single polynucleotide chains by this mechanism, and this suggests that this site requires a double-stranded structure if its substrate is to be orientated properly. The failure of the enzyme to degrade a single chain 5′⟶3′ may explain how normal replication proceeds with synthesis taking place in the absence of concomitant 5′⟶3′ hydrolysis. As the 3′-OH terminating chain is synthesized along the template, the 5′-P terminating chain may be displaced from the active site for a stretch of several nucleotides and thus be rendered insusceptible to hydrolysis. This sequence of events is illustrated in Figure 13.9.

The discontinuous synthesis of DNA

Replication of the bacterial chromosome occurs sequentially so that both strands of the duplex template are copied simultaneously from initiation at the origin to completion of synthesis at the terminus. Until fairly recently, it was assumed that the synthesis of new strands must be continuous; this would imply that one of the daughter strands is synthesized in the direction 5′⟶3′ while the other is assembled from 3′⟶5′. However, whilst the action of DNA polymerase demonstrated *in vitro* can account for growth of the new 5′⟶3′ strand, it does not explain the mode of assembly of the other new strand. No enzymic action for catalysing synthesis 3′⟶5′ has been found, and proposals to account for synthesis in this direction, such as the use of 3′-triphosphates, have proved unfounded.

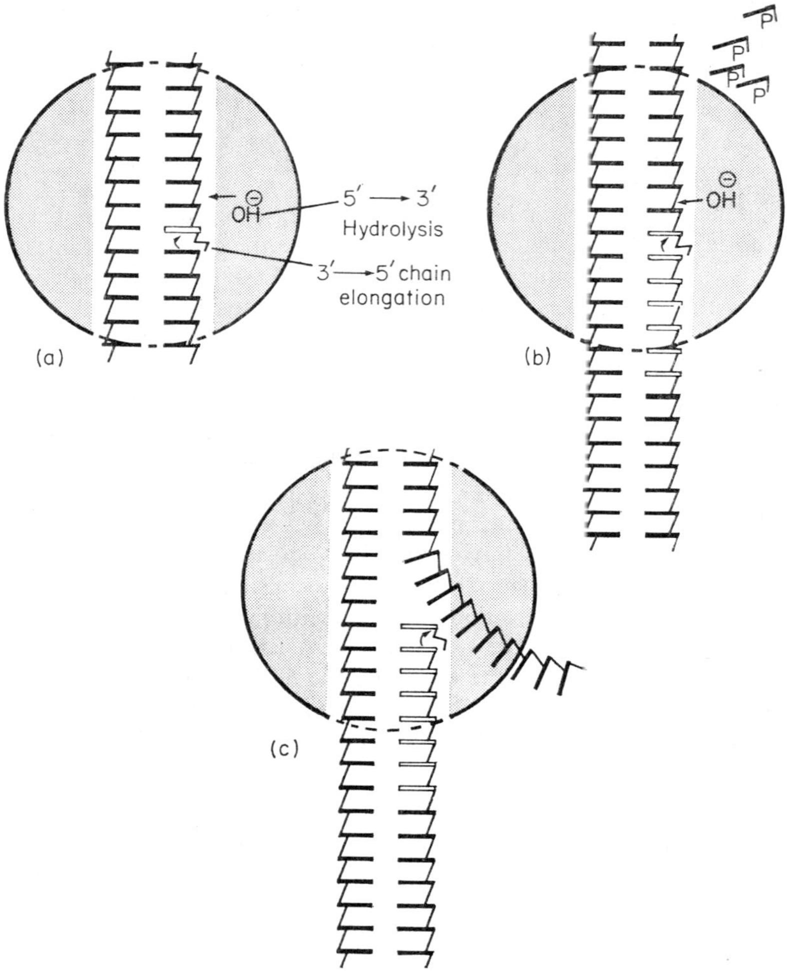

Figure 13.9. Chain elongation on a double-stranded template. (a) As the primer chain is extended by addition of nucleotides to its free 3′-hydroxyl end, the 5′-phosphate terminating chain above it is degraded from 5′ to 3′ by hydrolytic attack. (b) Chain elongation is continuing at the expense of degradation of the 5′-phosphate chain. (c) The 5′-phosphate chain has become displaced from the template strand and chain elongation continues without entailing its degradation

This difficulty was overcome by the finding that synthesis is discontinuous, that is the new DNA strands are synthesized in short stretches in a 5′⟶3′ direction by the polymerase. The short segments are not covalently linked to each other, and, subsequently, another enzyme is responsible for forging covalent links between them to produce a continuous daughter strand. It seems probable that both strands of DNA are synthesized in this manner, as illustrated in Figure 13.10.

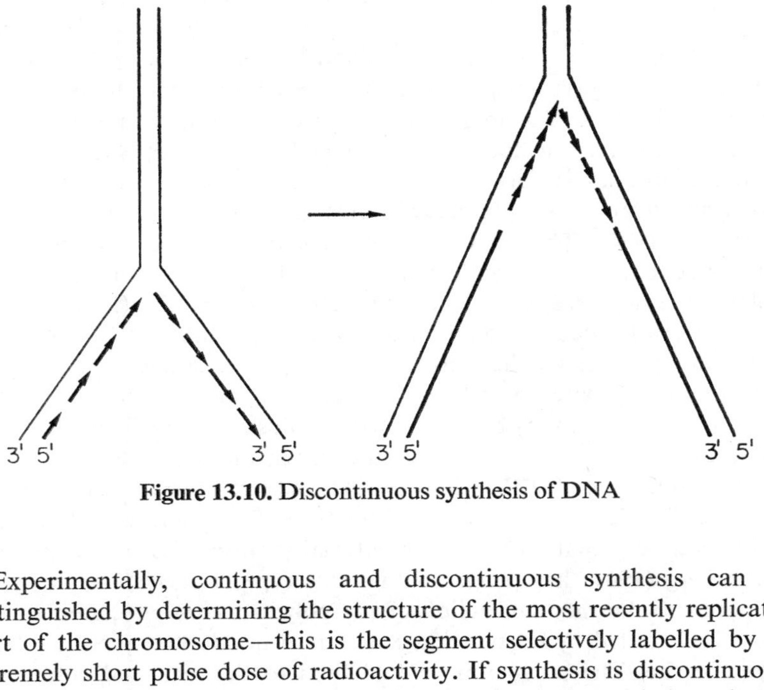

Figure 13.10. Discontinuous synthesis of DNA

Experimentally, continuous and discontinuous synthesis can be distinguished by determining the structure of the most recently replicated part of the chromosome—this is the segment selectively labelled by an extremely short pulse dose of radioactivity. If synthesis is discontinuous, the label should be found in many unconnected short chains, that is slowly sedimenting DNA, whereas if continuous it should be present in a large unit sedimenting much faster. Okazaki and coworkers (1968) used a technique in which cells synthesizing either bacterial or phage DNA were exposed to H^3-thymidine for a short period, after which the pulse labelling was terminated by the addition of KCN and ice. The DNA was then denatured and characterized by its sedimentation characteristics.

After a two-second pulse the radioactivity incorporated was found almost entirely in short fragments sedimenting slowly between 7S and 11S; this would suggest that the discontinuous unit is some 1000–2000 nucleotides long—about the order of size of a cistron. This mode of synthesis appears to apply to both newly synthesized strands, not just the 3′⟶5′, because virtually all the label was recovered in the slowly sedimenting fragments (if only one strand were involved half the label would be in fast-sedimenting DNA). When a two-minute pulse dose was used, the label was found in a fraction with an average sedimentation coefficient of 45S, that is, the fragments have become covalently linked into a continuous large strand. The experimental results obtained are shown in Figure 13.11. Recent work has shown that a similar mode of synthesis probably applies to both Chinese hamster and Hela cells (Painter and Schaeffer, 1969; Schandl and Taylor, 1969).

When *E. coli* polymerase is used in an *in vitro* system, the product of enzymic replication of a single-stranded DNA template is normal duplex DNA. However, when double-stranded DNA is used as a template, an abnormal product is formed; it is not easily denatured, it renatures readily and it appears as branched fibres under the electron microscope (Kornberg, 1966). This unusual DNA could be accounted for if the polymerase commences replication at the free 3′-OH termini of the duplex to synthesize complementary 5′⟶3′ strands (*in vitro*, the polymerase will attach at both ends of a linear DNA duplex). At some point during synthesis the unreplicated 5′ parent strand might act to compete as template for the polymerase enzyme so that the newly synthesized strand 'loops back' upon it, as illustrated in Figure 13.12. A second event of this nature might then occur in which the parental strand is displaced in favour of the newly synthesized strand to form the hairpin loops seen by electron microscopy. This type of structure would account for the unusual denaturation and renaturation properties of the product.

Kornberg (1969) has proposed that this type of mechanism may be analogous to the mode of action of the polymerase *in vivo*. Replication would be initiated by introducing a nick into the duplex, presumably at a specific site. The polymerase would bind to the nick and commence synthesis of a new 5′⟶3′ chain upon the parental 3′⟶5′ chain. At first, the 5′-P parental chain may be degraded by the enzyme exercizing its 5′⟶3′ exonuclease action; this 'clears the way' for synthesis of the new 5′⟶3′ chain. However, the parental chain is then displaced from its complement and thus becomes immune to further attack of this kind. Replication proceeds along the 3′-OH parental chain with the 5′-P parental chain continuing to 'peel back' until it competes with the template strand by attracting the polymerase. The enzyme then switches strands, thus producing a covalent fork in the growing chain; this must subsequently be cleaved by an endonuclease. A repetition of this sequence

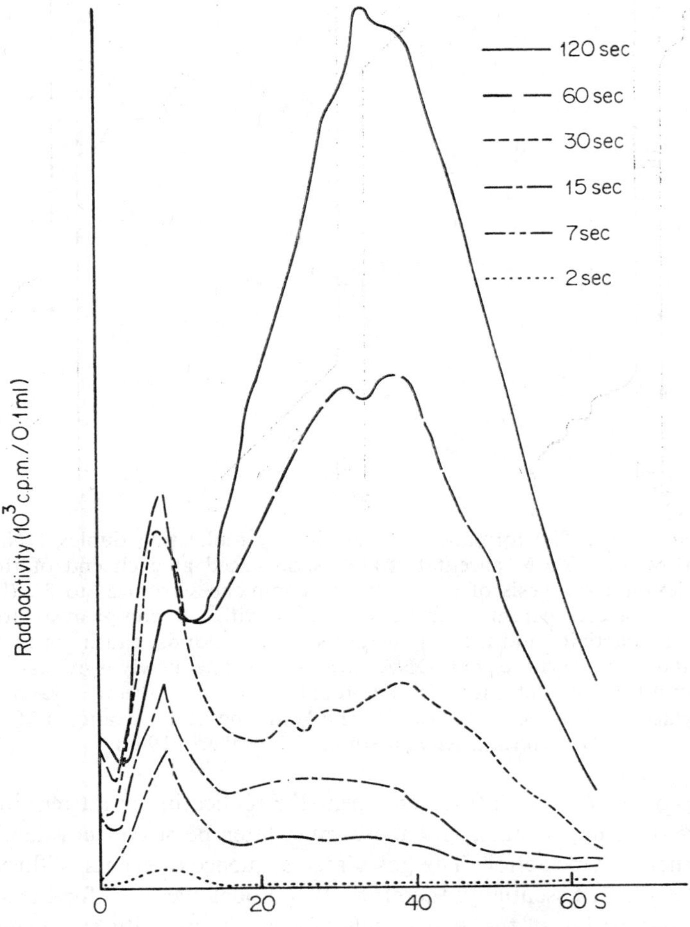

Figure 13.11. Sedimentation characteristics of DNA after a radioactive pulse label. At shorter pulse times, most of the label is found in a fraction sedimenting around 10S, but as the time of incubation is increased a greater proportion of the label is found in the heavy sedimenting (40S) peak. (After Okazaki and coworkers, 1968)

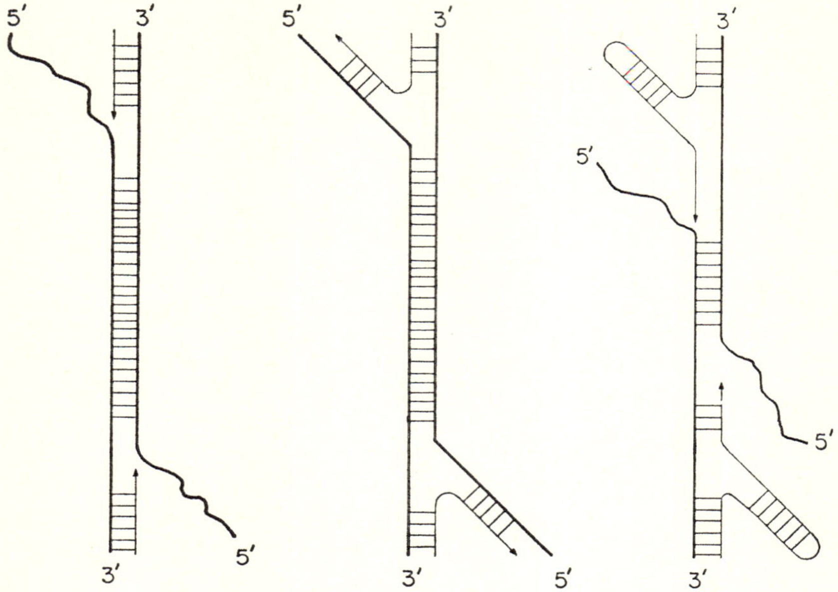

Figure 13.12. The formation of hairpins by replicating duplex DNA *in vitro*. (a) The 5′ parental strand is displaced at each end of the duplex and synthesis of a new strand commences from 5′ to 3′. (b) The displaced parental strand competes with its complement for template activity and the polymerase switches strands, causing a loop in the newly synthesized DNA strand. (c) The newly synthesized strand itself competes for template function and the parental strand is displaced. This generates a 'hairpin' in the new strand. (After Schildkraut, Richardson and Kornberg, 1964)

would produce the small fragments near the replicating point required by the pulse-labelling results, and these must later be sealed covalently by some other enzyme action. The postulated sequence of events is illustrated in Figure 13.13. Essentially, therefore, this model accounts for the simultaneous replication of the two strands of opposite polarity by a staggered action in which the polymerase switches from one parental strand to the other during synthesis. Although the model accounts for much of the data, there are difficulties in that the basis for the competitive action which would cause the polymerase to switch strands is not at all understood. Presumably, this would have to be a regular event in order to account for the consistent production of Okazaki fragments. It is worth noting also at this point that, although DNA polymerase can replicate DNA *in vitro*, there is no conclusive evidence that it does so *in vivo* (see also page 359).

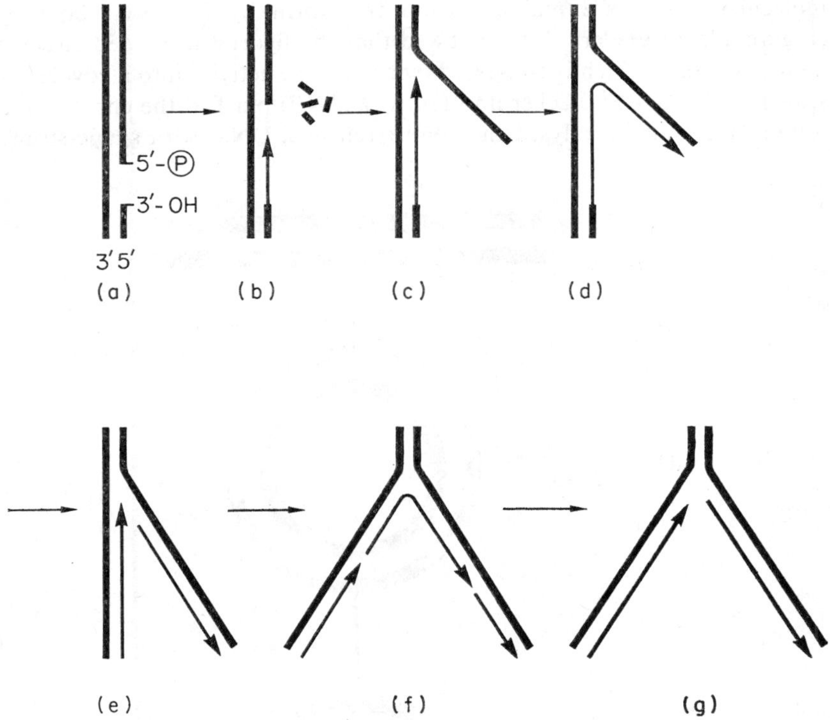

Figure 13.13. A model for discontinuous synthesis of DNA. (a) Initiation of synthesis at the nick. (b) As 5′→3′ chain elongation takes place, the 5′ parental strand is degraded. (c) The 5′ parental strand is displaced and immune to further degradation. (d) The polymerase switches template strands. (e) Okazaki fragments are created by nuclease action on the covalent fork. (f) Synthesis continues. (g) Ligase action joins the discontinuous fragments. (After Kornberg, 1969)

Polynucleotide ligase

Both *E. coli* and phage T4 synthesize an enzyme—polynucleotide ligase—which covalently links the discontinuous stretches of newly synthesized DNA into completed continuous strands. It was first detected as an activity necessary for the successful replication of phage λ. When *E. coli* cells are infected with this phage a large fraction of its linear molecules are converted into a covalently bonded circular form. λ DNA carries complementary single-stranded nucelotide sequences at its 5′ termini (known as *sticky ends*), whose cohesion produces circular molecules which are maintained in this form by hydrogen bonding (Figure 13.14). Gellert (1967) showed that the hydrogen-bonded circle is an intermediate which is converted by an enzyme activity into a covalently sealed circle. Gefter, Becker and Hurwitz (1967) purified the enzyme and found that it also catalysed the conversion of λ DNA with single-strand

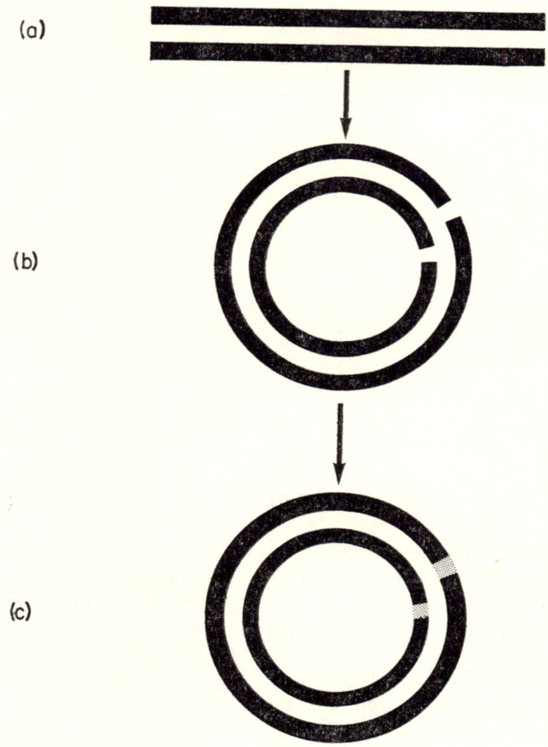

Figure 13.14. The formation of circular phage lambda: (a) linear form with sticky ends; (b) circular form maintained by hydrogen bonds; (c) covalent circle formed by ligase action

breaks to the duplex form, that is it can also rejoin internal breaks in the duplex structure.

Both the enzyme specified by phage T4 during its infection of bacteria, and the enzyme of *E. coli* itself have been studied in some detail (see

Gellert and coworkers; 1968; Olivera and coworkers, 1968; Olivera, Hall and Lehman, 1968; Richardson and coworkers, 1968). The enzyme induced by T4 in bacterial infection requires ATP as the cofactor whilst the *E. coli* enzyme requires NAD for activity. With both enzymes the reaction occurs in two phases. The first is formation of an enzyme–AMP complex; this is achieved by splitting out pyrophosphate from ATP with the T4 enzyme, and by the loss of NMN from NAD with the *E. coli* enzyme. The complex has been isolated in each case. The T4 enzyme was incubated with ATP bearing a P^{32} label in its γ phosphate and H^3 in its adenine moiety; after separation by gel filtration, the tritium was found to be present in fractions which contained the enzyme activity, but the P^{32} label was not. When the P^{32} was present in the α instead of the γ phosphate, an equal proportion of each radioactive isotope was associated with the enzyme. A similar complex was isolated for the *E. coli* enzyme by incubation with two NAD species, one with P^{32} in the adenylate, and one with H^3 in the nicotinamide moiety; a large fraction of the P^{32} was associated with the ligase activity but the H^3 was not bound to the enzyme.

Incubation of the ligase-AMP complex from either source with duplex DNA carrying singe-strand breaks results in a sealing of the breaks and concomitant release of 1 mole of AMP for every phosphodiester bond synthesized. It seems probable that each enzyme displays the same reaction in this phase. DNA preparations containing single-strand breaks with 3′-OH and 5′-P end groups are active in causing breakdown of the complex, but breaks with 3′-P and 5′-OH end groups are not repaired. This demand is very specific; the use of a phosphatase enzyme under conditions where it removes the internal phosphate from the 5′-P end group abolishes activity of the ligase.

The enzyme–AMP complex might bind to either the 3′-OH or the 5′-P terminus, and these possibilities were distinguished by use of the enzyme exonuclease I; this degrades one strand of DNA sequentially from the 3′-OH terminus to release monophosphates, but leaves the 5′ terminal dinucleotide intact. Thus if AMP is linked at the 5′-phosphate it will be released in the form of a trinucleotide (that is attached to the terminal dinucleotide), but if it is attached to a 3′-OH terminus it will be released as a mononucleotide (by the normal enzyme action). The intermediate actually isolated after treatment of the DNA–AMP complex with exonuclease proved to be the former; this implies that the subsequent bond formation must take place by a mechanism similar to that responsible for chain elongation, that is, attack by a free hydroxyl group on the phosphate of the incoming group. This is shown in Figure 13.15.

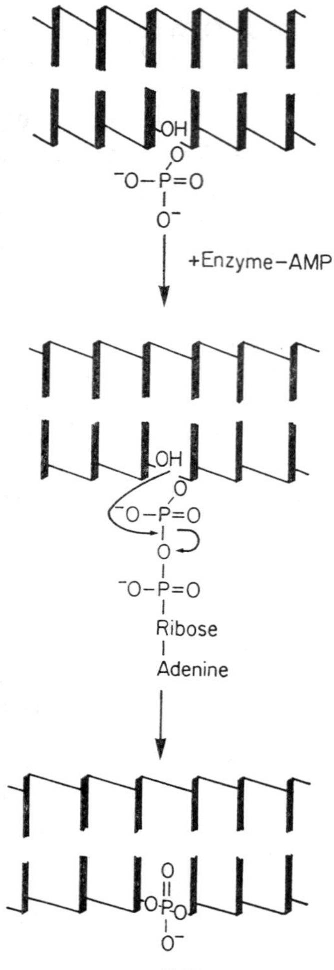

Figure 13.15. The action of polynucleotide ligase. The enzyme–AMP complex binds to a gap terminating in 3′-OH and 5′-P, and AMP reacts with the free phosphate group. Attack by the 3′-OH group on this moiety then forms a phosphodiester bond and seals the gap. (After Olivera, Hall and Lehman, 1968)

Gene 30 of phage T4 has been identified as the structural gene which codes for the ligase, and the use of temperature-sensitive mutants at this locus has suggested that the ligase is responsible for joining the discontinuous segments formed in the replication of a DNA duplex *in vivo* (Newman and Hanawalt, 1968; Okazaki and coworkers, 1968). Normally, the radioactive pulse label which first appears to sediment between 7S and 11S is rapidly transferred to much larger units sedimenting from 30S to 60S. If the ligase represents the activity which is responsible for linking the short segments, the impairment of a temperature-sensitive ligase by increase in temperature should prevent this transition and cause an accumulation of the short chains.

E. coli cells were infected with such a phage mutant, or with the wild-type species, and incubated for 70 minutes at 20°C to reach the stage at which phage DNA is being actively synthesized. Because the synthesis of host DNA is inhibited by T4 infection, only phage DNA is under synthesis at this time. The temperature was then raised to the non-permissive level of 43–44°C, and 1–2 minutes later cells were given a pulse label of H^3-thymidine. The rate of incorporation of the radioactive label into DNA was the same in cells infected with either mutant or wild-type phage. However, upon infection with the wild type an increase in the time of the pulse label to 40 or 60 seconds resulted in the transfer of almost all the radioactivity to large continuous units, whereas after infection with the mutant an increase in the time of incubation merely increased the amount of radioactivity in the 7–11S fraction and there was no transfer to a faster sedimenting fraction. When the entire procedure was performed at 20°C, the accumulation of short chains was not observed, and the mutant showed the same behaviour as the wild type. When mutant-infected cells were pulse labelled at 43°C and subsequently incubated at 30°C, the size of the unit bearing the radioactive label gradually increased after the temperature shift-down, and the small chains disappeared. This indicates that discontinuous segments can again be covalently linked when the ligase activity is restored.

Methylation of DNA

The DNA of *E. coli* contains small amounts of the methylated bases 6-methylaminopurine and 5-methylcytosine. Methylation is accomplished by enzymes which transfer methyl groups from the donor *s*-adenosylmethionine to bases in the DNA (for review see Gold, 1966; Srinivasan and Borek, 1966). Generally, DNA isolated from any particular source is fully methylated and cannot accept any further methyl groups through enzymes from the same source. Two techniques have therefore been used to study methylation. *In vitro*, DNA from one source can be used as the substrate for methylating enzymes from another; in this case, although the DNA has been fully methylated by its own enzymes, it can accept further methyl groups through catalysis by heterologous enzymes. The other technique, which has the advantage that it can also be used *in vivo*, is to subject methionine-requiring bacteria to starvation for this amino acid. Their DNA then becomes under-methylated, but upon addition of methionine is restored to its normal condition. Studies with cell free systems have shown that both the DNA and the methylating enzymes influence the pattern of methylation.

Lark (1968a) examined the timing of methylation with respect to replication by the procedure illustrated in Figure 13.16. A culture of *E. coli* $15T^-$ was pulse labelled for 5 minutes with C^{14}-thymine and then transferred to a medium with normal thymine but containing H^3-methionine. After 10 minutes the culture was transferred to a non-radioactive medium which contained 5BU in place of thymine. DNA was extracted at various times after the final transfer, and the fraction sedimenting at hybrid density, that is the newly replicated section comprising one normal and one heavy strand, was examined for radioactivity. Neither the C^{14} nor the H^3 label was observed at this density until almost one generation after the initial pulse dose, and the H^3 label was replicated immediately after the C^{14}. In view of the sequential mode of replication of DNA, this suggests that the methylated segment of the chromosome is located immediately after the region initially labelled with C^{14}; this suggests that DNA is methylated as or very shortly after it is synthesized.

However, although replication and methylation normally occur together, the two processes can be divorced. Methionine was removed from the culture medium and the DNA labelled with 5BU during the period of starvation. The cells were then cultured in a normal medium containing H^3-methionine. The tritium label was subsequently found in association with the density label, indicating that unmethylated DNA can be converted to its normal state upon the restoration of methionine. Thus replication and methylation can occur independently.

Lark (1968b) studied the effect of inhibiting methylation upon the subsequent replication of DNA. With *E. coli.* $15T^-$, starvation for methionine allows completion of synthesis of the chromosome, but does not permit initiation of the next round of replication. This means that it is only possible to study the replication of the DNA which was methylated in the previous replication cycle—replication of the unmethylated DNA would not occur until the next round of replication, that is until after the

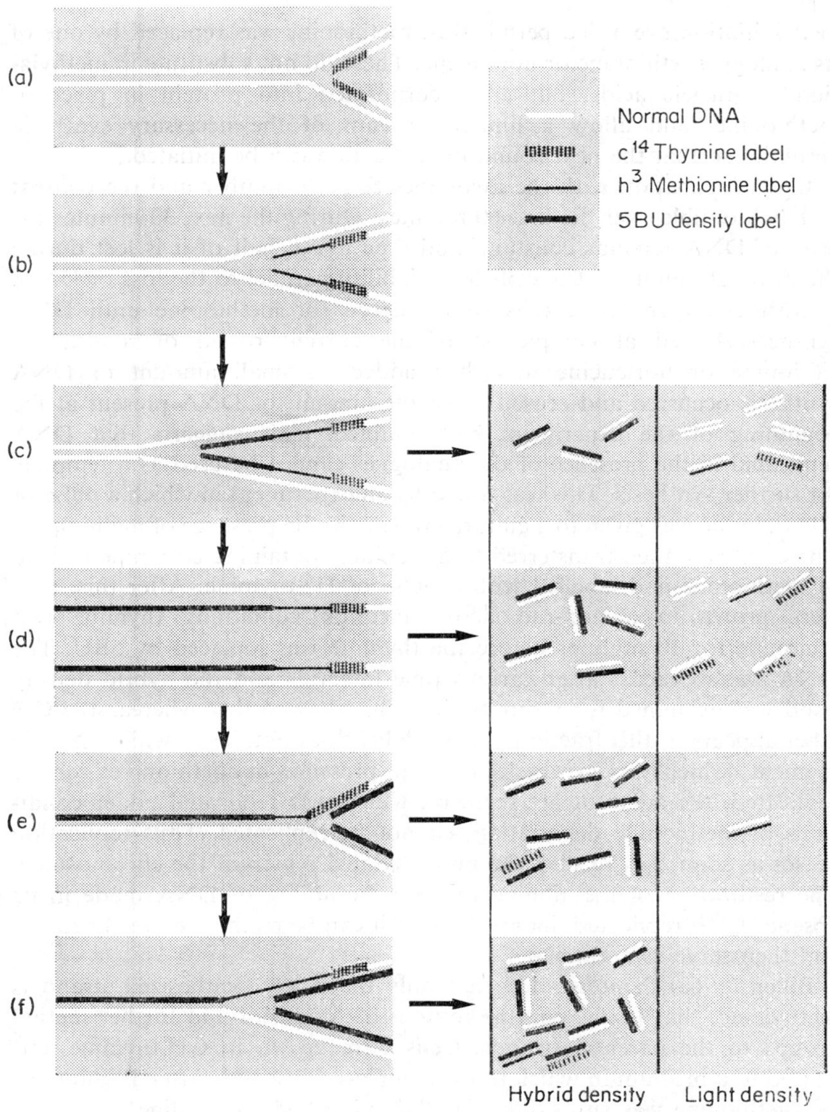

Figure 13.16. (a) Five-minute pulse incorporation of C^{14}-thymine. (b) Ten-minute pulse incorporation of H^3-methionine. (c) Switch to density label with 5BU. (d) Completion of round of synthesis. At this stage, the radioactive segments have not yet replicated again and both radiocative labels are found in light DNA. (e) Replication of the first radioactive label transfers the C^{14} isotope to hybrid density. (f) The H^3 label is found at hybrid density immediately after the transfer of the C^{14} label. This demonstrates that the tritium was originally incorporated adjacent to the C^{14}, that is the methylation event must be taking place at or immediately after the growing point

next initiation event. To permit this, methionine was replaced by one of its analogues, ethionine or norleucine. These do not substitute in methylation of nucleic acids, but are incorporated into protein in place of methionine, and allow a limited amount of the necessary synthetic activities so that the next round of replication can be initiated.

Upon incubation with the analogues the cell number and the content of DNA double over the first 60 minutes. During the next 30 minutes the level of DNA remains constant, and then about half of it is lost during the next 120 minutes. The cells lose viability parallel to the loss of DNA. In further experiments, cells were starved for methionine until DNA synthesis halted at completion of the current round of replication. Ethionine or norleucine was then added; a small amount of DNA synthesis occurred and ceased when the amount of DNA present at the beginning of the experiment had doubled. This suggests that DNA replicated in the presence of the analogues cannot be used as a template for further synthesis. This was tested by an experiment in which a pulse of C^{14}-thymine was given to a culture growing in the presence of methionine. The cells were then transferred to a medium containing ethionine instead of methionine and labelled with a pulse of H^3-thymine. After this, they were grown in a non-radioactive medium containing thymine and ethionine for 10 minutes before the thymine was replaced by 5BU. The DNA was extracted after various time intervals and the hybrid density band was examined for radioactivity. This showed that whereas the C^{14} label appears in this fraction, the H^3 label does not. Thus whilst the C^{14} segment, which was synthesized in the presence of methionine, can be replicated, the adjacent H^3 segment, which was replicated under conditions of methionine deprivation, cannot be replicated. Thus replication ceases as soon as it reaches the unmethylated region of the chromosome. The restoration of methionine restores the ability of DNA made in its absence to be replicated, indicating that it can be methylated retroactively and then serve as a template.

Billen (1968) demonstrated that only the newly synthesized strand is methylated, that is, the enzyme responsible does not add further methyl groups to the parental strands. Cells were grown in C^{14}-thymine and shifted to a medium in which this was replaced by 5BU. After 10 minutes, H^3-methionine was substituted for the unlabelled methionine previously present. The DNA was then extracted and heat denatured; the H^3 label appeared only in the heavy fraction corresponding to the newly synthesized strands.

The function of this methylation appears to be to provide a mechanism for the cell to distinguish its own genome from other DNA; DNA from different species has a different methylation pattern and is destroyed upon entry into the cell. This interpretation is borne out by the phenomenon of host modification and restriction (for review see Arber, 1968; Glover and Colson, 1969). Phage λ can be grown on any of several strains of *E. coli* and is subjected to a modification characteristic of its particular host; phage grown on K are modified in one way, those grown on B suffer a different modification, whilst *E. coli* C imposes no detectable modification. The modification resides in the pattern of methylation of the DNA; if the phage are grown on methionine-starved strains the usual host specificity is not conferred upon the progeny. As well as phage λ, other phages and the bacterial chromosome itself are substrates for this modifying methylation.

A host cell will only accept the entry of DNA bearing its own characteristic modification pattern; thus whilst λ.K (λ modified by *E. coli* K) can infect *E. coli* K successfully, if the host is strain B, the invading DNA is 'restricted'—it is degraded upon or very shortly after its entry into the cell, by the introduction of a number of double-stranded scissions. Restriction can thus be seen as a mechanism which guards against invasion by 'foreign' DNA. Meselson and Yuan (1968) identified and isolated an enzyme from *E. coli* K which proved to be an endonuclease specifically active against λ DNA grown in other *E. coli* strains. Joint renaturation of denatured DNA from strains K and C was used to produce hybrids with one strand from each parent; these were not attacked by the enzyme, which therefore presumably requires recognition of both strands of its substrate as 'foreign'. Mutant bacteria have been found which lack either the restriction capacity, so that they do not degrade foreign DNA, or both modification and restriction, in which case neither modify nor restrict their own or foreign DNA. Mutants lacking only modification have not been found; these would be lethal since they would be unable to methylate their own DNA, which would then suffer restriction, with consequent cell death. Restriction also explains the loss of DNA found by Lark (1968b) after replication during methionine deprivation; presumably, in the absence of its specific pattern of methylation groups, the DNA is (falsely) recognized as foreign and hence degraded.

It has proved difficult to analyse the genetic basis of restriction and modification because of the lack of satisfactory techniques for the selection of recombinants and the formation of diploids for the genetic region concerned. However, Boyer and Roulland-Dussoix (1969) made use of a new technique for the isolation of F-heterogenotes, and extended the genetic analysis by constructing permanent partial diploids of *E. coli* with various arrangements of mutant and wild-type restriction and modification alleles. The r^-m^+ and r^-m^- mutants (which arise with equal frequency) complement one another, so that at least two cistrons must be involved. Further, there are two complementing classes of r^-m^- mutants, so that at least three cistrons must be involved overall (all located close together). They proposed that these code for restriction, modification and recognition polypeptides. Either all three might form an oligomeric enzyme which has both restriction and modification properties, or there might be two separate enzymes which share a common recognition polypeptide.

The initiation of replication

Much work has been performed on *E. coli* strains derived from the $15T^-$ (thymine-requiring) mutant, which have two significant properties. First, DNA synthesis ceases if thymine is removed from the incubation medium. Second, the absence of thymine causes the cells to lose viability in a characteristic manner, as shown by the loss of colony-forming ability. This arises from either of two causes. If the bacteria carry lysogenic prophage, thymine starvation induces the prophage, thus destroying the cells. However, thymine-less death still occurs in strains 'cured' of prophage, this time as a result of their fragility with respect to the plating procedures used to grow the colonies (Donachie, 1967). (It is worth noting also that thymine starvation is highly mutagenic.)

Using certain derivatives of this strain, in particular the $T^-A^-U^-$ mutant (often termed TAU-bar), which is auxotrophic for thymine, arginine and uracil, it is possible to block selectively any or all of DNA, RNA or protein synthesis. Maaløe and Hanalwalt (1961) transferred an exponentially growing culture of these cells from a complete medium into one lacking thymine but containing arginine and uracil (—T, +AU). After a lag period there was an exponential loss of viability, with no survivors. However, if the culture was transferred instead to a medium lacking all three growth factors (—T, —AU), the survival curve was almost identical, but tailed off to a plateau corresponding to a level of about 3 per cent. survivors. Thus although most cells die when DNA synthesis is blocked, irrespective of whether RNA and protein synthesis can proceed or not, a small fraction live if these latter processes are prevented, but not if they are allowed. Maaløe and Hanawalt suggested that this immune fraction might represent cells at some particular stage of the growth cycle.

This theory was tested by removing arginine and uracil from exponentially growing cultures, so that RNA and protein synthesis was blocked, and then by allowing varying periods of time before the removal of thymine to block DNA synthesis. The immune fraction was then measured for each of the intervals used. The longer the preincubation with thymine which allowed DNA but *not* RNA and protein synthesis, that is the time in (+T, —AU) medium before DNA synthesis was also blocked by the removal of thymine, the greater the fraction of cells which was immune to thymine-less death. Eventually, the whole population was brought into the immune state (after 90 minutes pre-incubation). These results are shown in Figure 13.17.

It was suggested that cells are immune to thymine-less death when they are not engaged in DNA synthesis at the time of their transfer into the thymine-less medium. That is, they have just completed one replication cycle but have not yet commenced the next. RNA and/or protein synthesis must be necessary in between the rounds of replication for initiation of the second cycle; thus there are no survivors after transfer from (+T, +AU) to (—T, +AU) culture conditions because all the cells can undertake RNA and protein synthesis, commence another replication cycle and suffer thymine-less death. However, upon transfer from (+T, +AU) to (—T, —AU) medium, the necessary RNA and/or protein synthesis cannot occur, so that the 3 per cent. of cells in between replication cycles are unable to commence another cycle and therefore remain immune. If cells are pre-incubated in (+T, —AU) medium before their transfer to the (—T, —AU) conditions, they can complete the current replication cycle successfully but are unable to commence another. Thus as the preincubation period increases more cells complete their current replication cycles and remain in a state in which they are not synthesizing DNA; in this condition they are immune to thymine-less death. If immune cells are transferred into a culture in which RNA and protein synthesis is possible, as expected they lose their immunity by initiating another cycle of DNA synthesis.

Hanawalt and coworkers (1961) examined these predictions experimentally by autoradiographic analyses, and found that individual cells continue DNA synthesis for differing lengths of time after RNA and protein synthesis is halted. Whatever the fraction of immune cells, a similar fraction lacks the ability to synthesize DNA under conditions of inhibition of RNA and protein synthesis. Lark, Repko and Hoffman (1963), using the technique of introducing a radioactive label into a portion of the bacterial chromosome and following its next replication by density labelling, showed that after amino acid starvation a labelled segment is not transferred to hybrid density. Presumably, it cannot

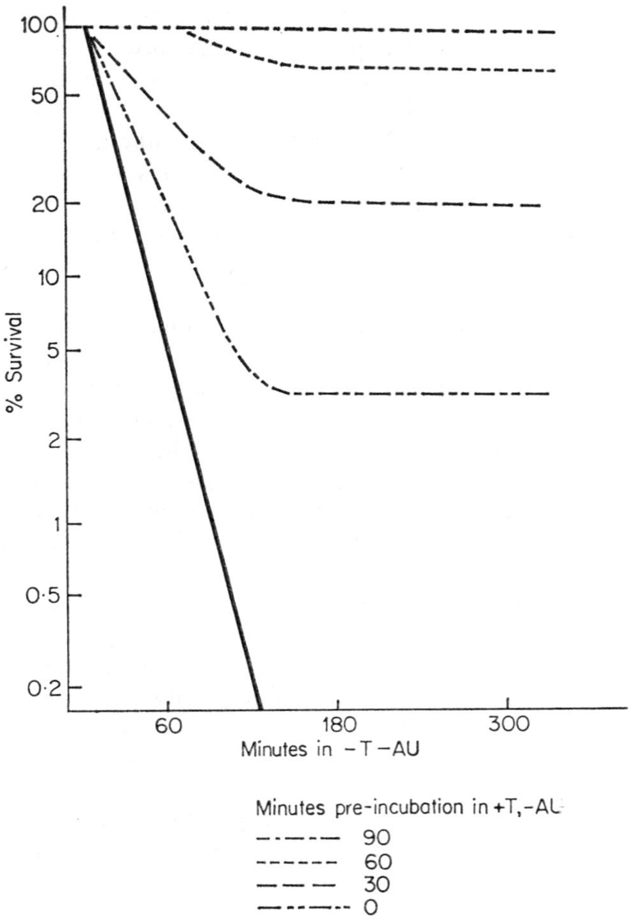

Figure 13.17. The thymine-less death of TAU-bar cells in (—T,+AU) medium after various periods of pre-incubation in (+T,—AU) medium. As the time of pre-incubation is increased, more cells can complete their current replication cycle but not commence another; in this state they are immune to thymine-less death and so the proportion of survivors increases. The heavy curve represents survival upon transfer to (—T,+AU) medium. (After Maaløe and Hanawalt, 1961)

replicate again because of an inability to initiate the next replication cycle.

The synthesis of specific proteins at certain times of the replication cycle is necessary for initiation (for review see Lark, 1969). When spores of *Bacillus subtilis* are germinated, DNA replication is initiated synchronously in the population. The addition of high concentrations of chloramphenicol to block protein synthesis after the round of replication has begun does not interfere with DNA synthesis; the amount of DNA doubles and then replication ceases. This corresponds to completion of the current round of replication, but inability to commence further rounds. Addition of the inhibitor at various times after spore germination has shown that synthesis of the protein(s) required for initiation is accomplished at a certain time during the replication cycle—after this time the addition of inhibitor does not prevent initiation.

In *E. coli* $15T^-$, high concentrations ($150\mu g/ml$) of chloramphenicol inhibit initiation as effectively as amino acid starvation, but low concentrations ($25\mu g/ml$) are somewhat less effective. Lark and Renger (1969) have shown that two separate processes requiring protein synthesis are necessary for initiation; one is inhibited by even the low concentration of chloramphenicol (the CM-sensitive process), but the other is only inhibited by the higher level (the CM-resistant step). These two events have been separated temporally in the replication cycle.

The procedure used was to subject cells to amino acid starvation so that they completed their current round of replication, but did not initiate the next. Amino acids were then restored in the presence of a pulse of radioactive thymine; this label is then specifically incorporated into the region of the chromosome origin. Subsequently, the cells were transferred to a medium containing the heavy isotopes C^{13} and N^{15}, DNA extracted after a period of growth, and subjected to centrifugation on a CsCl density gradient; the appearance of radioactivity in the fraction of hybrid density indicates that the origin has replicated again, that is, that there has been another initiation event. The administration of various inhibitory treatments during the period of growth in the heavy isotope medium was used to measure the time at which cells are able to initiate a new cycle of replication in the absence of protein synthesis. This can only occur after they have already synthesized the proteins required for initiation.

Thus after labelling the origin with tritium, cells were grown in a normal non-radioactive medium. Samples were removed at various times and transferred into a heavy medium containing $150\mu g/ml$ chloramphenicol. Growth was allowed to continue for 90 minutes in this medium before the DNA was examined for transfer of radioactivity to hybrid density. The results showed that cells become immune to the effect of inhibition of their protein synthesis some 15 minutes before the initiation event itself actually occurs. Once the initiation has taken place, they lose this immunity and again require protein synthesis before another cycle of replication can be initiated.

A similar experiment was performed with a lower concentration of chloramphenicol (25μg/ml). In this case, the potential to initiate replication in the heavy medium was acquired early in the replication cycle, 15 minutes before immunity was achieved in the experiments using the higher level of inhibitor. This suggests that two separate steps are involved, the CM-sensitive step taking place some 15 minutes before the CM-resistant step.

Phenethyl alcohol also inhibits the initiation of DNA synthesis in bacteria. During treatment with this inhibitor the CM-sensitive protein is synthesized, and sufficient accumulates to allow several subsequent rounds of replication even in the presence of chloramphenicol. This suggests that the phenethyl alcohol must be acting upon the CM-resistant protein, and this conclusion is supported by the finding that it can exert its inhibitory effect only if added during the period of the replication cycle preceding the synthesis of the CM-resistant species. Phenethyl alcohol is known to affect the cell membrane—*E. coli* mutants resistant to it exhibit an abnormal membrane function and fail to divide normally—so it seems likely that this initiation step may involve the cell membrane, in some manner.

Thus initiation of a new cycle of DNA replication requires the synthesis of at least two proteins, one sensitive to chloramphenicol and one sensitive to phenethyl alcohol. Each of these steps must occur once in each replication cycle. The data also suggests that a third step may be necessary since a period of 15 minutes intervenes between the second protein synthetic step and the actual initiation event. The cause of this lag is not known, but presumably whatever process is involved does not demand protein synthesis.

The isolation of mutants defective in replication has suggested that (at least) two gene functions must be involved in initiation in *B. subtilis*. Gross, Karamata and Hempstead (1968) isolated temperature-sensitive mutants by treating cells with nitrosoguanidine and then by measuring the ratio of protein to DNA synthesis at both low (30°c) and high (45°c) temperatures. The twenty-nine mutants found were mapped in four linkage groups by transformation, and the recombination frequencies within the four sites suggested that each probably comprises a single gene.

In two of the groups, DNA synthesis itself was prevented by a rise in temperature; since the activity *in vitro* of a crude DNA polymerase extract was normal at both temperatures, these (presumably) represent mutation in other features of the replication process. In the two other groups, the amount of DNA synthesis after an increase to the non-permissive temperature was the same as that achieved after the inhibition of protein synthesis; this suggests that cells completed their current round of replication, but were not able to initiate a new round.

Chromosome replication and the cell division cycle

It seems evident that there must be a correlation between replication of the chromosome and division of the cell such that rounds of replication occur with the same frequency as cycles of cell division. Bacteria grow at widely differing rates depending upon their conditions of culture (for example, *E. coli* may vary from doubling times of 20 to 180 minutes or more), and this raises the question of whether the rate of replication is proportional to the growth rate, or whether replication takes place at a constant rate which is independent of the doubling time. In the former case, it might be expected that cells would always be in a state of replicating their chromosomes, presumably so that one round of replication takes as long as the time between cell divisions. On the other hand, if the rate of replication is constant, at lower growth rates there should be gaps in the cell division cycle where no replication occurs.

Clark and Maaløe (1967), working with a synchronous culture of *E. coli* growing upon glucose medium with doubling times from 37–55 minutes, found that rounds of replication were initiated about half-way through the division cycle rather than at its beginning. However, despite the lack of coincidence between initiation and cell division, the periodicity between successive initiations of replication was the same as the generation time.

Helmstetter and Cooper (1968) measured the rate of DNA synthesis during the division cycle of *E. coli* B/r. The technique they used was to pulse label an exponentially growing culture with H^3-thymidine and to find the amount of label incorporated into cells of different ages by measuring the radioactivity in their progeny. This was done by binding the pulse-labelled culture to the surface of a millipore membrane filter; irrigating the surface of the membrane with medium allows the bound cells to grow and release progeny into the effluent after a cell division. The new-born cells were collected and their radioactivity measured as a function of the time at which they were eluted; this gives the rate of DNA synthesis in their ancestors as a function of the time in the division cycle when they were labelled—the sooner a new-born cell is eluted from the filter, the older its parent must have been when bound to the membrane.

The start of a round of replication is shown by a sudden increase in DNA synthesis in cells of that age, and completion of a round by a fall in the level of synthesis. The results showed that the time for a round of replication appears to be fairly constant at faster growth rates; for doubling times between 22 and 60 minutes one round of DNA synthesis takes about 41 minutes. This is in agreement with the finding of Clark and Maaløe (1967) that the rate of DNA synthesis was more or less constant per growing point within their range of generation times. At slower growth rates, the rate of replication is lowered, probably due to limitations on precursor or energy supplies.

On the basis of their findings, Cooper and Helmstetter (1968) proposed a model to link the replication cycle with the cell division cycle (for the faster growth rates). The replication cycle can be defined in terms of two constants. The rate of replication is the time which it takes for a replication point to proceed from the origin to the terminus of the chromosome; this constant, *C*, is about 40 minutes. They observed also that there is always a constant time, *D*, between the end of a round of chromosome replication, that is the moment at which the replication point arrives at the chromosome terminus, and the cell division resulting from this replication. This is about 20 minutes (more precisely, 22 minutes for doublings taking from 22–60 minutes).

The features of this model are illustrated in Figure 13.18. The left hand side shows the rate of DNA synthesis for a period of one division cycle in terms of the time before the next cell division. The initiations and completions of rounds of replication are observed respectively as increases or decreases in the amount of DNA synthesis, except for doubling times of 20 and 40 minutes when an initiation and termination occur simultaneously and obscure each other. Because of the constant value of *D*, rounds of replication are always completed 20 minutes before their ensuing cell division. The right hand side of the figure shows that state of the chromosomes according to the model.

A round of chromosome replication must be initiated a fixed time, $C+D=60$ minutes, before a cell division. Thus initiations do not coincide with division, but as the growth rate increases (doubling time decreases), initiation takes place earlier in the division cycle. If cells are growing faster than 60 minutes per doubling time, the next round of replication must be initiated before completion of the current division cycle. For example, in cells with a 50-minute doubling time, initiation of

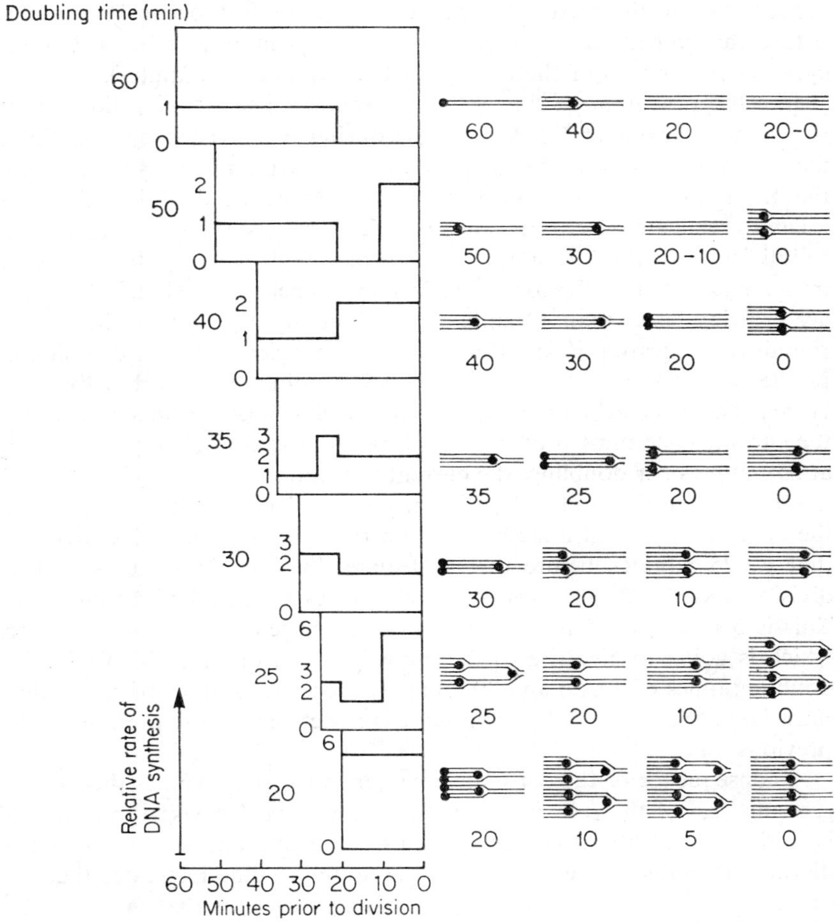

Figure 13.18. The rate of DNA synthesis in *E. coli* at various doubling times. Left: the rate of DNA synthesis during the division cycle in terms of the number of growing points. Increases represent the starts of rounds of replication and decreases their termination. At 40 and 20 minutes per doubling these occur simultaneously and obscure each other. At doubling times slower than 40 minutes there are gaps in the replication cycle when there is no DNA synthesis. Rounds of replication always take 40 minutes and terminate 20 minutes before a cell division. Right: the state of the chromosomes during the division cycle. Growing points are represented by black dots. At doubling times faster than 40 minutes chromosomes have multiple replication forks, that is another round of replication is started before the last has finished. (After Cooper and Helmstetter, 1968)

replication for the next but one cell division must occur 10 minutes before the immediate cell division (that is, 10 minutes of this division cycle + 50 minutes of the next gives the required 60 minutes).

In cells growing more slowly than 40 minutes per doubling time, there is a period devoid of DNA synthesis during the division cycle because the round of replication is completed in less time than it takes to complete the division cycle. In cells growing at 40 minutes per doubling, the time taken to replicate the chromosome is exactly the same as the time between cell divisions so that the starts and ends of rounds of replication coincide in the middle of the division cycle, 20 minutes before division. When cells are growing faster than 40 minutes per doubling, they must initiate new rounds of replication before there has been sufficient time for completion of the previous one. Thus a new growing point must start before the current one has reached the terminus of the chromosome; this means that the chromosome must bear multiple forks. For example, in cells growing at 35 minutes per doubling, the current replication cycle will end $D = 20$ minutes before the next cell division. However, the replication cycle for the division after next must be initiated 60 minutes prior to the division, that is, 25 minutes before the immediate division (25 minutes of this division cycle + 35 minutes of the next gives the required 60 minutes). This is 5 minutes before completion of the present round of replication which is going to give rise to the immediate cell division. Similarly, for doubling times of 30, 25 and 20 minutes, successive rounds of replication must be initiated 10, 15 and 20 minutes respectively before the end of the previous one.

A consequence of the constant time per replication cycle is that as the growing rate of the cells increases (doubling time decreases), replication is initiated earlier and earlier in terms of generations prior to the cell division D minutes after its end. If the doubling time is greater than 60 minutes, this initiation is in the same division cycle as that in which the replication is completed. Between 60 and 30 minutes doubling times, replication is initiated in the division cycle one cycle before that in which it is completed. Between 30 and 20 minutes, replication cycles are initiated two division cycles before their completion. This is illustrated in Figure 13.19 which plots the rate of DNA synthesis against cell age in terms of fractions of a generation rather than real time.

As can be seen from the right-hand side of Figure 13.18, cultures of cells growing at any particular rate cannot be characterized as one or two chromosomes per cell, since for growth rates faster than 40 minutes per doubling some cells have one replicating chromosome and some have two. The chromosome configurations for any particular doubling time

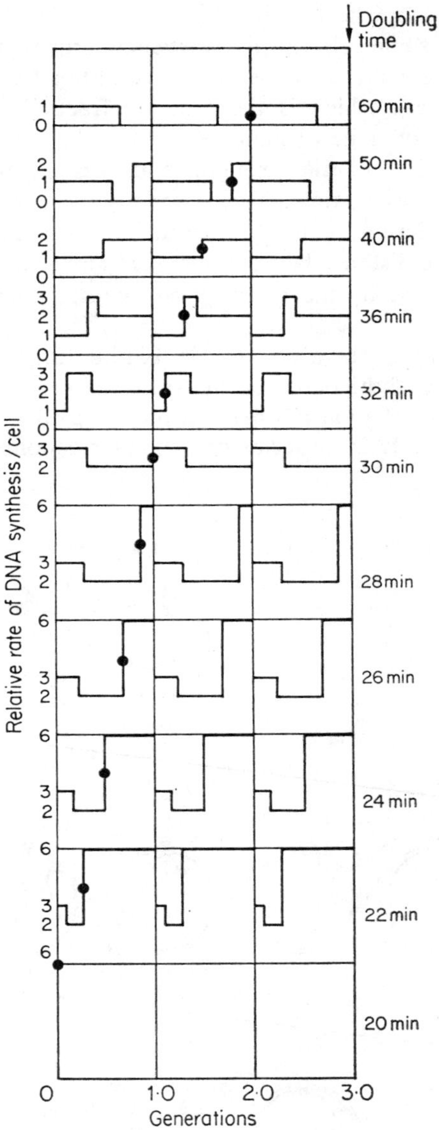

Figure 13.19. The rate of DNA synthesis in terms of growing points during three successive division cycles; the cell age is plotted in generation cycles instead of real time. The set number is the time from the filled circle to the final division. (After Cooper and Helmstetter, 1968)

were characterized by the introduction of the concept of 'chromosome sets'. The chromosome sets for any particular growth rate comprise the series of states of the chromosome illustrated in this figure, and is defined by the 'set number', which is the time, in fractions of a generation, between the start of a new round of replication and the cell division following completion of that round. Numerically, this is $(C+D)/T$, where T is the doubling time. This describes the configurations of the whole culture, obviating the need to characterize the number of chromosomes per cell. The distance between the filled circles of Figure 13.19 and the division indicated by the arrow gives the set numbers for various doubling times.

The chromosome states illustrated in Figure 13.18 do not show the true physical form of the chromosome since the genome is circular, not linear. The process of replication for a doubling time of 30 minutes is illustrated in Figure 13.20 in terms of a circular genome. The final stage

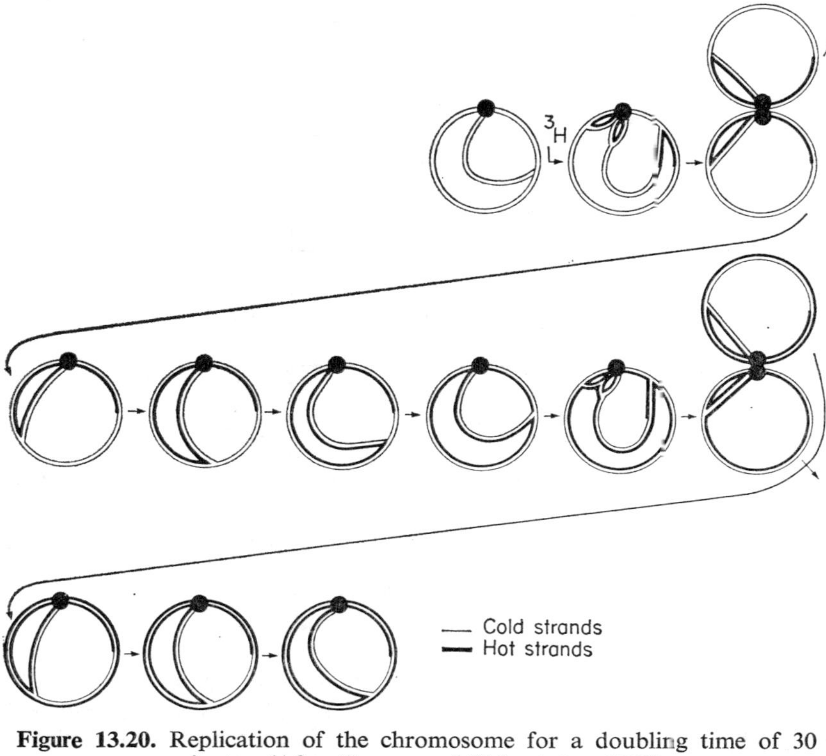

Figure 13.20. Replication of the chromosome for a doubling time of 30 minutes. (After Cooper and Helmstetter, 1968)

corresponds to that detected autoradiographically by Cairns (1963)—see Figure 13.2.

The model does not appear to hold for slower growth rates. For doubling times greater than 60 minutes, there should be a gap in DNA synthesis at the beginning of the division cycle before initiation, and another gap of 20 minutes after the completion of replication before cell division occurs. However, for generation times greater than one hour per doubling, *C* and *D* are no longer constant in time, but instead occupy fixed proportions of the division cycle. Replication is initiated at the beginning of a division cycle, and the chromosome is synthesized during the first two-thirds of the generation, with a gap devoid of DNA synthesis occupying the one-third remaining to cell division.

The Helmstetter–Cooper model differs from that previously proposed by Lark (1966) for *E. coli* $15T^-$. The amount of DNA synthesis in growing cells was measured by labelling a culture with radioactive thymine for one-tenth of a generation and subsequently examining the cells by autoradiography. This showed that the faster the growth rate of the cells, the greater the amount of DNA synthesis occurring in any particular cell. At the longer doubling times, there were gaps in the division cycle during which no replication occurs. To account for this data, Lark proposed that cells may possess different numbers of chromosomes according to their growth conditions. In glucose cultures (the fastest growing), cells would possess two chromosomes which are replicated simultaneously, whereas in succinate medium, although there would be two chromosomes per cell and their replication would occupy the whole generation period, only one would be replicated at a time. In aspartate medium, only one chromosome would be present, and would suffer replication throughout the division cycle. In acetate cultures (the slowest growing), only half the cells would be in the act of replication at any instant, with a single chromosome replicated during the latter half of the division cycle.

This model is incompatible with the Helmstetter–Cooper proposal that cells should contain multifork chromosomes. However, they observed that the pattern of grain counts found by Lark and interpreted in terms of different numbers of chromosomes differed very little from the pattern expected on their model, and indeed, cannot be distinguished from it. Further observations by Lark (see Bird and Lark, 1968) suggest that $15T^-$ shows the same pattern of replication as B/r in that there is a constant rate for chain elongation, sequential initiation of rounds of replication, and multiple replication forks at faster growth rates. However, in faster growing cells Lark also observed a specific pattern of segregation of parental chromosomes into progeny cells which can be accounted for by the presence of multiple chromosomes, but cannot be explained on the Helmstetter–Cooper model. Thus Bird and Lark suggested that the precise mechanisms relating chromosome replication and segregation to cell division may differ in the two species. This could be due to physiological differences between the two cell types; for example, $15T^-$ cells growing at 40 minutes per doubling have about the same size and DNA content as B/r cells growing at 27 minutes per doubling.

Replication must be linked in some manner to the apparatus responsible for division if the cell is to divide only when replication of its chromosome is completed. The large number of mutants defective in cell division which have been found confirms that it must involve many processes. Clark (1968b) used two approaches to determine the timing of events in cell division. Killing by infection with phage T4 mutants unable to multiply in their host was used to test for septum formation. If a host cell retains its integrity, infection is lethal, but if a septum has formed, the distal end of a cell infected with a single phage particle is immune from the effects of infection. The time of formation of a rigid cross wall was measured by subjecting cells to sonication. When an osmotically rigid wall has developed, there is a sudden increase in the number of survivors from sonication because one end of the cell can survive when the other is damaged by the treatment. These experiments showed that with a culture of *E. coli* doubling every 45 minutes, a round of replication was completed at 20 minutes, septum formation was achieved at 30 minutes and a rigid cross wall formed at 37 minutes. Thus the *physiological division* resulting from septum formation, in which the cytoplasmic events at opposite ends of the cell become independent of each other, occurs well after the completion of replication, and well before the physical separation of a parent into two progeny cells. So far as the regulation of cell division is concerned, it is probably only the events prior to this compartmentation which are relevant.

The most likely form for the connection between replication and division would be for completion of a round of replication to generate a signal which activates the first in a series of steps leading to division, the whole process requiring *D* minutes for conversion. There is now considerable evidence that the completion of a round of replication is a necessary prerequisite for division, although it has not been formally proven that it is sufficient and provides the sole trigger. Helmstetter and Pierucci (1968) and Clark (1968a, 1968b), have shown that once a round of replication has been terminated, further DNA synthesis is not necessary for cell division. The former workers inhibited DNA synthesis by the use of ultraviolet irradiation, treatment with mitomycin C, or the addition of nalixidic acid. They found that cell division continued for 20 minutes after DNA synthesis ceased; this corresponds to cells which have completed rounds of replication undertaking the subsequent cell division. Clark, using *E. coli* B/r cells in a synchronous culture, showed that whilst inhibition of DNA synthesis by nalixidic acid prevented cell division if added before completion of a round of replication, addition of the inhibitor after completion had no effect. The same results were achieved when DNA synthesis was blocked instead by inactivation of a temperature-sensitive mutant in replication itself by an increase in temperature. Further support is provided by the finding of Gross, Karamata and Hempstead (1968) that temperature-sensitive mutants in initiation can continue to divide at high temperature, whereas mutants in the replication process itself are not able to do so.

The results of shift-up experiments (see Clark, 1968a; Cooper, 1969; Lark, 1966) are also consistent with the proposal that termination of replication triggers cell division. After a shift, cells appear to complete the currently proceeding round of replication as characteristic of their old medium, and only show a change to the new conditions in the next round. When culture conditions are changed from a slow to a faster growth rate, an initiation of replication takes place shortly after the shift-up. However, the next cell division takes place after exactly the same time as it would have upon the slower medium. This suggests that it is dependent upon the currently active DNA polymerase completing replication and triggering the *D* minute wait for division. (Under the conditions used, there was only one growing point at the time of transfer.) It is the second division which follows more rapidly as characteristic of the new medium; that is, it takes place *D* minutes after the enzyme which initiated replication upon the shift-up has reached the chromosome terminus.

Replication and the cell membrane

Although bacteria do not possess a system of the complexity of the mitotic apparatus which ensures the even segregation of chromosomes in higher organisms, some system must be responsible for the distribution of newly replicated chromosomes into the two daughter cells of a division. It seems probable that this system must involve an association between the chromosome and the cell surface—most likely the membrane. Various models, differing in their exact details, have been proposed upon this basis, and there is now evidence in favour of a connection between the genome and the cell membrane.

Morphological studies (Ryter, Hirota and Jacob, 1968) on both *E. coli* and *B. subtilis* have shown that there is indeed an association between the bacterial nuclear complex and the cell membrane. The isolation of DNA has often produced fractions which are associated with membrane material, although it is is usually difficult to exclude the possibility that their association may have taken place during the extraction procedures. However, Tremblay, Daniels and Shaechter (1969) have fractionated cell components by a method which depends upon the ability of membrane material to attach to crystals of detergent. The fractions isolated by this method has the properties expected of a bacterial 'nuclear complex'—it consists of a portion of the cell membrane, virtually all the DNA of the cell, nascent RNA and ribosomes. Since DNA alone does not interact with the detergent, its presence in this fraction must be due to its attachment to the membrane. The RNA of the complex appears to be newly synthesized molecules (as shown by a pulse label), and the ribosomes are probably attached to it (they are released from the complex by treatment with ribonuclease).

If gently lysed preparations of *E. coli* or *B. subtilis* are subjected to sucrose density gradient centrifugation, most of the cellular DNA sediments relatively slowly, but a small fraction sediments much faster and forms a pellet. It seems likely that the DNA in this fraction is membrane-bound since treatment with ionic detergents converts it to the more slowly sedimenting form. If H^3-thymidine is added to cultures in exponential growth, it first appears in the DNA of this membrane fraction. The addition of a chase of cold thymidine causes the tritium to assume the same profile as the bulk of the DNA. This suggests that the membrane fraction comprises the newly synthesized DNA at the replication point. Smith and Hanawalt (1967) refined this extraction technique by adding a 'shelf' of very high concentration sucrose to the bottom of the density gradient. The membrane-bound DNA then bands at the shelf and is thus separated from both the bulk of the DNA sedimenting on the gradient, and from the cellular debris which pellets at the bottom of the tube. The DNA of the fraction bonding at the shelf was isolated from the other components present in the fraction at the interface, and then subjected to CsCl density gradient centrifugation. When a pulse of 5BU was used to label the growing point, the shelf DNA showed as a peak bonding between the unreplicated parental DNA and the newly replicated hybrid density species. This is what would be expected of a fork-shaped growing structure in which the forks have been replicated and are thus of hybrid density, and the 'handle' comprises unreplicated parental DNA.

Fairly drastic treatment was required to free the DNA of the shelf fraction, suggesting that it is likely to be in some form of a complex with the other components. The complex from *E. coli* was not affected by digestion with pronase, so that it does not contain protein, but it was sensitive to disaggregation by deoxycholate, suggesting that the DNA is bound to lipid. On the other hand, pronase digestion of the *B. subtilis* complex released the DNA from association with the other components. It seems likely, therefore, that the shelf complex arises from the attachment of the chromosomal growing point to the cell wall, and the different chemical nature of the cell surface in different bacterial species accounts for the different treatments required to release the DNA.

Sueoka and Quinn (1968) used this technique to examine the possibility that the DNA might be attached to the membrane at other points in addition to the growing region. They extracted the DNA of the membrane fraction and tested it in transformation studies to determine which genetic markers were represented. A marker mapping very near to the origin was increased in frequency relative to other markers, and there was also a higher level of a marker mapping near to the chromosome terminus. The possibility that the origin might be attached to the membrane was then examined by germinating *B. subtilis* spores in the presence of H^3-thymidine; after germination, the first round of replication takes place synchronously in a spore population, so this specifically labels the origin region. The tritium label appeared in the shelf fraction, and in contrast to pulse labels of the growing point, was not chased into the slowly sedimenting DNA by allowing a subsequent growth of the cells in cold thymidine. This suggests that the origin is permanently attached to the membrane, the attachment site possibly extending past the terminus region. (It should be noted that Rosenberg and Cavalieri (1968), on the basis of physico-chemical studies with *E. coli* DNA, have suggested that there is no single and unique membrane attachment site on the genome. They suggest that the DNA is associated with the membrane at some 30 points, either randomly or at fixed places; the growing point and origin may comprise two of these, but if so should constitute only a small proportion of the total attachment regions.)

Sueoka and Quinn proposed a model for chromosome replication in which they suggested that the origin–terminus region and replication forks are bound at one area—the replication apparatus—on the cell membrane. These regions must be together at the beginnings and ends of rounds of replication, and the model shown in Figure 13.21 suggests that they remain together during replication instead of migrating apart and later returning. In this case, the origins of the two daughter chromosomes must be attached in some manner to the old terminus during replication.

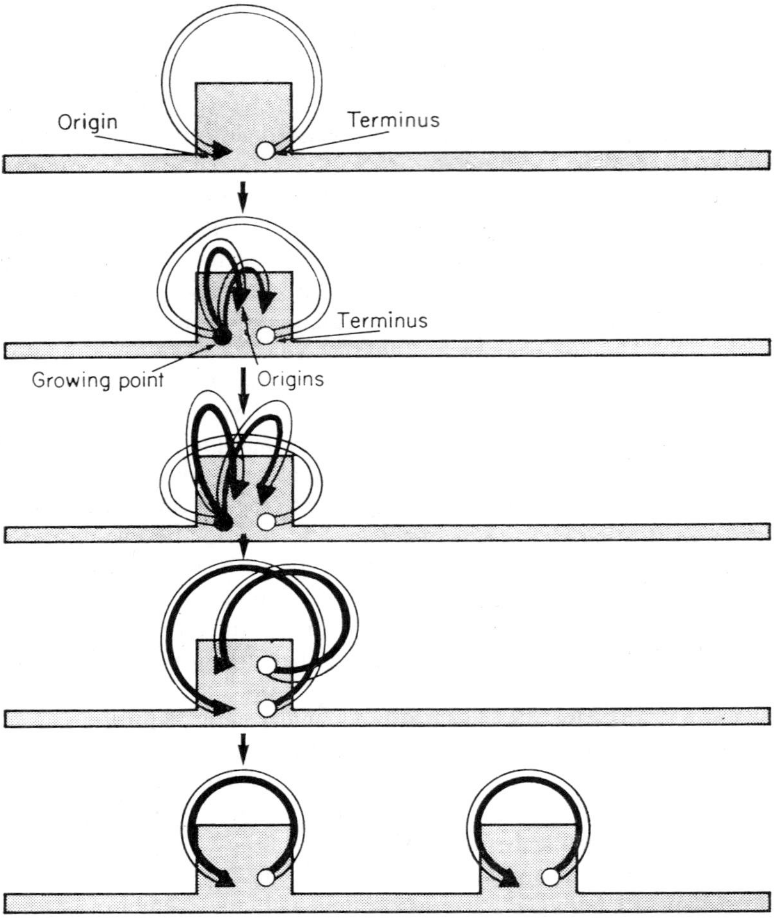

Figure 13.21. A model for DNA replication in which the origin, terminus and growing point are all linked to the membrane through a replication apparatus. (After Sueoka and Quinn, 1968)

A suggestion by Yoshikawa (1967) could provide a mechanism for this. This was based upon the finding that, when *B. subtilis* spores are germinated in the presence of 5BU, the newly synthesized material is covalently linked to the preexisting DNA. This can be accounted for if initiation of a replication cycle consists of extending the terminal nucleotide sequence of the old strands by adding more bases, as shown in Figure 13.22. When replication is completed, the product is two covalently-linked daughter chromosomes whose separation must require the introduction of two breaks, one in each strand.

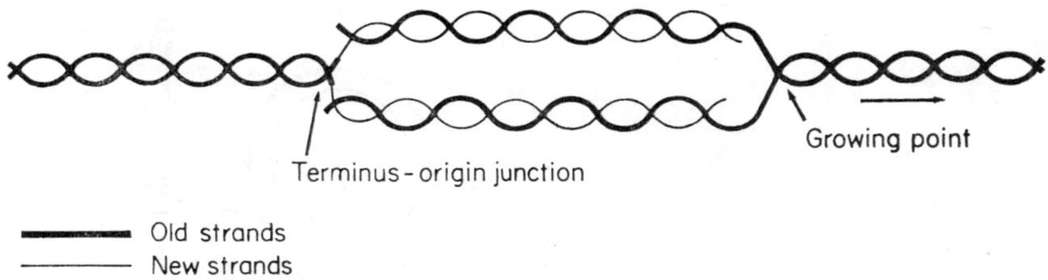

Figure 13.22. Initiation of replication by covalent extension of old strands

Gilbert and Dressler (1968) have proposed a general model—the rolling circle—to account for the features of both bacteriophage and bacterial replication. This demands a circular DNA template, and ensures that all its genetic information is preserved by always copying more than one genome's worth of information from the circle. Synthesis begins by opening one strand of the original circular duplex (in phage systems this would be the positive strand) and attaching the newly exposed 5′-phosphate end to the membrane. Chain elongation then begins at the free 3′-hydroxyl end, using the closed strand remaining circular as the template. As synthesis proceeds, it may displace the strand which has been opened from the circular template strand, and this strand will continue to 'peel off' ahead of the growing point for the new strand, as illustrated in Figure 13.23. The 'tail' attached to the membrane could thus contain several genomes' worth of information. Since synthesis on the circular template is 5′⟶3′ in direction, it can be continuous. A new (negative) strand would be synthesized on the tail by discontinuous synthesis, and circular progeny molecules could be constructed by recombinational events between homologous sequences one genome apart in length on the tail.

A difficulty in constructing models for the replication of a circular duplex is that if the strands are to be able to unwind, either the DNA must possess a 'swivel' mechanism of some nature, or a break must be introduced. This is accounted for on the rolling circle model since one parental strand is always open. Gilbert and Dressler suggested that synthesis might be driven either by a torque exerted round the circle, or more likely, by the displacement of the old tail strand by the newly

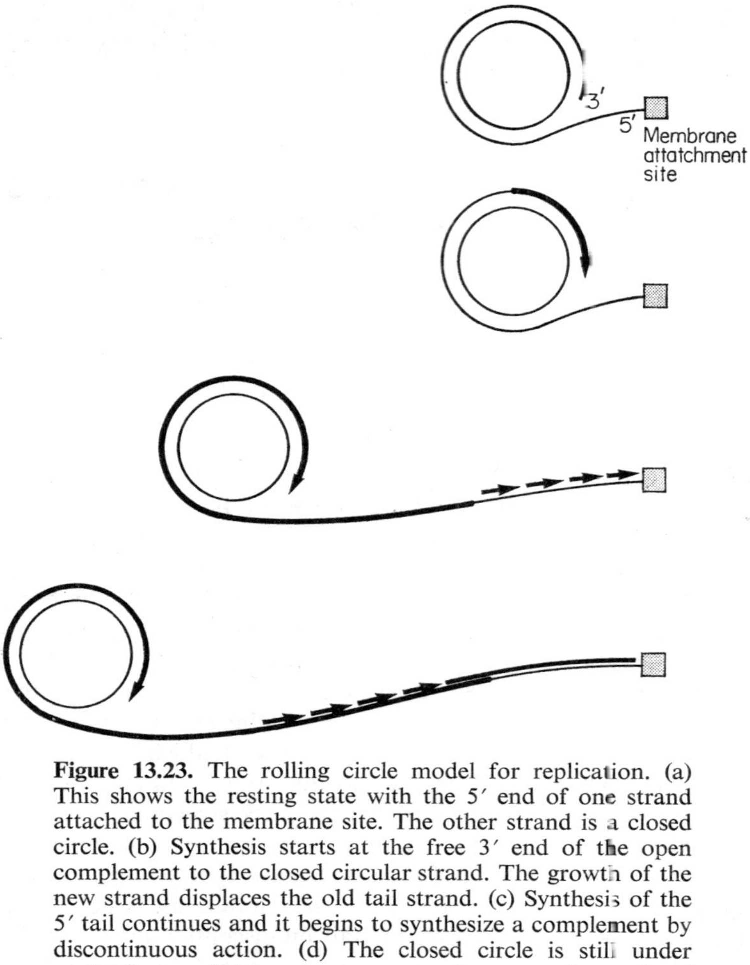

Figure 13.23. The rolling circle model for replication. (a) This shows the resting state with the 5′ end of one strand attached to the membrane site. The other strand is a closed circle. (b) Synthesis starts at the free 3′ end of the open complement to the closed circular strand. The growth of the new strand displaces the old tail strand. (c) Synthesis of the 5′ tail continues and it begins to synthesize a complement by discontinuous action. (d) The closed circle is still under replication and there is now several genomes' worth of genetic information in the tail. (After Gilbert and Dressler, 1968)

growing strand. The former would require integrity of the circle if synthesis is to occur, and any break would halt replication. The latter only requires integrity of the strands ahead of the growing point. If the enzyme meets a nick in the circular template strand, it will become detached from its template, whilst a nick in the tail strand before its complement has been synthesized will result in displacement from the membrane.

This form of the model implies that replication occurs continually, and whilst this may be desirable for phage replication, in bacteria some mechanism must be responsible for terminating DNA synthesis after one genome length has been copied. The model shown in Figure 13.24 incorporates a single control to achieve this end. In its resting state, one strand of the chromosome comprises a closed circle, and the other strand is open with its 5′ end attached to the membrane. Because of the displacement of the tail strand, there is a region of single-stranded DNA on the closed

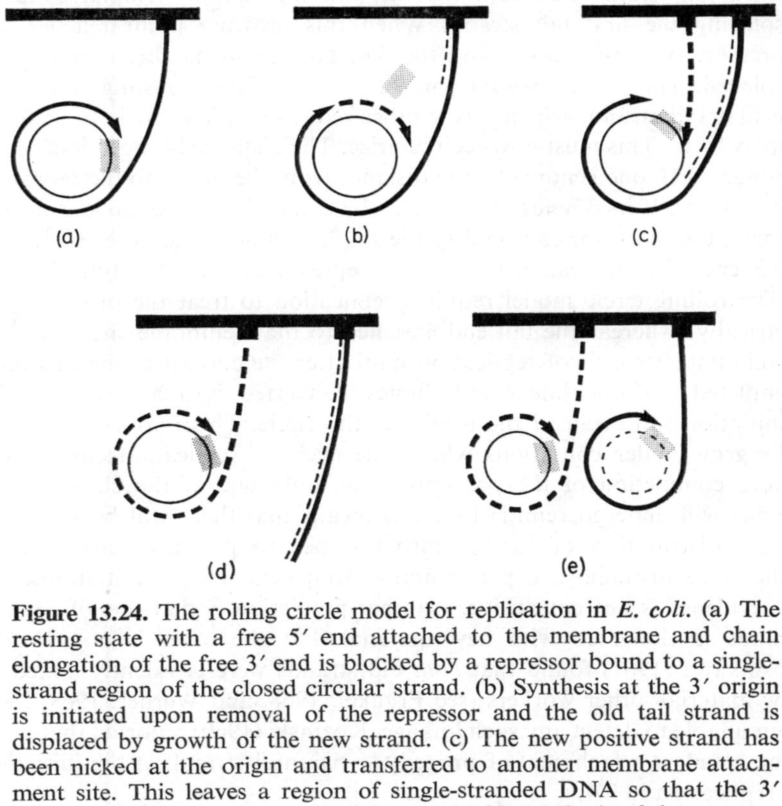

Figure 13.24. The rolling circle model for replication in *E. coli*. (a) The resting state with a free 5′ end attached to the membrane and chain elongation of the free 3′ end is blocked by a repressor bound to a single-strand region of the closed circular strand. (b) Synthesis at the 3′ origin is initiated upon removal of the repressor and the old tail strand is displaced by growth of the new strand. (c) The new positive strand has been nicked at the origin and transferred to another membrane attachment site. This leaves a region of single-stranded DNA so that the 3′ origin can elongate only as far as the block. (d) Synthesis of the two new strands continues until the tail is released completely from its old partner and a new complement has been synthesized for the closed circle. (e) The 3′ ends of the displaced tail are digested, and then associate to form a circle. The 3′ end of the tail strand then grows up to the repressor to leave the chromosome in its resting state. (After Gilbert and Dressler, 1968)

circle, to which a 'repressor' molecule binds; this prevents the progress of the polymerase from the free 3′ end at the origin beyond this point. When the repressor is removed, synthesis continues, displacing the tail as it progresses. Upon the appearance of a new membrane attachment site, a nick is made at the origin on the circular duplex, and the 5′ end thus freed is transferred to it. This displacement again creates a single-stranded region on the circle, and the repressor binds to it; synthesis from the newly freed 3′ end at the origin on the duplex is thus halted when it reaches this point. However, synthesis from the growing point of the second tail strand then continues around the circular template strand, displacing the first tail strand. When this growing point reaches the repressor, synthesis ceases, and the first tail strand has been completely displaced. This leaves one chromosome in its circular resting state, and the first tail strand, which has synthesized a complement, is in a linear duplex form. This must now recircularize. The distance between its 5′ ends is longer than one genome by the distance from the origin to the repressor binding site. If its 3′ ends are digested by exonuclease, the circle can then reform, the repressor can bind to the single-stranded region revealed, and the 3′ end of the tail can grow up to the repressor to yield a resting circle.

The rolling circle model requires replication to treat the origin asymmetrically; whereas the tail end attached to the membrane site is unable to reinitiate a round of replication until after the current round has been completed and the linear tail duplex converted into a resting circle, reinitiation can occur at the origin on the circle. This implies that when cells grow under conditions where one cycle of replication commences before completion of the previous cycle, only one of the chromosome origins will undergo reinitiation; this means that there will be an asymmetric distribution of forked chromosomes to progeny cells. This is difficult to distinguish experimentally from symmetric reinitiation, and the evidence is not conclusive. Another prediction is that one of the old strands is elongated with new material, but the other is not. Whilst Yoshikawa (1967) found that *both* old strands were covalently linked to new material, other workers (see Frankel, 1968, and Werner, 1968) have been unable to detect any such linkage. Kozinski (1969) reported that there is no covalent attachment of progeny to parental strands at the initiation of replication in T4.

Jacob, Ryter and Cuzin (1966) suggested that if each of the two daughter chromosomes resulting from a round of replication is attached to the membrane, the growth of a septum in between their attachment sites could automatically segregate them into different progeny cells, as illustrated in Figure 13.25. Ryter, Hirota and Jacob (1968) observed that in this case the attachment might have an effect upon the pattern of their segregation at cell division. This possibility was analysed experimentally by labelling *E. coli* or *B. subtilis* DNA with H^3-thymidine, growing the cells for several generations in non-radioactive medium, and determining the position of the label in the progeny by autoradiography. If the DNA strands are distributed at random, the pattern of labelling in the second generation must be as shown in Figure 13.26. If there is a definite form of segregation, only one of the possibilities shown in this figure would be found. The results suggested that segregation is random, that is at each cell division each of the old strands has the same probability of distribution into either of the progeny. (It should be noted that Lark (1966) obtained different results which suggested a definite pattern of segregation in the formation of daughter cells.)

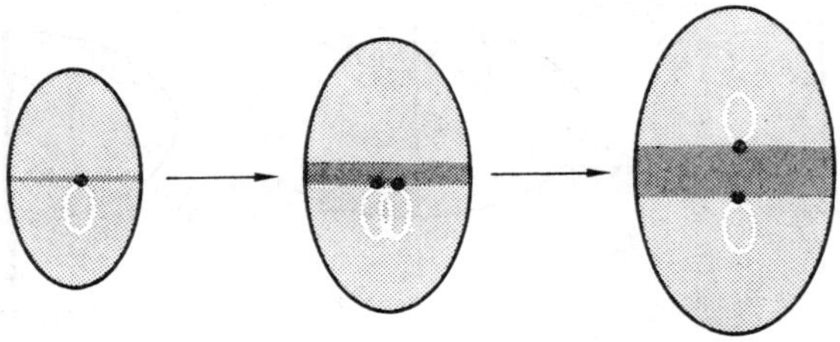

Figure 13.25. The automatic segregation of daughter chromosomes into the two progeny cells by growth of a septum between their membrane attachment sites. (After Jacob, Brenner and Cuzin, 1966)

The control of replication

Donachie (1968) reported that the ratio of the cell mass to the number of chromosome origins tends to be constant at the time when initiation occurs. He was able to derive the relationship

$$\log_e M_d = \log_e M_i/N_i + (C+D)/T$$

where M_d and M_i are the cell masses at division and initiation of replication, and N_i is the number of chromosome origins at initiation. The mass of cells at division was found experimentally, and the mass at initiation can be calculated from this as the mass $C+D = 60$ minutes earlier. The equation above holds experimentally if the ratio M_i/N_i is constant; this

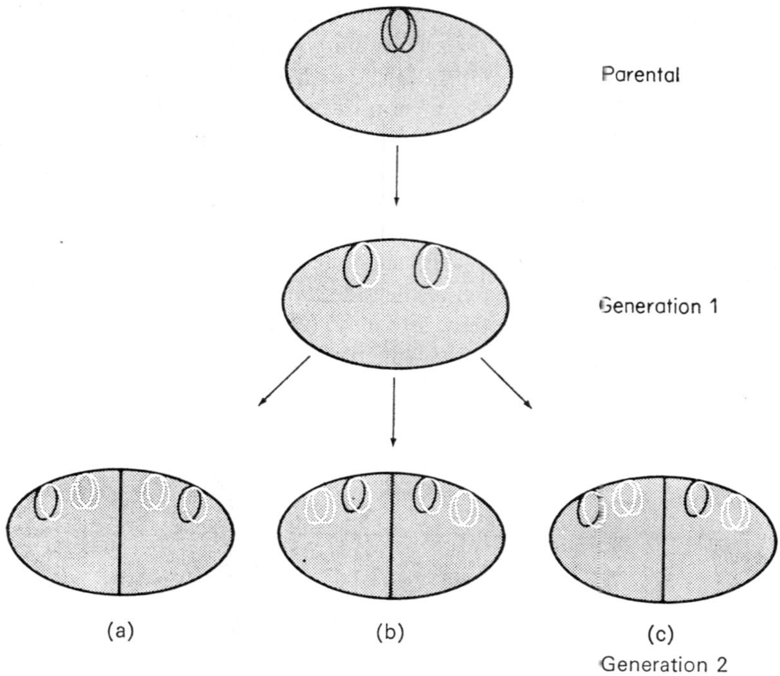

Figure 13.26. Patterns for chromosome segregation at cell division. (a) The parental strands segregate to the far ends of the cell. (b) The parental strands remain near the septum. (c) Both parental strands segregate to the left (or, not illustrated, to the right). If the segregation of chromosomes is random, these should be found with the respective frequencies of 25, 25 and 50%; this agrees with the experimental results. (After Ryter, Hirota and Jacob, 1968)

implies that the cell can titrate its mass in some way and judge when it is ready for initiation. Two models have been proposed to account for this ability. One suggests that an initiator of replication is accumulated during growth, and the other that an inhibitor of initiation is diluted out.

Jacob, Brenner and Cuzin (1963) observed that when chromosomal segments are introduced into a recipient bacterium by conjugation or transformation, they cannot multiply as such, but only do so after they have been inserted by recombination into the host genome. This can be explained by a positive control system in which a chromosome must be switched on if it is to replicate. They suggested that a cytoplasmic initiator protein is responsible for the control of replication by its action on an operator-like region (the replicator). Each genetic element, such as a bacterial chromosome, bacteriophage or episome, would constitute a unit of replication—the replicon—characterized by possession of the structural gene responsible for initiator synthesis and the specific receptor site (replicator) for its action. When an episome or lysogenic bacteriophage is introduced into a bacterial chromosome, it comes under the control of the latter replicon. Any segment of DNA lacking these control functions would be unable to replicate.

Helmstetter and coworkers (1968) and Pierucci (1969) proposed that an initiator of replication accumulates in the cell during growth, and the presence of a critical amount (a unit of initiator) triggers replication. The initiator is synthesized continually, and the same one unit of initiator per chromosome origin is required for initiation at all growth rates. The generation time of cells growing exponentially in any particular medium reflects the time required to synthesize sufficient units of initiator at that growth rate. The constant cell mass per chromosome origin can be accounted for if the synthesis of initiator is related in some way to the mass or volume of the cell, and the theoretical quantity of initiator which this model requires to be synthesized at various growth rates does appear to be proportional to the experimentally determined cell volume.

The results of shift-up experiments are consistent with this interpretation. After an increase in growth rate, the time of the next initiation can be calculated as the time when sufficient initiator will have accumulated. This can be tested experimentally, since a cell division must occur $C+D$ minutes later. For example, suppose cells growing at 40 minutes per doubling are shifted to a growth rate of 20 minutes per doubling. A cell which had initiated a round of replication 10 minutes before the shift-up will have synthesized $10/40=\frac{1}{4}$ of a unit of initiator per origin at the time of the change in its growth conditions. It will synthesize the rest of the unit in $\frac{3}{4}\times 20=15$ minutes, whereas in the absence of the shift-up this event would not have taken place for another 30 minutes. Different cells will be at different stages of their replication cycles when transferred, but the distribution of the ensuing cell divisions with time can be calculated from this model and measured experimentally. The theoretical predictions proved to be in agreement with the experimental results.

This model also predicts that if DNA synthesis is inhibited selectively in a culture in which RNA and protein synthesis remains unaffected, initiator protein should continue to accumulate; thus the subsequent restoration of DNA synthesis should result in a period of accelerated initiation of replication, and consequently cell division. After DNA synthesis has been stopped, cell division should continue normally for D minutes, as cells which have completed their current rounds of replication undertake the consequent division. Division must then stop and initiator accumulate. After the restoration of DNA synthesis, the rate of division should be the same as its previous rate for $C+D$ minutes, but after this there should be a wave of cell divisions due to the multiple initiations resulting from the accumulation of initiator during the block to DNA synthesis. Experimental results obtained by either stopping DNA synthesis completely through the withdrawal of thymidine from the medium, or slowing the rate of movement of a replication point by the substitution of 5BU for thymine, gave results in agreement with these predictions.

Pritchard, Barth and Collins (1969) have proposed an alternative scheme in which an inhibitor of initiation is diluted in concentration as the cell grows. The model postulates that an initiator protein is produced continually and constitutes a constant fraction of the total protein at all growth rates. An inhibitor protein is coded by a gene located adjacent to the chromosome origin (or which is part of it), and is transcribed only at the time of its replication. Each such gene, and thus each chromosome origin, is responsible for the synthesis of a fixed number of molecules of inhibitor protein irrespective of the growth rate. The inhibitor interacts with either the chromosome origin itself, or with the initiator protein (formally, this is the same model), in such a way that a two-fold change in its concentration effects a transition between complete and no inhibition of initiation.

When the cellular concentration of inhibitor falls below a critical value, initiation is triggered. Initiation itself causes the production of more inhibitor protein as the relevant gene is replicated and transcribed, thus raising the level of inhibitor above the critical value. As the cell grows, its volume increases, and since no more inhibitor is synthesized its concentration falls due to dilution. When it passes below the critical value the next round of initiation commences. In an equilibrium situation, the concentration of inhibitor will oscillate over a two-fold range, irrespective of the absolute number of inhibitor molecules produced at each pulse. The frequency of initiation will thus be determined by the dilution rate of the inhibitor, that is the reciprocal of the growth rate. This results in successive acts of initiation occurring after successive doublings of the cell. Since initiation is achieved at a critical value of inhibitor concentration, that is at a constant value of a/V_i ($a=$ the number of inhibitor molecules, $V_i=$ the volume of cell at initiation), and since the number of inhibitor molecules depends upon the number of chromosome origins, V_i/a, the cell volume per number of origins is constant at initiation for all growth rates.

An important feature of this model is that it is self-regulatory; it requires that the *mean* time interval between successive acts of initiation equals the doubling time, although individual cells can show a distribution about this mean. Thus if initiation takes place before the inhibitor concentration has fallen to its critical value, the daughter cells will be smaller than usual, and the pulse of inhibitor synthesized at the initiation event will produce a higher concentration than usual. Thus it will take longer before the next initiation occurs, restoring the average between initiation events.

CHAPTER 14

Repair of DNA

Ultraviolet irradiation damage

Ultraviolet irradiation exerts a mutagenic effect upon bacteria, and depending upon the dose and the bacterium, may be lethal. The irradiation treatment damages DNA (there is no evidence to suggest damage to RNA or protein) by covalently linking adjacent pyrimidine residues on the same strand to form the cyclobutane ring shown in Figure 14.1. This results in a distortion of the structure of the duplex since the two bases are no longer connected only through its phosphodiester backbone. The main product of the reaction is thymine–thymine dimers, but some cytosine–cytosine and cytosine–thymine species are also produced. The dimers exert an inhibitory effect upon both replication and transcription; the kinetics of replication of such DNA suggest that the replicating enzyme is held up when it arrives at a dimer. Recent work has suggested that it may be able to pass the distortion, but that this process may be error-prone and induce mutation. RNA polymerase ceases transcription at ultraviolet damaged sites, and is released from the template, resulting in the production of shorter RNA molecules.

Photoreactivation

A resistance to the damage caused by ultraviolet irradiation may be conferred by the presence of *repair* systems with the ability to remove pyrimidine dimers in either of two ways. The degree of resistance varies widely with the organism, depending upon the efficiency of its particular repair processes. These systems have been studied mostly in bacteria, notably with *Micrococcus radiodurans*, which is especially efficient in removing dimers, and in *E. coli*, where a large number of mutants lacking the ability to overcome ultraviolet damage have been isolated and can be compared with relatively more resistant strains. There is also evidence that there may be similar repair systems in higher organisms.

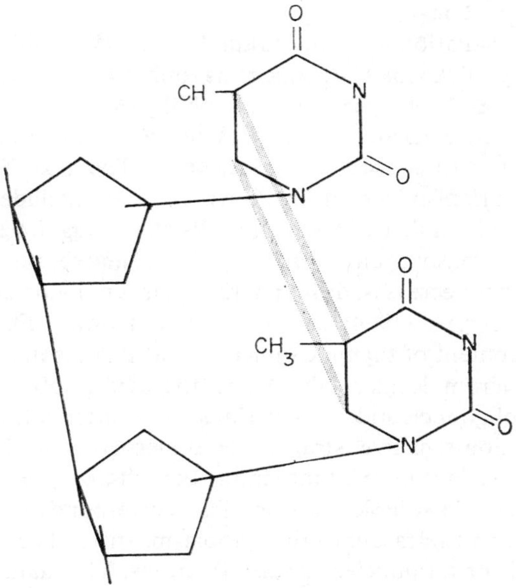

Figure 14.1. Thymine dimer formation through the cyclobutane ring structure

The dimers are stable to acid and enzymic hydrolysis and can be determined quantitatively if the DNA is labelled with H^3-thymidine, so that it is possible to follow their fate after the cell has been irradiated. Both *E. coli* and yeast are *photoreactivable*; after exposure to a source of visible light above 330mμ (most effectively at about 400mμ), the damage to the DNA is reversed. This ability can be lost by mutation in a single cistron; *phr*$^+$ cells have an enzyme which monomerizes the dimers by splitting the bonds joining them, and this is missing in *phr*$^-$ bacteria. Such repair can be exercized by a cell extract *in vitro*; this can restore biological activity by splitting dimers present in transforming DNA from an organism lacking the enzyme (Setlow and Setlow, 1963). The extract from *phr*$^-$ cells lacks this activity. The chemical specificities of the extract systems from *E. coli* and yeast are identical (for review see Hanawalt, 1968). The enzyme binds specifically to ultraviolet-irradiated but not to unirradiated DNA, in the dark, to form a stable complex. After exposure to light, the complex dissociates to release the active enzyme and a repaired DNA which lacks the dimers.

Excison-repair systems

In relatively radiation-resistant strains of *E. coli* the block to replication caused by ultraviolet treatment is only temporary, and after a period in the dark cells recover their ability to synthesize DNA and form colonies. The efficiency of this *dark reactivation* can be measured by the time taken before synthesis is resumed. The possibility that this system might work also by splitting dimers has been excluded by measurements of the total number of dimers in cells at various times after irradiation (Setlow, Swenson, and Carrier, 1963); unlike photoreactivation where the number decreases, during dark repair it remains unaffected.

Setlow and Carrier (1964) and Boyce and Howard-Flanders (1964a) measured the content of thymine dimers in the acid-insoluble cell fraction (containing macromolecular DNA) and the acid-soluble fraction (containing small oligonucleotides) at various times after ultraviolet irradiation. In radiation-sensitive strains the dimers remained in the acid-insoluble fraction, but in resistant strains they disappeared from this and appeared in the acid-soluble fraction. This corresponds to their excision from the DNA and release into the cytoplasm, still linked as dimers, and generally in tri- or tetranucleotide size fractions. This data does not give the size of the actual polynucleotide length removed from the DNA because this may have been further degraded after its excision. However, the relative loss of labelled H^3 dimers to thymine monomers would roughly correspond to the removal of from twenty to fifty nucleotides for each dimer. Although several resistant strains achieve removal of dimers at the same rate, they differ in the efficiencies of the subsequent steps, as is shown by their different rates of reappearance of colony formation ability. Pettijohn and Hanawalt (1964) examined the steps subsequent to excision by replacing the thymine of the culture medium with 5BU after ultraviolet irradiation; the use of thymine-requiring bacteria ensured that they could not produce endogenous thymine but were dependent upon the supply in the medium. This means that any DNA synthesized during recovery is density labelled. The arrangement of the 5BU on the chromosome was determined by thermal denaturation of the DNA, fragmentation of the single strands and sedimentation. Whereas normal replication occurs sequentially from a single growing point, in dark repair the 1 per cent. of thymine sites occupied by 5BU were distributed in random regions over the entire chromosome. Some DNA molecules were repaired in only one strand, and others had been labelled in both. This has been termed *non-conservative* replication, and suggests that after the region around a thymine dimer has been removed, complementary base pairing is used to replace it and restore the original structure of the damaged strand (see Figure 14.2). Although both strands of DNA can undertake repair, it seems likely that at any given time only one strand is repaired at any given point, since otherwise an overlap of the deleted regions on complementary strands would result in chromosome breakage and permanent loss of genetic information. A recent review is by Hanawalt (1968).

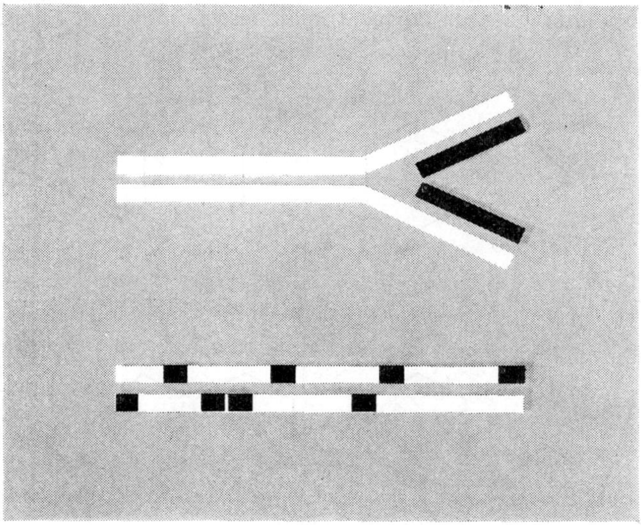

Figure 14.2. Normal sequential replication, in which replication is from one growing point only, and repair replication, when it takes place simultaneously at many locations on the chromosome

Further experiments (Hanawalt, 1967; Hanawalt and coworkers, 1968) were used to test whether the DNA of cells which have undergone non-conservative repair replication is normal and can subsequently undertake cycles of semi-conservative replication. The processes of repair and normal replication were separated by ensuring that cells were unable to perform normal replication whilst the repair activities were proceeding. Cultures of *E. coli* TAU-bar were starved for amino acids for 90 minutes so that the entire population of cells had completed their normal replication cycles. After ultraviolet irradiation, non-conservative repair replication made good the damage. Upon the restoration of a medium containing the auxotrophic requirements, a normal round of semi-conservative replication followed. Thus the DNA resulting from repair replication is biologically normal.

The same repair system which is responsible for acting on ultraviolet damage can also repair other types of damage to the DNA. Hanawalt and Haynes (1965) observed that the relative sensitivities of the ultraviolet-resistant *E. coli* B/r and sensitive B_{s-1} are similar for both ultraviolet inactivation and nitrogen mustard treatment. The latter reagent introduces cross links between adjacent guanines, as shown in Figure 14.3.

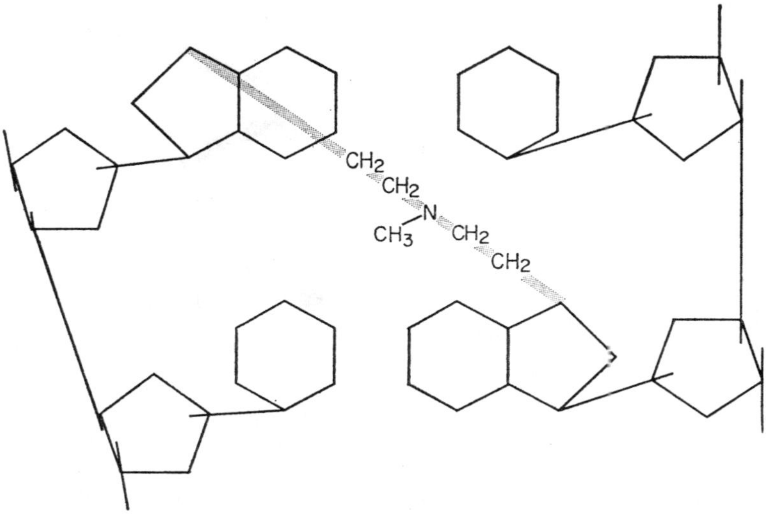

Figure 14.3. The formation of inter-strand cross-links between adjacent guanines by nitrogen mustard

After treatment with nitrogen mustard, 5BU labelling of a thymine-requiring strain shows the same pattern of non-conservative repair replication as after ultraviolet treatment. Further, Boyce and Howard-Flanders (1964b) showed that mitomycin C, which acts by the introduction of cross links between DNA strands, causes damage which is repaired by ultraviolet-resistant but not by ultraviolet-sensitive strains of *E. coli.* This suggests that the excision system does not recognize the exact chemical nature of the damage, but rather, is a general error-correcting mechanism which recognizes some feature of the distortion of the secondary structure of the double helix. Hanawalt (1968) has suggested that the damage recognition step may be formally equivalent to 'threading' DNA through a close fitting 'sleeve' which gauges the closeness of fit to the Watson–Crick structure.

Several enzymic processes are involved in recovery from ultraviolet damage, although the order in which they act is not known. Two models have been proposed (see Hanawalt and Haynes, 1967). The 'cut and patch' model suggests that an endonuclease excises a short three to four nucleotide segment containing the dimer. The resulting gap is then enlarged by exonuclease attack on the exposed 3′-hydroxyl terminus, and a polymerase undertakes repair replication by inserting complementary nucleotides. Finally, polynucleotide ligase connects the last newly inserted nucleotide to the exposed 5′-phosphate terminus of the original strand.

The 'patch and cut' scheme suggests that repair is initiated by a single-strand break introduced near the dimer. Repair replication commences immediately, and the defective strand 'peels back' simultaneously with the insertion of complementary nucleotides. Repair is terminated by a second nick, which releases the old damaged segment. Finally, the ligase rejoins the newly synthesized strand segment to the old strand by covalent linkage. This model has the advantage that it does not involve the introduction of a long and possibly vulnerable stretch of single-stranded DNA before repair replication begins, and also that it would be possible for a single enzyme complex to catalyse all the processes as it moves along the DNA. Of course, there are other possible models between these two extremes.

Some of the enzymes involved in the steps of repair replication have been identified and their properties studied *in vitro*. The bacterium *Micrococcus lysodeikticus* has been used as a source for repair nucleases because it has only low levels of other nuclease activities, and this assists identification of the enzymes involved in repair. A crude extract from this bacterium degrades ultraviolet-irradiated DNA much more rapidly than unirradiated DNA (Takagi and coworkers, 1968; Grossman and coworkers, 1968). Two enzyme activities are involved, an endonuclease and an exonuclease, and these have been purified and their catalytic properties studied in detail (Kaplan, Kushner and Grossman, 1969). The first step in repair is the introduction of a single-strand break adjacent to the dimer—this is probably induced by a recognition of the distorted structure around the dimer (for review see Setlow, 1968). The ultraviolet endonuclease from *M. lysodeikticus* is completely specific for ultraviolet-irradiated duplex DNA, and there is a linear relationship between the dose of irradiation and the number of phosphodiester bonds broken. Its mode of action was identified by comparing the results of neutral and

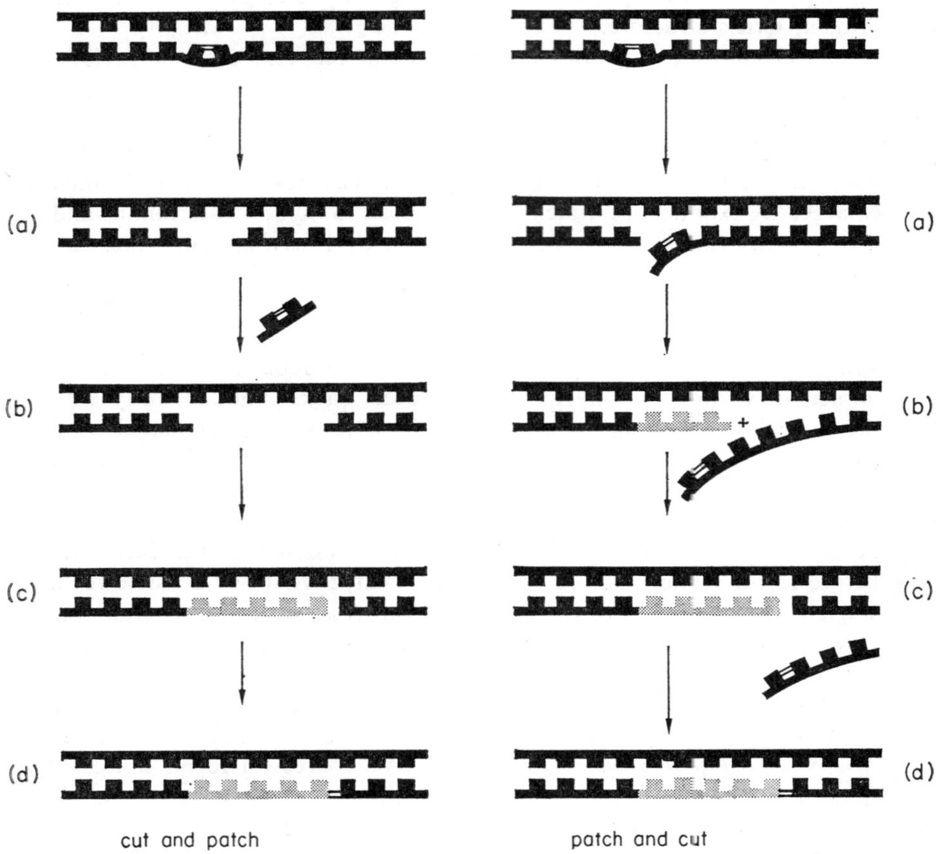

Figure 14.4. Repair of ultraviolet damage to DNA. Cut and patch: (a) An endonuclease excises the thymine dimer from the DNA. (b) Exonuclease attack extends the region of single-strand degradation. (c) Repair replication fills the gap. (d) The newly synthesized segment is joined covalently to the old strand by ligase action. Patch and cut: (a) A nick is made adjacent to the dimer. (b) Repair replication commences, and as it proceeds the damaged strand is displaced. (c) A second nick releases the segment bearing the thymine dimer and terminates repair replication. (d) Ligase action joins the new to the old segments. (Note: the polynucleotide length excised is rather larger than illustrated, usually twenty to fifty nucleotides)

alkaline sucrose density gradient centrifugations. The former showed no difference in the sedimentation characteristics of the DNA, but under the latter conditions, in which the DNA is denatured to single strands, a new peak appeared, representing molecules of about one-tenth their previous length. This indicates that only single-strand breaks are made, and simultaneous scission of both strands does not occur.

Although it seems probable, there is as yet no direct evidence that the degradation of cellular DNA which is observed after ultraviolet irradiation is due to a specific enlargement of the region around the site of the thymine dimers (for review see Hanawalt, 1968). The second enzymes activity found in *M. lysodeikticus* represents an exonuclease which undertakes a limited degradation of DNA *in vitro* from the position of a single-strand break introduced near a thymine dimer by the first enzyme. The ultraviolet exonuclease only attacks denatured DNA (either native or irradiated), but does not attack native or even irradiated DNA *per se*. However, irradiated DNA which has been subjected to the ultraviolet endonuclease is also susceptible as a substrate. The combined action of the enzymes is to excise five nucleotides for each single-strand break. Since considerably more degradation takes place *in vivo*, the excess may be due to other nucleases in the cell and is not necessarily essential for ultraviolet repair.

In *E. coli*, none of the enzymes which can synthesize or degrade DNA *in vitro* has as yet been implicated in repair *in vivo*, although there are indications that different enzymes are responsible for normal and repair replication. Couch and Hanawalt (1967) examined the ability to perform repair replication of temperature-sensitive mutants of *E. coli* which cannot undertake normal semi-conservative replication above 42°C; the extent of repair replication was not affected by temperature, suggesting that different enzyme systems must be involved. Another factor in support of this is the finding reported by Hanawalt and coworkers (1968) that the normal and repair systems exhibit different relative affinities for thymine and 5BU; the repair system shows a more stringent demand for thymine rather than its analogue. However, in lieu of a demonstration that the Kornberg DNA polymerase does undertake normal replication *in vivo*, it is tempting to speculate that this enzyme may be involved in repair, with some other enzyme as yet unidentified responsible for normal replication. Indeed, it is easier to understand the many catalytic activities exhibited *in vitro* by DNA polymerase (see pages 305–311) in terms of an enzyme involved in repair rather than normal replication. And certainly, this enzyme can perform *in vitro* the insertion of nucleotides to replace an excised region of DNA; duplex DNA subjected to the enzyme exonuclease III suffers a sequential hydrolysis of one strand in the 3′ to 5′ direction, and can be restored to its double-stranded form by the action of DNA polymerase.

Indeed, Kornberg has recently found (see Kelly and coworkers, 1969) that the *E. coli* DNA polymerase can excise thymine dimers from DNA. This would support the idea that the role of this enzyme *in vivo* may be concerned with repair. The enzyme appears to catalyse both the excision and replacement steps of repair, and this has allowed formulation of a simplified model for the repair of irradiated DNA. After an endonuclease recognizes a distorted region, it nicks the DNA near the 5′ side of the dimer, and the polymerase then binds at the nick. This model requires that the break should introduce a free 3′-OH terminus; if this is so, the polymerase can extend the chain from the nick by adding nucleotides to this terminus. Because of its inability to H-bond to the complementary strand, the thymine dimer will cause a region of denaturation; so that as the polymerase engages in chain extension, it is easy for it to displace the damaged region. When it has reached the far side of the distortion, the enzyme excises the damaged region as an oligonucleotide. It is then only necessary for the ligase to seal the newly synthesized segment to the previously existing strand. This is essentially the patch and cut scheme of Figure 14.4, with the DNA polymerase operating both steps (b) and (c). It is perhaps worth noting that this behaviour implies that the 5′-3′ nuclease activity of the enzyme cannot be properly characterized as exclusively either endonucleolytic or exonucleolytic.

Repair systems exercising similar functions to those found in bacteria are also present in a variety of other organisms. Davies (1965, 1967) reported that thymine dimers induced in the green alga *Chlamydomonas reinhardii* are subject to either photoreactivation or dark repair. Mutants with enhanced sensitivity to radiation have been isolated in the yeast *Saccharomyces cerevisiae* and in the fungus *Ustilago maydis*, and these fall into phenotypic categories similar to those of bacterial mutants (Snow, 1967; Nakai and Matsumoto, 1967; Holliday, 1965). There is also evidence that such repair systems may exist in mammalian cells; non-conservative repair replication has been demonstrated after ultraviolet irradiation or nitrogen mustard treatment of Hela cells in tissue culture (Cleaver and Painter, 1968; Roberts, Crathorn and Brent, 1968).

The genetics of radiation resistance

A number of genetic loci are concerned in determining the degree of sensitivity of *E. coli* to radiation, and mutations have been identified and mapped in both the K12 and B strains (summarized by Witkin, 1967, and Hanawalt, 1968). The various mutants show a range of phenotypic effects upon the characteristics of sensitivity to ultraviolet irradiation, sensitivity to X-rays, genetic recombination, and the ability to undertake cell division after irradiation. They fall into the following five classes:

(i) *hcr*$^-$ mutants show a ten- to twenty-fold increase in sensitivity to both the lethal and the mutagenic effects of ultraviolet irradiation. They lack the ability to repair damage to either the DNA of their own genome, or to the DNA of an infecting phage; there is a high correlation between the level of survival of colony-forming ability in ultraviolet-irradiated bacterial cells and the level of plaque-forming ability when the cells are used as hosts for ultraviolet-irradiated phages. (Many bacteriophages depend upon their bacterial host for repair systems, and *hcr* refers to this bacterial ability, host cell reactivation.) Howard-Flanders, Boyce and Theriot (1966) reported that this phenotype can be achieved through mutation in any one of the three loci in *E. coli* K12; impairment of any of the cistrons *uvr*A, *uvr*B or *uvr*C (which are not clustered together) results in inability to remove pyrimidine dimers from DNA.

(ii) A second class of mutants is ultraviolet sensitive in that they cannot undertake the repair of damage to their own genomes, but they retain the ability of host cell reactivation. These have been termed *uvr*$^-$. (The nomenclature for radiation-sensitive mutants is confused; the three *uvr*A, B and C mutants described in the previous paragraph do *not* fall into the *uvr*$^-$ phenotypic category, but are *hcr*$^-$). Neither the *hcr*$^-$ nor the *uvr*$^-$ classes of mutant show an enhanced sensitivity to X-irradiation.

(iii) Increased sensitivity to X-rays is conferred by mutation at the *exr* locus; *exr*$^-$ mutants are also sensitive to ultraviolet irradiation.

(iv) Mutation in the *rec* system causes an increased sensitivity to both u.v.- and X-irradiation, and such cells are also deficient in genetic recombination. Several such loci have been identified. The *exr* and *rec* mutations are discussed in detail in the next section.

(v) Another class of mutant affects the ability of cells to divide after u.v.-irradiation or X-ray treatment; these have been termed *lon*$^-$ (in *E. coli* K12) or *fil*$^-$ (in *E. coli* B). The cell division mechanisms of *lon*$^+$ and *fil*$^+$ strains are vulnerable to irradiation, and their cells cannot produce septa and divide if their DNA contains unrepaired radiation damage; however, the mutants can do so. The mutation affects only the division ability, and not the capacity to excise dimers or overcome their effects.

The actions of some, and possibly all, of these mutants are additive; for example, the highly sensitive *E. coli* B_{s-1} is a double mutant which is both *hcr*$^-$ and *exr*$^-$, and as such is about fifty to a hundred times more sensitive than its parent *E. coli* B/r. The additive effect suggests that the various classes of mutant represent impairment of different repair systems.

The induction of mutation by ultraviolet irradiation

The presence of unexcised pyrimidine dimers exerts an inhibitory effect upon the ability of DNA to act as a template for further DNA or RNA synthesis *in vitro* or *in vivo*, and until recently it was thought that this inhibition could account for the lethal effect of ultraviolet irradiation, but that some other form of damage must be responsible for its mutagenic effects. However, the product which *is* synthesized *in vitro* is mutant through possession of guanines in place of adenines, and in particular there is a lack of the AA sequences complementary to thymine dimers (see Setlow, 1968). Bridges and Munson (1968) found that an *E. coli* strain lacking the ability to excise dimers was nevertheless viable and could undertake replication and cell division in the presence of a moderate number of dimers. These cells showed a low level of mutation in each replication cycle. If this is caused by the unexcised dimers, mutation should be prevented by subjecting the cells to photoreactivation. Following the photoreversibility of mutation showed that the dimers persist for about four generations, and have a low probability of giving rise to mutation in each replication cycle. This suggests that dimers *can* be replicated, but that this process is error-prone and induces mutation.

Rupp and Howard-Flanders (1968) and Howard-Flanders and co-workers (1968) investigated this problem by studying replication after UV-irradiation of a *uvr*A$^-$ mutant unable to excise thymine dimers. They found that replication *in vivo* continued at a normal rate until a dimer was reached, when there was a delay of about 10 seconds. The DNA synthesized was labelled by a ten-minute pulse of H^3-thymidine and subjected to a sedimentation analysis. Under alkaline conditions, when DNA is converted to single-stranded species, it sedimented more slowly than usual, corresponding to a reduction in its molecular weight. In fact, the molecular weight distribution agreed with the calculated distribution of the molecular weights of the DNA between randomly spaced thymine dimers. Further incubation of the cells before extraction of the DNA showed that the slowly sedimenting form was converted to the usual faster sedimenting species after some 15–40 minutes. The simplest interpretation of these results is that DNA polymerase leaves a gap in the newly synthesized complementary strand opposite a parental dimer, and this gap is later filled by some repair system other than the excision–repair system.

Two other systems, the *exr* and the *rec* have been attributed with the role of providing repair mechanisms for the production of viable DNA from DNA containing unexcised ultraviolet damage. The *exr* system is concerned with the repair of X-irradiation damage, which takes two forms. One is the creation of single-strand breaks in the DNA, and these are repaired by a fast process, probably just a simple ligase action. In addition, however, a considerable degradation of DNA occurs in response to the irradiation; this must be due to a system distinct from that involved in pyrimidine dimer excision since mutants which are unable to excise their dimers are still capable of degrading their genomes in response to X-irradiation (Billen and coworkers, 1967; Emmerson and Howard-Flanders, 1964). This degradation is fairly considerable; Emmerson and Howard-Flanders suggested that some 6×10^3 nucleotides are excised from the genome for each irradiation event. In *exr*$^-$ mutants there is increased DNA degradation, and a depression of the incorporation of nucleotides into DNA after either u.v.- or X-irradiation.

A connection between the *exr* system and the repair of ultraviolet-induced damage is indicated by the finding that although mutants to ultraviolet sensitivity do not seem to be particularly sensitive to X-rays, all the X-ray sensitive mutants so far isolated are also ultraviolet-sensitive. Futher, as well as causing an increased fatality rate in response to ultraviolet-irradiation, the *exr* locus has a profound effect upon the induction of mutations by ultraviolet. If *exr*$^-$ strains are given doses of ultraviolet which are mutagenic in *exr*$^+$ strains, ultraviolet-induced mutations are not found amongst the survivors. Witkin (1967) has suggested that the *exr* system provides a repair mechanism which is error-prone and this tends to induce mutation; *exr*$^-$ mutants lack the repair system and so there are fewer survivors after irradiation, but the bacteria that do survive must have overcome their damage by some other, more accurate system and there are no ultraviolet-induced mutations. An obvious possibility for the role of the *exr* system is to fill gaps, such as those left by the replication of DNA containing unexcised thymine dimers, by some error-prone mechanism, possibly the random insertion of nucleotides. In this case, the mutation from *exr*$^+$ to *exr*$^-$ should interfere with the mutagenic effect of other agents which act by single-strand gaps. Bridges,

Law and Munson (1969) used both ionizing gamma radiation and thymine deprivation, which induce single-strand gaps, to test this. They found that the mutagenic effect of either agent was severely curtailed by loss of the *exr*$^+$ function.

Another mechanism which allows the cell to survive even when its DNA contains unexcised thymine dimers is provided by the *rec* system. Rupp and Howard-Flanders (1968) observed that the *uvr*A$^-$ (excision-defective) mutant can survive with as many as fifty unexcised thymine dimers in its genome, but the *uvr*A$^-$ *rec*A$^-$ double mutant can only survive one to two such species. Ganesan and Smith (1968) observed that the *rec* system cannot exert its effect until after the irradiated chromosome has replicated; this implies that it may be concerned with the repair of gaps in DNA, and this would also explain the increased sensitivity of *rec*$^-$ mutants to X-irradiation (which produces both single- and double-strand breaks in DNA). *Rec*$^-$ mutants are also deficient in genetic recombination (see next chapter), and Howard-Flanders and coworkers (1968) suggested that this system may operate by the retrieval of genetic information between sister duplexes after replication. There are several possibilities for such action, as shown in Figure 14.5. Pyrimidine dimers might induce two-strand crossing-over so that the radiation damage all accumulates in one duplex, or alternatively, single sister strand exchange could take place, either extensively (in which case both conserved and semi-conserved stretches of duplex result), or for only short stretches (in which case the chromosome remains largely semi-conserved).

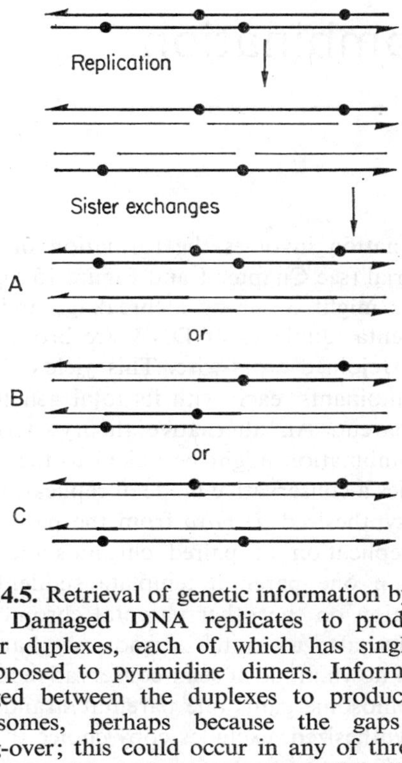

Figure 14.5. Retrieval of genetic information by the *rec* system. Damaged DNA replicates to produce two daughter duplexes, each of which has single-strand gaps opposed to pyrimidine dimers. Information is exchanged between the duplexes to produce viable chromosomes, perhaps because the gaps induce crossing-over; this could occur in any of three ways: (A) two-strand crossing-over causes the dimers to accumulate in one duplex and the other is then normal; (B) single-strand exchanges take place between opposite strands, which causes the formation of both conserved and semi-conserved regions of each duplex; (C) the dimers induce only short lengths of single strand exchange, so that the chromosomes remain predominantly semi-conserved. (After Howard-Flanders and coworkers, 1968)

CHAPTER 15

Recombination

Hybrid DNA

Genetic recombination involves the formation of new arrangements of the genetic material (see Chapter 1 and Figure 15.1), and early theories postulated that this might occur by a 'breakage and reunion' event in which the two parental duplexes of DNA are broken at corresponding positions and then rejoined cross-wise. This yields chromosomes which are reciprocal recombinants, each with its total genetic material derived some from each parent. An alternative theory—termed *copy-choice*—suggested that recombination might be linked to the replication process. However, this implies a conservative mode of replication in which progeny chromosomes are synthesized *de novo* from the parental species; in this case, during the replication of paired chromosomes, synthesis of the daughter genome on one parental template suddenly switches to the corresponding position on the other parental chromosome. This causes the reciprocal replica also to switch strands, so that reciprocal recombinants result. The known mode of semi-conservative replication, in which each daughter chromosome gains one parental strand and only its complement is newly synthesized, excludes copy-choice theories as originally formulated, although more recent modifications overcome this difficulty. However, these neccessitate a series of complex, and rather unlikely, assumptions. Further, an important distinction between the two models (see Figure 15.1) is that according to breakage and reunion theory, recombinant chromosomes possess physical material from both parental chromosomes, whilst copy-choice yields recombinant genomes synthesized from new material and inheriting only genetic information from the parental structures. The former expectation has been confirmed experimentally.

Experimental work has proceeded largely in two systems. In eucaryotic organisms, chromosomes comprise complex nucleoprotein structures which are not at all well understood, precluding a direct approach. Recombination takes place at the four-strand stage of meiosis, and

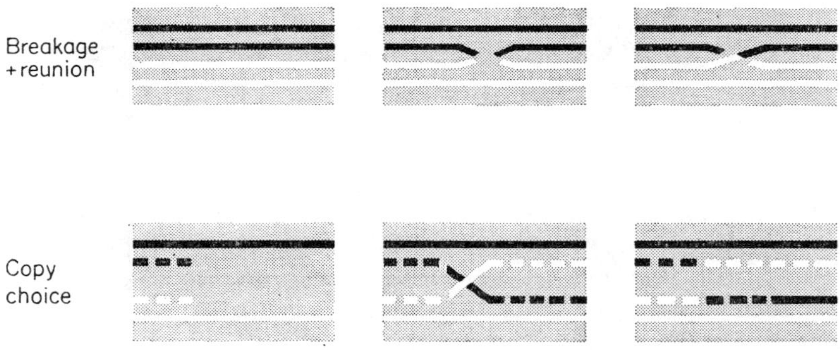

Figure 15.1. Models for recombination. Breakage and reunion: two of the four strands suffer breakage at the same place and rejoin cross-wise. Each recombinant gains physical material from each parent. Copy-choice: during synthesis of new chromosomes, the new strands switch templates. Each of the resultant recombinants has genetic information from each parent, but has physical material derived from only one parent comprising one chromosome, and newly synthesized material comprising the other

genetic analysis has been used in fungal systems where the four chromatids of one meiosis segregate into a linear order in a single spore. This has permitted a detailed analysis of the process of crossing-over between alleles, where a recombination event has occurred in a single specified cistron. A more direct approach has been possible with bacterial and bacteriophage systems. Studies with bacterial mutants defective in recombination have suggested that the enzyme systems involved may have features in common with those responsible for the correction of damage to DNA, and work with phages has enabled isolation of intermediate species in the recombination process.

Since recombinants rarely contain additions or deletions to the genetic material, the recombination event must occur between precisely the same nucleotides in each parental DNA molecule. Homologous pairing requires that identical parts of the genome are brought into apposition at synapsis, although Thomas (1966) has observed that the physical nature of DNA crystals suggests that the exterior of any duplex segment is structurally equivalent to any other. In higher organisms, where the DNA is associated with protein, synapsis is a function of chromosomes rather than DNA molecules, but phenomena such as lysogeny demonstrate that DNA molecules can have a specific affinity for each other, although the molecular basis for this is not yet clear. However, there is a precise mechanism for the association of single polynucleotide chains by base pairing between complementary sequences. Recent models for recombination have suggested that it occurs in stages, initially only one strand being broken in each parental duplex. Reunion does not take place by an end-to-end association of broken duplexes, but by base pairing between complementary sequences of the different parental single strands. This creates recombinant duplexes which have one chain from each parent over the region around the crossover event—this has been termed *hybrid DNA*. Nevertheless, although this accounts for the precision of recombination at the molecular level once two homologous regions of DNA are near each other, it is still necessary to account for the pairing of these regions in the first instance.

Two models have been proposed, differing in the pattern of the initial breakages required. Holliday (1964) envisaged these as occurring in strands of the same polarity. After breakage at defined points in each chromosome, the single strands separate from their partners along the chromosomes in the same direction. Cross-wise annealing forms the hybrid DNA. If recombinants are to result, a second break must be introduced, this time in the hitherto unbroken strands; if breakage takes place again in the strands which have broken previously, the original chromosomes can separate without recombination, although each will contain a region of hybrid DNA. Figure 15.2 illustrates this model.

Previously, Whitehouse (1963) had proposed that the initial breakages would occur at the same site on each of the parental DNA duplexes, but in chains of opposite polarity, as illustrated in Figure 15.3. The broken ends then fall away in the same direction from their partner strands and new strand segments are synthesized along the unbroken strands to replace them. These newly synthesized strands then fall away in turn to undergo complementary base pairing with the corresponding parental stretches which have broken away on the other DNA duplex. Again, this forms a sequence of hybrid DNA in which each of the two strands of the duplex carries genetic information from one of the parents. The final step is to delete segments of the hitherto unbroken chains so that there has been no *net* synthesis of DNA, and to join the ends of the hybrid DNA segments covalently to the new parental ends exposed by this degradation. The single parental strands can fall away from the initial break in either direction, but if they fall away in both there is no recombination, although a stretch of hybrid DNA results.

Studies on recombination between alleles support the concept that recombination takes place through the formation of hybrid DNA. In the *Ascomycetes* fungi, the products of a single meiosis are held together in a single large cell (the ascus), and the four haploid nuclei produced by the

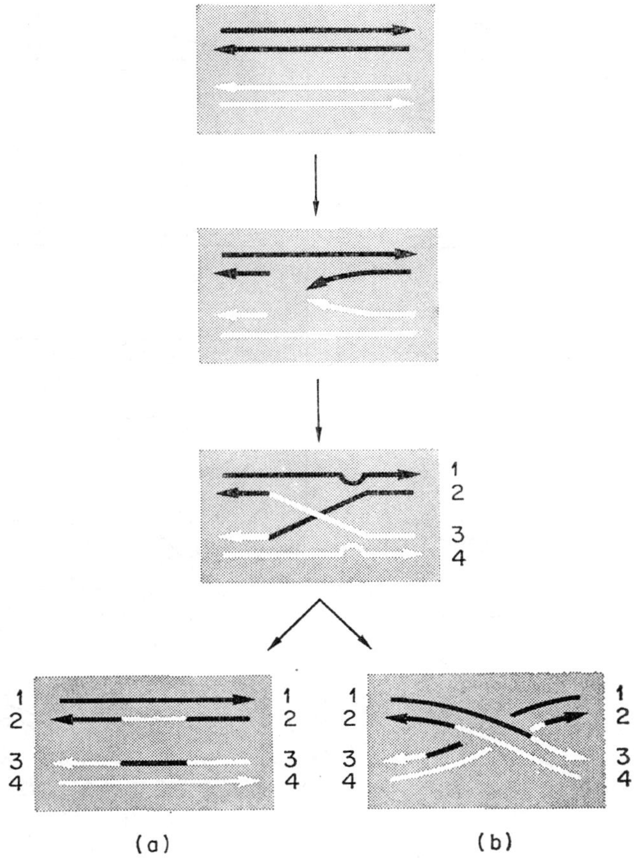

Figure 15.2. The formation of hybrid DNA by breakage in strands of the same polarity. After breakage, the strands pair with their complement in the other parent and reunite cross-wise. If the second breakage occurs in the same strands as involved in the first (that is in strands 2 and 3 as illustrated in (a)) there is no recombination, but each chromosome bears a region of hybrid DNA. If breakage is in the other strands (1 and 4) as illustrated in (b), recombinants result. (After Holliday, 1964)

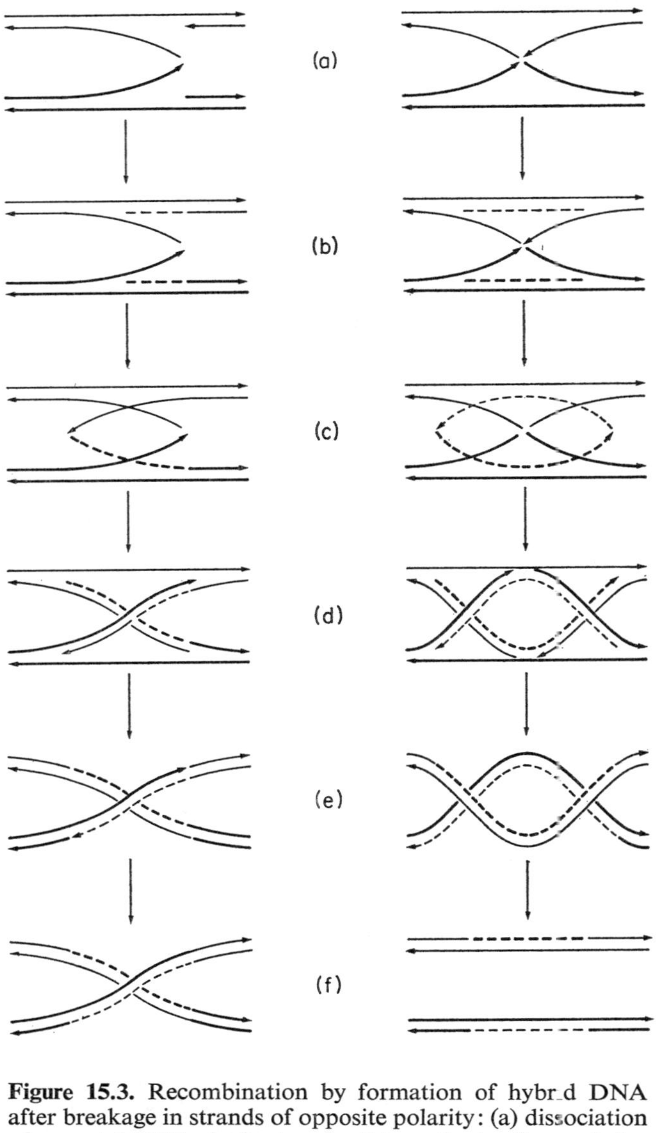

Figure 15.3. Recombination by formation of hybrid DNA after breakage in strands of opposite polarity: (a) dissociation of parental DNA single strands; (b) synthesis of new chains to replace dissociated regions; (c) dissociation of newly synthesized chains; (d) formation of hybrid DNA by complementary base pairing between newly synthesized chains of one parent and old strands of the other; (e) degradation of

meiosis are arranged in a linear order. A mitosis occurs after completion of the meiosis, and this results in production of a spore with eight haploid nuclei in linear order. Effectively, each of the chromosomes in these represents the genetic character of one of the eight single strands of the four chromosomes resulting from the meiosis. Thus if the chromosomes resulting from the meiosis are genetically homogeneous, that is both the strands of each chromosome carry the same information, the characters should be arranged in pairs in the final ascus, as shown in Figure 15.4.

A system that has been widely used for such analysis is that of spore color mutants in *Sordaria*; this has the advantage that the conflicting parental characters can be observed directly. If meiosis is normal, each ascus should have four haploid nuclei with the colour of one parent and four with the colour of the other. However, on occasion there is an abnormal segregation in which there is two of one type to six of the other. This has been termed *gene conversion.* A further type of abnormality has been found with 5 : 3 or 3 : 5 ratios of mutant to wild-type colour. This implies that the DNA resulting from the meiosis must be genetically heterozygous, that is the two strands of one duplex carry different information. This has been termed *post-meiotic segregation* since the characters must then segregate at the mitosis succeeding the meiosis. As well as the asci with abnormal proportions of the two segregating characters, there are classes with the normal 4 : 4 ratio, but with an abnormal sequence in that the spores are not in pairs. Figure 15.5 shows the unusual arrangements found in *Sordari fimicola* for segregation after a cross between black-spored and grey-spored strains.

The occurrence of aberrant segregation at the spore colour locus shows a high correlation with the occurrence of crossing-over in the immediate neighbourhood, suggesting that the unusual ratios found reflect the processes involved in recombination (an extensive review is by Whitehouse, 1969). When hybrid DNA is formed, the two chromosomes resulting each have one strand from each parent over the hybrid region. If one parent is mutant for a site located within this segment of DNA, the

---

unpaired parental single strands; (f) joining of breaks by covalently linking newly synthesized segments to old strand regions. Left: recombinants are formed from dissociation of parental strands in only one direction from the break. Right: the formation of hybrid DNA without recombination by dissociation in both directions from the initial break. (After Whitehouse and Hastings, 1965)

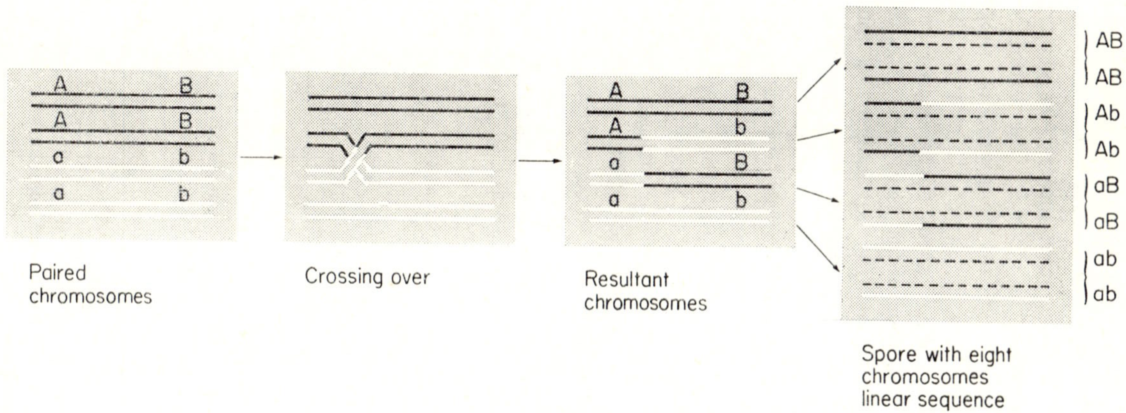

Figure 15.4. Spore formation in *Ascomycetes*. The eight chromosomes in the spore are arranged in linear sequence, and each represents the genetic characteristics of one of the eight single DNA strands in the four chromosomes produced by meiosis. Thus the chromosomes of the spore comprise four pairs. Recombination has taken place between the genetic markers A/a and B/b; the A/a markers lie in the parental order AAAAaaaa, and the B/b markers in the order BBbbBBbb

chromosome will be heterozygous for it and may have mispaired bases. Correcting enzymes, similar to those which recognize and repair ultraviolet damage, could act upon the mispaired bases to restore normal Watson–Crick hydrogen bonding between the two complementary DNA strands; after excision of one of the strand segments, repair replication would insert bases correctly hydrogen-bonded to the bases on the unexcised strand. Since hybrid DNA tends to occur with crossovers there is a correlation between crossing-over and the correction of mispairing.

If the process of excision and repair fails to take place, the abnormal situation of a normal 4 : 4 ratio, but with unpaired sequences in the ascus, results from the separation of the mispaired strands at the replication prior to the post-meiotic mitosis. If only one of the two mispaired chromosomes is corrected, a 5 : 3 or 3 : 5 ratio results, depending upon whether the correction is to wild type by excision of a mutant strand, or vice versa. If both chromosomes are corrected in the same direction, that is either both are converted to the wild type or both to the mutant sequence, a 6 : 2 or 2 : 6 ratio is given. As well as these non-reciprocal events, reciprocal recombination could also result if both chromosomes are corrected, but in opposite directions. Figure 15.5 illustrates these various possibilities.

Whitehouse (1966) has modified his original theory to account for the finding that there is a polarity in the occurrence of gene conversion events; the frequency of conversion has been found to fall in a gradient along the gene. He suggested that recombination can be initiated only at one end of a gene, and hybrid DNA then extends from this site for a variable distance along the chromosome, and on occasion may continue for a short distance into the adjacent gene. Polarity results because the nearer a mutant site is to the end of the gene where recombination is initiated, the greater the likelihood that it will fall into a stretch of hybrid DNA and hence be available for conversion.

Recombination in bacteriophages

Wild type (r^+) T-even phage are characterized by small plaques with a turbid halo, whereas r^- mutants exhibit larger and clearer plaques. Hershey and Chase (1951) discovered that some 2 per cent. of the plaques arising from the individual particles produced by an $r^+ \times r^-$ cross have the mottled appearance which is normally obtained after mixed infection of a bacterium by both species. When the particles from these mottled plaques were used for infection, they gave about equal numbers of pure

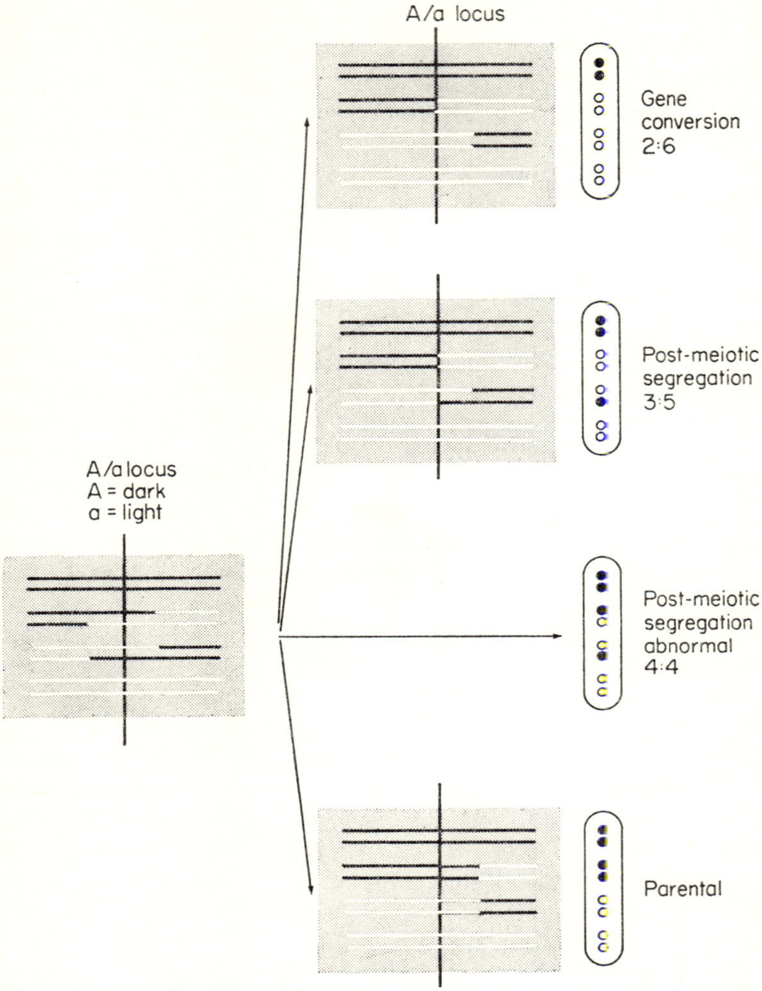

Figure 15.5. Segregation of spore colour mutants in *Sordaria*. Gene conversion: both of the regions of hybrid DNA are converted to the mutant characteristic to give a 2 : 6 ratio. If both converted to wild type, a 6 : 2 would result. Post-meiotic segregation: if only one of the hybrid regions is corrected a 3 : 5 or 5 : 3 ratio results; the diagram shows conversion of one wild strand to mutant. If no correction takes place, there are equal numbers of both mutant and wild type characters, but their arrangement is abnormal. Parental: one duplex is converted so that both strands are mutant, and the other is converted so that both strands are wild type

r^+ and pure r^- plaques, and the same 2 per cent. of mottled plaques. Since the mottled plaques arose as a result of infection by only a single phage particle, the implication of this result is that this particle must have been heterozygous and carried both the r^+ and r^- alleles. Such heterozygotes have since been found to arise with equal frequency for any genetic marker used, and the same characteristic 2 per cent. frequency is found under a range of conditions of infection. The average length of the heterozygous region is quite short and less than the size of the *r*II cistrons. The most obvious interpretation is that such phages are intermediates in the process of reproduction. The replication and recombination of phages takes place in a single pool from which genomes are randomly withdrawn for insertion into mature particles; the constant frequency of the heterozygotes can be accounted for if they represent random withdrawal of an intermediate which is continually generated in the pool, but is lost at the same rate due to maturation. In fact, phage heterozygosis arises from two causes. One corresponds to the production of a region of hybrid DNA in which the two strands of the duplex are derived from different parents; this is an intermediate and would disappear by segregation at the next round of replication if it had not been trapped in a phage head by its withdrawal from the pool. The second class is a consequence of the phenomenon of terminal redundancy (Chapter 3); phages with duplicated regions at the two ends of their chromosomes may carry two different alleles if these ends come from different parents. A full discussion is given by Hayes (1968b).

Experiments using mixed infection by two phages have demonstrated that recombination must occur by a breakage and reunion process rather than by a copy-choice mechanism. Meselson and Weigle (1961) showed that recombination between phage λ can occur independently of replication. Genetic crosses were performed between two phage λ species distinguished by different mutant characteristics. Both the infecting phages were grown on bacteria cultured in heavy medium so that the parental DNA was extensively labelled with C^{13} and N^{15} isotopes and could be identified by density gradient centrifugation. After mixed infection of bacteria growing on a normal (C^{12}, N^{14}) medium, three classes of phage particle were found, sedimenting at the heavy, hybrid and light positions corresponding to 0, 1 and 2 rounds of replication. Genetic recombinants were present in all three classes, including the unreplicated heavy phages. Meselson (1964) confirmed this finding by crossing two phage λ species each of which contained different radioactive labels, so that only physical incorporation from both parent types could produce recombinants containing both types of isotope. A further finding was that a small amount of DNA synthesis also occurred during recombination; this appeared to represent the removal and resynthesis of some 5–10 per cent. of the genome.

Tomizawa and Anraku (1965) investigated the effects of inhibiting DNA synthesis after the infection of bacterial cells with phage T4. Inhibition was achieved either by treatment of the cells with fluorodeoxyuridine, or by the use of phage *amber* mutants unable to replicate in non-permissive hosts. This resulted in a several-fold increase in the incidence of r^+/r^- heterozygotes, suggesting that DNA synthesis is required at some stage of the recombination process subsequent to their formation. After joint infection with two phage species, one labelled with the radioactive isotope P^{32} and the other with a 5BU density label, they found linear phage genomes comprised of the two types of parental material joined end to end, but not covalently linked. They suggested that these *joint molecules* represent the primary product of the recombination process and are later converted to covalently linked phage recombinants by a process which requires DNA synthesis.

A series of experiments were performed by Kozinski (Kozinski, 1968; Kozinski, Kozinski and James, 1967) to study the effects of inhibiting protein synthesis by the addition of chloramphenicol at various times after infection. Essentially, the method is to grow *E. coli* B in a medium containing heavy isotopes and infect the bacteria with light P^{32}-labelled phage particles. Parental phage DNA can then be identified by its radioactive label, and newly synthesized progeny phage DNA by its density label. After the addition of chloramphenicol, incubation of the infected cells is continued for 45 minutes, after which the phage DNA is extracted and analysed on both CsCl density gradients (when it remains in duplex form) and on alkaline sucrose density gradients (when it is denatured to single strands).

In the first 3–10 minutes after infection, there is extensive replication of the injected parental phage DNA, during which the radioactive label is mostly found associated with a hybrid density band. Subsequently (around 20 minutes after infection) recombination between the replicated parental DNA and the entirely heavy progeny strands occurs, and the radioactive label is displaced toward a heavier location on the density gradient. If chloramphenicol is added during the first 5 minutes after infection, the injected DNA is unable to replicate. Addition of the inhibitor after 5 minutes allows the replication of the phage DNA, but it does not subsequently undertake recombination. If protein synthesis is not inhibited until 7 minutes after infection, molecular exchanges take place to form joint molecules, but the integrity of the polynucleotide chain is not restored—denaturation of the recombinant species releases the pure parental fragments. The addition of chloramphenicol between 7 and 9 minutes allows the parental fragment to become covalently bonded to the adjacent progeny strand, and it can no longer be released by denaturation.

This demonstrates that the expression of phage-coded genes is necessary if recombination is to occur, and this appears to involve both degradative and repair enzymes. Concomitant with the inhibitory effect of chloramphenicol on recombination, there is an inhibition of endonucleolytic activity directed against the phage DNA. The addition of the inhibitor at various times after infection allows degradation to proceed to different extents, and suggests that at least three nuclease activities must be involved. About 5 minutes after infection, the DNA suffers single-strand breakages, probably at non-random locations, with an average distance between breaks of 20–25 per cent. of the total length of the phage molecule. Their formation is inhibited if chloramphenicol is added within 3 minutes after infection. Further nicks are introduced into the DNA about 2 minutes later (inhibited by CM added at 5 minutes after infection) opposite to the primary nicks; this causes double-strand breaks and yields a group of stable fragments which are capable of replication as such. Finally, about 15 minutes after infection (inhibited by CM added at 7 minutes) single-strand breaks are introduced randomly into these fragments, and it is at this time that parent-progeny recombination begins to be expressed.

The obvious candidate for the enzyme required to catalyse the repair of gaps remaining in polynucleotide strands after joint molecule formation is the T4 polynucleotide ligase. However, the use of mutants which are defective in ligase activity has shown that nevertheless their ability to form covalent recombinants is unimpaired (Kozinski and Kozinski, 1969; Richardson and coworkers, 1968). The possibility that this action is mediated by the *host* ligase is excluded by the finding that addition of chloramphenicol up to 7 minutes after infection can prevent recombination from proceeding beyond the joint molecule stage. It seems likely, therefore, that some other 'repairase', coded by the phage, is responsible for this activity.

Recombination and repair in bacteria

There are many similarities between the enzyme activities required for recombination and those known to be concerned in the repair of damage to DNA, and this raises the possibility that some of the enzymes involved may play a role in both processes. Both require an endonucleolytic enzyme to nick the DNA; repair demands the excision of damaged regions, and recombination may involve the degradation of any single-stranded regions which remain uninvolved in the cross-wise complementary pairing; and resynthesis to fill any gaps in the DNA and the covalent joining of newly synthesized to old strands are common requirements. In addition, systems which recognize distortions in the hybrid DNA segment and correct mispairing may be similar to those involved in the excision–repair process for overcoming ultraviolet damage.

The possibility that the excision enzymes themselves might be used in recombination has been excluded by the finding that bacterial mutants which are defective in excision are able to support normal recombination. However, mutants have been isolated in *E. coli* F^- cells with the two phenotypic properties of increased sensitivity to ultraviolet irradiation, and the inability to form recombinants with DNA introduced from an *Hfr* strain; thus they appear infertile in such crosses. Several of these *rec*$^-$ mutants have been isolated (see Clark and coworkers, 1966; Howard-Flanders and Theriot, 1966; Van de Putte, Zwenck and Rorsch, 1966). They fall into two classes, the 'reckless' and the 'cautious'. The former release an abnormally large amount of radioactivity from their genomes after ultraviolet irradiation if they have been labelled with H^3-thymidine during growth, and also are unable to incorporate thymine into their DNA after irradiation. It seems likely that they are defective in the ability to terminate degradation of DNA during excision and initiate repair synthesis, that is, excision starts but does not stop. The second class of mutants show a much lower degradation of DNA after irradiation, presumably due to interference with an excision process.

The increased ultraviolet sensitivity of these strains results from their reduced capacity to cope with thymine dimers which are not excised but remain in the DNA. As discussed in Chapter 14, the most probable explanation is that the *rec* system is involved in promoting genetic exchanges, either single strand or double strand, between homologous segments of DNA. The exact details of the enzyme activities involved remain to be found.

References

The italic numbers following each reference indicate the pages upon which it is cited.

Adams, A., T. Lindahl, and J. R. Fresco (1967). Conformational differences between the biologically active and inactive forms of a *t*RNA. *Proc. Nat. Acad. Sci. USA*, **57**, 1684–1691.(*196*)

Adams, J. M. (1968). On the release of the formyl group from nascent protein. *J. Mol. Biol.*, **33**, 571–590.(*119*)

Adams, J. M., and M. R. Capecchi (1966). *N*-formylmethionyl-*t*RNA as the initiator of protein biosynthesis. *Proc. Nat. Acad. Sci. USA*, **56**, 147–155.(*119*)

Adams, J. M., P. G. V. Jeppesen, F. Sanger, and B. G. Barell (1969). Nucleotide sequence from the coat protein cistron of R17 bacteriophage RNA. *Nature*, **223**, 1009–1015.(*82*, *289*)

Adelberg, E. A., and S. N. Burns (1960). Genetic variations in the sex factor of *E. coli*. *J. Bacteriol.*, **79**, 321–330.(*66*)

Ahmed, A. (1968). Organisation of the histidine-3 region of *Neurospora*. *Mol. Gen. Gen.*, **103**, 185–193.(*259*)

Ahmed, A., M. E. Case, and N. H. Giles (1964). The nature of complementation among mutants in the histidine-3 region of *N. crassa*. *Brookhaven Symp. Biol.*, **17**, 53–65.(*259*)

Alberghina, F. A. M., and S. R. Suskind (1967). Ribosomes and ribosomal protein from *N. crassa*. I. Physical, chemical, and immunochemical properties. *J. Bacteriol.*, **94**, 630–649.(*159*)

Allen, D. W., and P. C. Zamecnik (1962). The effect of puromycin on rabbit reticulocyte ribosomes. *Biochim. Biophys. Acta*, **55**, 865–874.(*107*)

Allende, J. E., and H. Weissbach (1967). GTP interaction with a protein synthesis initiation factor preparation from *E. coli*. *Biochem. Biophys. Res. Commun.*, **28**, 82–88.(*110*)

Amaldi, F., and G. Attardi (1968). Partial sequence analysis of ribosomal RNA from Hela cells. I. Oligonucleotide pattern of 28S and 18S RNA after pancreatic ribonuclease digestion. *J. Mol. Biol.*, **33**, 737–756.(*176*)

Ames, B. N., and P. E. Hartman (1963). The histidine operon. *Cold Spring Harbor Symp. Quant. Biol.*, **28**, 349–356.(*245*, *263*, *266*, *292*)

Anderson, J. S., M. S. Bretscher, B. F. C. Clark, and K. A. Marcker (1967a). A GTP requirement for binding initiator *t*RNA to ribosomes. *Nature*, **215**, 490–492.(*121*, *122*, *129*)

Anderson, J. S., J. E. Dahlberg, M. S. Bretscher, M. Revel, and B. F. C. Clark (1967b). GTP stimulated binding of initiator *t*RNA to ribosomes directed by f_2 bacteriophage RNA. *Nature*, **216**, 1072–1076.(*121*, *122*)

Anton, D. N. (1968). Histidine regulatory mutants in *S. typhimurium*. V. Two new classes of histidine regulator mutants. *J. Mol. Biol.*, **33**, 533–546.(*246*)

Apirion, D. (1966). Altered ribosomes in a suppressor strain of *E. coli*. *J. Mol. Biol.*, **16**, 285–301.(*216*)

Apirion, D. (1967). Three genes that affect *E. coli* ribosomes. *J. Mol. Biol.*, **30**, 255–275.(*215*)

Apirion, D., and D. Schlessinger (1967). Reversion from streptomycin dependence in *E. coli* by a further change in the ribosome. *J. Bacteriol.*, **94**, 1275–1276. (*216*)

Arber, W. (1968). Host controlled restriction and modification of bacteriophage. *Symp. Soc. Gen. Microbiol.*, **18**, 293–314.(*325*)

Arditti, R. R., J. G. Scaife, and J. R. Beckwith (1968). The nature of mutants in the lac promotor region. *J. Mol. Biol.*, **38**, 421–426.(*233*)

Attardi, G. S., S. Naono, J. Rouviere, F. Jacob and F. Gros (1963). Production of *m*RNA and regulation of protein synthesis. *Cold Spring Harbor Symp. Quant. Biol.*, **28**, 363–374.(*271*)

Aubert, M., J. F. Scott, M. Reynier, and R. Monier (1968). Rearrangement of the conformation of *E. coli* 5S RNA. *Proc. Nat. Acad. Sci. USA*, **61**, 292–299. (*157*)

Avery, O. T., C. M. Macleod, and M. McCarty (1944). Studies on the chemical nature of the substance inducing transformation of *Pneumococcal* types, I. Induction of transformation by a DNA fraction isolated from *Pneumococcus* type III. *J. Exp. Med.*, **79**, 137–158.(*157*)

Baglioni, C., and T. Campana (1967). α-chain and globin: intermediates in the synthesis of rabbit hemoglobin. *Eur. J. Biochem.*, **2**, 480–492.(*294*)

Baguley, B. C., and M. Staehelin (1968). The specificity of *t*RNA methylases from rat liver. *Biochemistry, NY.*, **7**, 45–50.(*184*)

Baich, A., and M. Johnson (1968). Evolutionary advantage of control of a biosynthetic pathway. *Nature*, **218**, 464–465.(*261*)

Baker, R. F., and C. Yanofsky (1968a). Direction of *in vivo* degradation of an *m*RNA. *Nature*, **219**, 26–29.(*230, 279*)

Baker, R. F., and C. Yanofsky (1968b). The periodicity of RNA polymerase initiations: a new regulatory feature of transcription. *Proc. Nat. Acad. Sci. USA*, **60**, 313–320.(*274*)

Balbinder, E., A. J. Blume, A. Weber and H. Tamaki (1968). Polar and antipolar mutants in the *tryptophan* operon of *S. typhimurium*. *J. Bacteriol.*, **95**, 2217–2229.(*266*)

Baldwin, A. N., and P. Berg (1966a). Purification and properties of isoleucyl *t*RNA synthetase from *E. coli*. *J. Biol. Chem.*, **241**, 831–837.(*202*)

Baldwin, A. N., and P. Berg (1966b). *t*RNA induced hydrolysis of valyl adenylate bound to isoleucyl *t*RNA synthetase. *J. Biol. Chem.*, **241**, 839–845.(*205*)

Barnett, L., S. Brenner, F. H. C. Crick, R. G. Schulman, and R. J. Watts-Tobin (1967). Phase shift and other mutants in the first part of the *r*II B cistron of bacteriophage T4. *Phil. Trans. Roy. Soc. Ser. B*, **252**, 487–560.(*75*)

Bauerle, R. H., and P. Margolin (1966). The functional organisation of the tryptophan gene cluster in *S. typhimurium*. *Proc. Nat. Acad. Sci. USA*, **56**, 111–118.(*282*)

Baumberg, S., D. F. Bacon, and H. J. Vogel (1965). Individually repressible enzymes specified by clustered genes of the arginine system. *Proc. Nat. Acad. Sci. USA*, **53**, 1029–1032.(*257*)

Baydasarian, M., and Z. Cièsla (1968). Detection of an enzyme of the histidine operon on large polyribosomes of *S. typhimurium*. *Abstracts, Fed. Eur. Biochem. Socs., Madrid*, 308.(*230*)

Beadle, G. W., and E. L. Tatum (1941). Genetic control of biochemical reactions in *Neurospora*. *Proc. Nat. Acad. Sci. USA*, **27**, 499–506.(*20*)

Beckwith, J. R. (1964). A deletion analysis of the lac operator region in *E. coli*. *J. Mol. Biol.*, **8**, 427–430.(*232*)

Bell, D., and G. J. Russell (1967). Physical studies on mixed *t*RNAs from yeast. *Biochemistry, NY*, **6**, 3363–3368.(*193*)

Bennett, T. P. (1969). Evidence for one leucyl-*t*RNA synthetase with specificity for leucine *t*RNAs with different coding characteristics. *J. Biol. Chem.*, **244**, 3182–3187.(*202*)

Benzer, S., and S. P. Champe (1961). Ambivalent *r*II mutants of phage T4. *Proc. Nat. Acad. Sci. USA*, **47**, 1025–1038.(*82*)

Benzer, S., and S. P. Champe (1962). A change from nonsense to sense in the genetic code. *Proc. Nat. Acad. Sci. USA*, **48**, 1114–1121.(*82, 211*)

Berberich, M. A., J. S. Kovach, and R. F. Goldberger (1967). Chain initiation in a polycistronic messenger: sequential versus simultaneous derepression of the enzymes for histidine biosynthesis. *Proc. Nat. Acad. Sci. USA*, **57**, 1857–64. (*284*)

Berberich, M. A., P. Venetianer, and R. F. Goldberger (1966). Alternative modes of derepression of the histidine operon observed in *S. typhimurium*. *J. Biol. Chem.*, **241**, 4426–4433.(*284*)

Berg, C. M., and L. G. Caro (1967). Chromosome replication in *E. coli*. I. Lack of influence of the integrated F factor. *J. Mol. Biol.*, **29**, 419–431.(*301*)

Berg, P., and E. J. Ofengand (1958). An enzymatic mechanism for linking amino acids to RNA. *Proc. Nat. Acad. Sci. USA*, **44**, 78–85.(*102*)

Berger, H., W. J. Brammar, and C. Yanofsky (1968). Analysis of amino acid replacements resulting from frameshift and mis-sense mutations in the tryptophan synthetase A gene of *E. coli*. *J. Mol. Biol.*, **34**, 219–238.(*82*)

Bergman, F. H., P. Berg, and M. Dieckmann (1961). The enzymic synthesis of aminoacyl derivatives of RNA. II. The preparation of leucyl-, valyl-, isoleucyl-, and methionyl-*t*RNA synthetases from *E. coli*. *J. Biol. Chem.*, **236**, 1735–1740. (*102*)

Billen, D. (1968). Methylation of the bacterial chromosome: an event at the 'replication point'? *J. Mol. Biol.*, **31**, 477–486.(*324*)

Billen, D., R. R. Hewitt, T. Lapthisophon, and P. M. Achey (1967). DNA repair replication after UV light or X ray exposure of bacteria. *J. Bacteriol.*, **94**, 1538–1545.(*363*)

Bird, R., and K. G. Lark (1968). Initiation and termination of DNA replication after amino acid starvation of *E. coli* 15T$^-$. *Cold Spring Harbor Symp. Quant. Biol.*, **33**, 799–808.(*337*)

Birge, E. A., G. R. Craven, S. J. S. Hardy, C. G. Kurland, and A. Voynow (1969). Structural determinant of a ribosomal protein. *Science*, **164**, 1285–1286. (*160*)

Birnsteil, M. L., M. I. H. Chipchase, and B. B. Hyde (1966). The nucleolus as a source of ribosomes. *Biochim. Biophys. Acta*, **76**, 454–462.(*171*)

Birnstiel, M., J. Spiers, I. Purdom, K. Jones, and U. E. Loening (1968). Properties and composition of the isolated ribosomal DNA satellite of *X. laevis. Nature*, **219**, 454–463.(*173*)

Bjork, G. R., and I. Svensson (1969). Studies on microbial RNA; fractionation of *t*RNA methylases from *S. cerevisia. Eur. J. Biochem.*, **9**, 207–215.(*184*)

Bladen, H. A., R. Byrne, J. G. Levin, and M. W. Nirenberg (1965). An electron microscopic study of a DNA-ribosome complex formed *in vitro. J. Mol. Biol.*, **11**, 78–83.(*150*)

Bollen, A., J. Davies, M. Ozaki, and S. Mizushima (1969). Ribosomal protein conferring sensitivity to the antibiotic spectinomycin in *E. coli. Science*, **165**, 85–86.(*215*)

Bollum, F. J. (1967). Enzymatic replication of polydeoxynucleotides. In D. Shugar (Ed.), *Genetic Elements* 3–16, Academic Press, London.(*302*)

Bolton, E. T., and B. J. McCarthy (1962). A general method for the isolation of RBA complementary to DNA. *Proc. Nat. Acad. Sci. USA*, **48**, 1390–1397.(*41*)

Bonhoeffer, F., and W. Vielmetter (1968). Conjugational DNA transfer in *E. coli. Cold Spring Harbor Symp. Quant. Biol.*, **33**, 623–628.(*65,66*)

Bourgeios, S., M. Cohn, and L. Orgel (1965). Suppression of and complementation among mutants of the regulatory gene of the *lactose* operon of *E. coli. J. Mol. Biol.*, **14**, 300–302.(*231*)

Boyce, R. P., and P. Howard-Flanders (1964a). Release of ultraviolet light induced dimers from DNA in *E. coli* K12. *Proc. Nat. Acad. Sci. USA*, **51**, 293–300.(*354*)

Boyce, R. P., and P. Howard-Flanders (1964b). Genetic control of DNA breakdown and repair in *E. coli* treated with mitomycin C or UV. *Z. Vererbungsl.*, **95**, 345–350.(*356*)

Boyer, H. W., and D. A. Roulland-Dussoix (1969). A complementation analysis of the restriction and modification of DNA in *E. coli. J. Mol. Biol.*, **41**, 459–472.(*325*)

Bremer, H., and M. W. Konrad (1964). A complex of enzymatically synthesised RNA and template DNA. *Proc. Nat. Acad. Sci. USA*, **51**, 801–810.(*146*)

Brenner, S., L. Barnett, F. H. C. Crick, and A. Orgel (1961). The theory of mutagenesis. *J. Mol. Biol.*, **3**, 121–124.

Brenner, S., L. Barnett, E. R. Katz, and F. H. C. Crick (1967). UGA: a third nonsense triplet in the genetic code. *Nature*, **213**, 449–450.(*86*)

Brenner, S., and J. R. Beckwith (1965). *Ochre* mutants, a new class of suppressible nonsense mutants. *J. Mol. Biol.*, **13**, 629–637.(*84*)

Brenner, S., F. Jacob, and M. Meselson (1961). An unstable intermediate carrying information from genes to ribosomes for protein synthesis. *Nature*, **190**, 576–580.(*98*)

Brenner, S., A. O. W. Stretton, and S. Kaplan (1965). Genetic code: the nonsense triplets for chain termination and their suppression. *Nature*, **206**, 994–998.(*84, 212*)

Bretscher, M. S. (1965). Fractionation of oligolysyl-adenosine complexes derived from polylysine attached to *t*RNA. *J. Mol. Biol.*, **12**, 913–919.(*132*)

Bretscher, M. S. (1966). Polypeptide chain initiation and the characterisation of ribosomal binding in *E. coli. Cold Spring Harbor Symp. Quant. Biol.*, **31**, 289–296.(*121, 128*)

Bretscher, M. S. (1968a). Translocation in protein synthesis in a hybrid structure model. *Nature*, **218**, 675–677.(*110, 114*)

Bretscher, M. S. (1968b). Polypeptide chain termination: an active process. *J. Mol. Biol.*, **34**, 131–136.(*132*)

Bretscher, M. S. (1968c). Direct translation of a circular *m*RNA. *Nature*, **220**, 1088–1091.(*127*)

Bretscher, M. S. (1969). Direct translation of bacteriophage fd DNA in the absence of neomycin B. *J. Mol. Biol.*, **42**, 595–598.(*127*)

Bretscher, M. S., H. M. Goodman, J. R. Menninger, and J. D. Smith (1965). Polypeptide chain termination using synthetic polynucleotides. *J. Mol. Biol.*, **14**, 634–639.(*132*)

Bretscher, M. S., and K. A. Marcker (1966). Polypeptidyl-*t*RNA and aminoacyl-*t*RNA binding sites on ribosomes. *Nature*, **211**, 380–384.(*128*)

Bridges, B. A., J. Law, and R. J. Munson (1969). Mutagenesis in *E. coli*. II. Evidence for a common pathway for mutagenesis by ultraviolet light, ionizing radiation, and thymine deprivation. *Mol. Gen. Gen.*, **103**, 266–273.(*364*)

Bridges, B. A., and R. J. Munson (1968). The persistance through several replication cycles of mutation-producing pyrimidine dimers in a strain of *E. coli*. *Biochim. Biophys. Res. Commun.*, **30**, 620–624.(*362*)

Brimacombe, R., J. Trupin, M. Nirenberg, P. Leder, M. Bernfield, and T. Jaouni (1965). RNA codewords and protein synthesis. VIII. Nucleotide sequences of synonym codons for arginine, valine, cystein, and alanine. *Proc. Nat. Acad. Sci. USA*, **54**, 954–960.(*80*)

Brinton, C. C. Jr., P. Gemski Jr., and J. Carnahan (1964). A new type of bacterial pilus genetically controlled by the fertility factor of *E. coli* K12 and its role in chromosome transfer. *Proc. Nat. Acad. Sci. USA*, **52**, 776–783.(*64*)

Brown, D. B., and J. B. Gurdon (1964). Absence of ribosomal RNA synthesis in the anucleolate mutant of *X. laevis*. *Proc. Nat. Acad. Sci. USA*, **51**, 139–146. (*172*)

Brown, D. D., and C. S. Weber (1968a). Gene linkage by RNA-DNA hybridisation. I. Unique DNA sequences homologous to 4S RNA, 5S RNA, and *r*RNA. *J. Mol. Biol.*, **34**, 661–680.(*171, 172*)

Brown, D. D., and C. S. Weber (1968b). Gene linkage by RNA-DNA hybridisation. II. Arrangement of the redundant gene sequences for 21S and 18S *r*RNA. *J. Mol. Biol.*, **34**, 681–698.(*172*)

Brown, G. M., and G. Attardi (1965). Methylation of nucleic acids in Hela cells. *Biochem. Biophys. Res. Commun.*, **20**, 298–302.(*155*)

Brown, J. C., and P. Doty (1968). Protein factor requirement for binding of *m*RNA to ribosomes. *Biochem. Biophys. Res. Commun.*, **30**, 284–291.(*129, 151*)

Brownlee, G. G., F. Sanger, and B. G. Barrell (1967). Nucleotide sequence of 5S *r*RNA from *E. coli*. *Nature*, **215**, 735–736.(*156*)

Brownlee, G. G., F. Sanger, and B. G. Barrell (1968). The sequence of 5S *r*RNA. *J. Mol. Biol.*, **34**, 379–412.(*156*)

Bruening, G., and R. M. Bock (1967). Covalent integrity and molecular weights of yeast *r*RNA components. *Biochim. Biophys. Acta*, **149**, 377–386.(*154*)

Burgess, R. P., A. A. Travers, J. J. Dunn, and E. K. F. Bautz (1969). Factor stimulating transcription by RNA polymerase. *Nature*, **221**, 43–47.(*141*)

Byrne, R., J. G. Levin, H. A. Bladen, and M. W. Nirenberg (1964). The *in vitro* formation of a DNA-ribosome complex. *Proc. Nat. Acad. Sci. USA*, **52**, 140–148.(*150*)

Cairns, J. (1963). The chromosome of *E. coli. Cold Spring Harbor Symp. Quant. Biol.*, **28**, 43–46.(*296*, *337*)

Campbell, A. (1962). Episomes. *Advan. Genet.*, **11**, 101–146.(*56*)

Campbell, A. (1964). Transduction. In I. C. Gunsalus and R. Y. Stainer (Eds.), *The Bacteria*, Vol. 5. Academic Press, London.(*58*)

Cantor, C. R. (1967). Possible conformations of 5S *r*RNA. *Nature*, **216**, 513–514. (*157*, *158*)

Cantor, C. R. (1968). The extent of base pairing in 5S *r*RNA. *Proc. Nat. Acad. Sci. USA*, **59**, 478–483.(*157*)

Capecchi, M. R. (1967a). Polypeptide chain termination *in vitro*: isolation of a release factor. *Proc. Nat. Acad. Sci. USA*, **58**, 1144–1151.(*132*)

Capecchi, M. R. (1967b). Polarity *in vitro*. *J. Mol. Biol.*, **30**, 213–218.(*286*)

Capecchi, M. R., and G. N. Gussin (1965). Suppression *in vitro*: identification of a serine *t*RNA as a 'nonsense' suppressor. *Science*, **149**, 416–422.(*212*)

Caro, L. G., and C. M. Berg (1968). Chromosome replication in some strains of *E. coli* K12. *Cold Spring Harbor Symp. Quant. Biol.*, **33**, 559–576.(*301*)

Carter, T., and A. Newton (1969). Messenger RNA stability and polarity in the *lac* operon of *E. coli. Nature*, **223**, 707–710.(*272*)

Case, M. E., and N. H. Giles (1968). Evidence for nonsense mutants in the *arom* gene cluster of *N. Crassa. Genetics*, **60**, 49–58.(*260*)

Cashel, M., and J. Gallant (1968). Control of RNA synthesis in *E. coli*. I. Amino acid dependence of the synthesis of the substrates of RNA polymerase. *J. Mol. Biol.*, **34**, 317–330.(*291*)

Caskey, C. T., A. Beaudet, and M. Nirenberg (1968). Codons and protein synthesis. XV. Dis-similar responses of mammalian and bacterial *t*RNA fractions to *m*RNA codons. *J. Mol. Biol.*, **37**, 99–118.(*206*, *207*)

Cerda-Olmeda, E., P. C. Hanawalt, and N. Guerola (1968). Mutagenesis of the replication point by nitrosoguanidine: map and pattern of replication of the *E. coli* chromosome. *J. Mol. Biol.*, **33**, 705–720.(*301*)

Chae, Y–B., R. Mazumder, and S. Ochoa (1969). Polypeptide chain initiation in *E. coli:* isolation of homogeneous initiation factor f_2 and its relation to ribosomal protein. *Proc. Nat. Acad. Sci. USA*, **62**, 1181–1188.(*130*)

Chamberlin, M., R. L. Baldwin, and P. Berg (1963). An enzymically synthesised RNA of alternating base sequence: physical and chemical characterisation. *J. Mol. Biol.*, **7**, 334–349.(*146*)

Chambers, D. A., and G. Zubay (1969). The stimulatory effect of cyclic 3′5′ AMP on DNA directed synthesis of β-galactosidase in a cell-free system. *Proc. Nat. Acad. Sci. USA*, **63**, 118–122.(*238*)

Chapeville, F., F. Lipmann, G. von Ehrenstein, B. Weisblum, W. J. Ray, and S. Benzer (1962). On the role of *t*RNA in coding for amino acids. *Proc. Nat. Acad. Sci. USA*, **48**, 1086–1092.(*102*)

Chuang, H. Y. K., A. G. Atherly, and F. E. Bell (1967). Protection of the proline and valine activating enzymes by their amino acid substrates against thermal inactivation. *Biochem. Biophys. Res. Commun.*, **28**, 1013–1018.(*203*)

Church, R. B., and B. J. McCarthy (1968). Related base sequences in the DNA of simple and complex organisms. II. The interpretation of DNA/RNA hybridisation studies with mammalian nucleic acids. *Biochem. Genet.*, **2**, 55–74.(*41, 97*)

Clark, A. J., M. Chamberlin, R. P. Boyce, and P. Howard-Flanders (1966). Abnormal metabolic response to ultraviolet light of a recombination deficient mutant of *E. coli* K12. *J. Mol. Biol.*, **19**, 442–454.(*378*)

Clark, B. F. C., and K. A. Marcker (1966). The role of *N*-formyl methionyl-*t*RNA in protein biosynthesis. *J. Mol. Biol.*, **17**, 394–406.(*120, 121*)

Clark, D. J. (1968a). Regulation of DNA replication and cell division in *E. coli* B/r. *J. Bacteriol.*, **96**, 1214–1225.(*338*)

Clark, D. J. (1968b). The regulation of DNA replication and cell division in *E. coli* B/r. *Cold Spring Harbor Symp. Quant. Biol.*, **33**, 823–838.(*338*)

Clark, D. J., and O. Maaløe (1967). DNA replication and the division cycle in *E. coli*. *J. Mol. Biol.*, **23**, 99–112.(*331, 332*)

Cleaver, J. E., and R. B. Painter (1968). Evidence for repair replication of Hela cell DNA damaged by u.v. light. *Biochim. Biophys. Acta*, **161**, 552–554.(*362*)

Cline, A. L., and R. M. Bock (1966). Translational control of gene expression. *Cold Spring Harbor Symp. Quant. Biol.*, **31**, 321–331.(*293*)

Colombo, B., C. Vesco, and C. Baglioni (1968). Role of ribosome subunits in protein synthesis in mammalian cells. *Proc. Nat. Acad. Sci. USA*, **61**, 651–658.(*125*)

Comb, D. G., and N. Sarkar (1967). The binding of 5S *r*RNA to ribosomal subunits. *J. Mol. Biol.*, **25**, 317–330.(*156*)

Cooper, S. (1969). Cell division and DNA replication following a shift to richer medium. *J. Mol. Biol.*, **43**, 1–12.(*338*)

Cooper, S., and C. E. Helmstetter (1968). Chromosome replication and the division of *E. coli* B/r. *J. Mol. Biol.*, **31**, 519–540.(*332, 333, 335, 336*)

Cory, S., K. A. Marcker, S. K. Dube, and B. F. C. Clark (1968). Primary structure of a methionine *t*RNA from *E. coli*. *Nature*, **220**, 1039–1040.(*186*)

Cotter, R. I., and W. B. Gratzer (1969). Conformation of *r*RNA of *E. coli:* an infra-red analysis. *Nature*, **221**, 154–157.(*162*)

Cotter, R. I., P. McPhie, and W. B. Gratzer (1967). Internal organisation of the ribosome. *Nature*, **216**, 864–868.(*162, 163*)

Couch, J. L., and P. Hanawalt (1967). DNA repair replication in temperature sensitive DNA synthesis deficient bacteria. *Biochem. Biophys. Res. Commun.*, **29**, 779–784.(*359*)

Cox, E. C., J. R. White, and J. G. Flaks (1964). Streptomycin action and the ribosome. *Proc. Nat. Acad. Sci. USA*, **51**, 703–709.(*214*)

Cramer, F., H. Doepner, F. V. D. Haar, E. Schimme, and H. Seidel (1968). On the conformation of *t*RNA. *Proc. Nat. Acad. Sci. USA*, **61**, 1384–1391. (*196–200*)

Cramer, F., and V. A. Erdman (1968). Amount of adenine and uracil base pairs in *E. coli* 23S, 16S, and 5S *r*RNA. *Nature*, **218**, 92–93.(*157*)

Creighton, H. B., and B. McClintock (1931). A correlation of cytological and genetical crossing over in *Zea mays*. *Proc. Nat. Acad. Sci. USA*, **17**, 492–497. (*14*)

Crick, F. H. C. (1958). On protein Synthesis. *Symp. Soc. Exp. Biol.*, **12**, 138–163. (*99*)

Crick, F. H. C. (1963). The recent excitement in the coding problem. *Prog. Nuc. Acid Res.*, **1**, 164–218.(*71*)

Crick, F. H. C. (1966). Codon anticodon pairing: the wobble hypothesis. *J. Mol. Biol.*, **19**, 548–555.(*194, 206*)

Crick, F. H. C. (1968). The origin of the genetic code. *J. Mol. Biol.*, **38**, 367–380. (*92*)

Crick, F. H. C., L. Barnett, S. Brenner, and R. J. Watts-Tobin (1961). General nature of the genetic code for proteins. *Nature*, **192**, 1227–1232.(*75*)

Crick, F. H. C., J. Griffith, and L. Orgel (1957). Codes without commas. *Proc. Nat. Acad. Sci. USA*, **43**, 416–421.(*74, 99*)

Dahlberg, J. E. (1968). Terminal sequences of bacteriophage RNAs. *Nature*, **220**, 548–552.(*139*)

Dalgarno, L., and F. Gros (1968). RNA synthesis in *relaxed* and *stringent E. coli*. Breakdown of preformed ribonucleoprotein particles and subsequent RNA synthesis. *Biochim. Biophys. Acta*, **157**, 64–75.(*183*)

Davern, C. I., and M. Meselson (1960). The molecular conservation of RNA during bacterial growth. *J. Mol. Biol.*, **2**, 153–160.(*97, 98*)

Davies, D. R. (1965). Repair mechanisms of variations in UV sensitivity within the cell cycle. *Mut. Res.*, **2**, 477–486.(*360*)

Davies, D. R. (1967). UV sensitive mutants of *Chalamydomonas reinhardii*. *Mut. Res.*, **4**, 765–770.(*360*)

Davies, J. (1964). Studies on the ribosomes of streptomycin sensitive and resistant strains of *E. coli*. *Proc. Nat. Acad. Sci. USA*, **51**, 659–663.(*214*)

Davies, J. (1966). Streptomycin and the genetic code. *Cold Spring Harbor Symp. Quant. Biol.*, **31**, 665–670.(*216, 219*)

Davies, J., D. S. Jones, and H. G. Khorana (1965). A further study of misreading of codons induced by streptomycin and neomycin using ribopolynucleotides containing nucleotides in alternating sequence as templates. *J. Mol. Biol.*, **18**, 48–57.(*216*)

Deutscher, M. P., and A. Kornberg (1969a). Enzymatic synthesis of DNA. XXVIII. The pyrophosphate exchange and pyrophosphorolysis reactions of DNA polymerase. *J. Biol. Chem.*, **244**, 3019–3028.(*310*)

Deutscher, M. P., and A. Kornberg (1969b). Enzymatic synthesis of DNA. XXIX. Hydrolysis of DNA from the 5′ terminus by an exonuclease function of DNA polymerase. *J. Biol. Chem.*, **244**, 3029–3037.(*309*)

Dickermann, H. W., and B. C. Smith (1967). Aminoacyl-*t*RNA$_{met}$ transformylase. II. Evidence for an enzyme: *t*RNA complex. *Arch. Biochem. Biophys.*, **122**, 105–110.(*120*)

Dickermann, H. W., E. Steers Jr., B. G. Redfield, and H. Weissbach (1966). Formylation of *E. coli* methionyl-*t*RNA. *Cold Spring Harbor Symp. Quant. Biol.*, **31**, 287–288.(*120*)

Di Mauro, E., L. Snyder, P. Marino, A. Lamberti, A. Coppo, and G. P. Tocchini-Valenti (1969). Rifampicin sensitivity of the components of DNA dependent RNA polymerase. *Nature*, **222**, 533–537.(*141*)

Dintzis, H. M. (1961). Assembly of the peptide chains of hemoglobin. *Proc. Nat. Acad. Sci. USA*, **47**, 247–260.(*93, 95*)

Donachie, W. D. (1967). Recovery from thymineless death in *E. coli* 15T$^-$. *Biochem. Biophys. Res. Commun.*, **29**, 172–177.(*326*)

Donachie, W. D. (1968). Relationship between cell size and time of initiation of DNA replication. *Nature*, **219**, 1077–1079.(*347*)

Doty, P., J. Marmur, J. Eigner, and C. Schildkraut (1960). Strand separation and specific recombination in DNAs: physical studies. *Proc. Nat. Acad. Sci. USA*, **46**, 461–476.(*35*, *38*, *39*)

Douglas, H. C., and D. C. Hawthorne (1966). Regulation of genes controlling synthesis of the galactose pathway enzymes in yeast. *Genetics*, **54**, 911–916. (*258*)

Drake, J. W., E. F. Allen, S. A. Forsberg, R–M. Preparata, and E. O. Greening (1969). Genetic control of mutation rates in bacteriophage T4. *Nature*, **221**, 1128–1131.(*304*)

Dube, S. K., K. A. Marcker, B. F. C. Clark, and S. Cory (1968). Nucleotide sequence of *N*-formyl methionyl-*t*RNA. *Nature*, **218**, 232–234.(*186*, *209*)

Dubnau, D., I. Smith, and J. Marmur (1965). Gene conservation in *Bacillus* species. II. the location of genes concerned with the synthesis of ribosomal components and *t*RNA. *Proc. Nat. Acad. Sci. USA*, **54**, 727–730.(*171*)

Echols, H., H. Garen, S. Garen, and A. Torriani (1961). Genetic control of repression of alkaline phosphatase in *E. coli*. *J. Mol. Biol.*, **3**, 425–438.(*252*)

Economou, A., and A. Nakamoto (1967). Further studies on the initiation of protein synthesis with *N*-formylmethionine in *E. coli* extracts. *Proc. Nat. Acad. Sci. USA*, **58**, 1033–1037.(*122*)

Edgar, R. S., and W. B. Wood (1966). Morphologies of bacteriophage T4 in extracts of mutant infected cells. *Proc. Nat. Acad. Sci. USA*, **55**, 498–505.(*50*)

Edlin, G., and P. Broda (1968). Physiology and genetics of the 'RNA control' locus in *E. coli*. *Bact. Rev.*, **32**, 206–226.(*291*)

Edlin, G., G. S. Stent, R. F. Baker, and C. Yanofsky (1968). Synthesis of a specific messenger RNA during amino acid starvation of *E. coli*. *J. Mol. Biol.*, **37**, 257–268.(*291*)

Eggen, K., and D. Nathans (1969). Regulation of protein synthesis directed by coliphage MS2 RNA. II. *In vitro* repression by phage coat protein. *J. Mol. Biol.*, **39**, 293–306.(*287*)

Eidlic, L., and F. C. Neidthardt (1965). Role of valyl-*t*RNA synthetase in enzyme repression. *Proc. Nat. Acad. Sci. USA*, **53**, 539–543.(*248*)

Eisenstadt, J. M., and G. Brawerman (1967). The role of the native subribosomal particles of *E. coli* in polypeptide chain initiation. *Proc. Nat. Acad. Sci. USA*, **58**, 1560–1565.(*136*)

Elseviers, D., R. Cunin and N. Glansdorff (1969). Reactivation of arginine genes under the influence of polar mutations. *Febs. Lett.*, **3**, 18–20.(*257*)

Emmerson, P. T., and P. Howard-Flanders (1964). Post irradiation degradation of DNA following exposure of u.v.-sensitive and resistant bacteria to X-rays. *Biochem. Biophys. Res. Commun.*, **18**, 24–29.(*363*)

Engelhardt, D. L., R. E. Webster, and N. D. Zinder (1967). *Amber* mutants and polarity *in vitro*. *J. Mol. Biol.*, **29**, 45–58.(*286*)

Englesberg, E., J. Irr, J. Power, and N. Lee (1965). Positive control of enzyme synthesis by gene C in the L-arabinose system. *J. Bacteriol.*, **90**, 946–957.(*249*)

Englesberg, E., D. Sheppard, C. Squires, and F. Meronk (1969). An analysis of 'revertants' of a deletion mutant in the C gene of the L-arabinose gene complex in *E. coli* B/r: isolation of initiator constitutive mutants. *J. Mol. Biol.*, **43**, 281–298.(*249*)

Englesberg, E., C. Squires, F. Meronk (1969). The L-arabinose operon in *E. coli* B/r: a genetic demonstration of two functional states of the product of a regulator gene. *Proc. Nat. Acad. Sci. USA*, **62**, 1100–1107.(*249, 250*)

Englund, P. T., M. P. Deutscher, T. M. Jovin, R. B. Kelly, N. R. Cozzarelli, and A. Kornberg (1968). Structural and functional properties of *E. coli* DNA polymerase. *Cold Spring Harbor Symp. Quant. Biol.*, **33**, 1–10.(*306*)

Englund, P. T., J. A. Huberman, T. M. Jovin, and A. Kornberg (1969). Enzymatic synthesis of DNA. XXX. Binding of triphosphates to DNA polymerase. *J. Biol. Chem.*, **244**, 3038–3044.(*308*)

Englund, P. T., R. B. Kelly, and A. Kornberg (1969). Enzymatic synthesis of DNA. XXXI. Binding of DNA to DNA polymerase. *J. Biol. Chem.*, **244**, 3045–3052.(*308*)

Epstein, C. J. (1966). Role of the amino acid 'code' and of selection for conformation in the evolution of proteins. *Nature*, **210**, 25–28.(*90*)

Epstein, W., and J. R. Beckwith (1968). Regulation of gene expression. *Ann. Rev. Biochem.*, **37**, 411–436.(*267*)

Erbe, R. W., M. N. Nau, and P. Leder (1969). Translation and translocation of defined RNA messengers. *J. Mol. Biol.*, **39**, 441–460.(*117*)

Ertel, R., B. Redfield, N. Brot, and H. Weissbach (1968). Role of GTP in protein synthesis: interaction of GTP with soluble transfer factors from *E. coli*. *Arch. Biochem. Biophys.*, **128**, 331–338.(*110*)

Ezekial, D. H., and B. Valulis (1966). Control of RNA synthesis in *E. coli* cells with altered *t*RNA concentration. *Biochim. Biophys. Acta*, **129**, 1839–1843. (*290*)

Fellner, P., and F. Sanger (1968). Sequence analysis of specific areas of the 16S and 23S *r*RNAs. *Nature*, **219**, 236–238.(*155*)

Fincham, J. R. S. (1966). *Genetic Complementation*, Benjamin, New York.(*21*)

Fink, G. R. (1966). A cluster of genes controlling three enzymes in histidine biosynthesis in *S. cerevisiae*. *Genetics*, **53**, 445–459.(*259*)

Fink, G. R., T. Klopotowski, and B. N. Ames (1967). Histidine regulatory mutants in *S. typhimurium*. IV. A positive selection for polar histidine requiring mutants from histidine operator constitutive mutants. *J. Mol. Biol.*, **30**, 81–96.(*273*)

Fink, G. R., and R. G. Martin (1967). Translation and polarity in the histidine operon. II. Polarity in the histidine operon. *J. Mol. Biol.*, **30**, 97–108.(*268, 273, 274*)

Fink, G. R., and J. R. Roth (1968). Histidine regulatory mutants in *S. typhimurium*. VI. Dominance studies. *J. Mol. Biol.*, **33**, 547–558.(*246, 247*)

Flaks, J. G., P. S. Leboy, E. A. Birge, and C. G. Kurland (1966). Mutations and genetics concerned with the ribosome. *Cold Spring Harbor Symp Quant. Biol.*, **31**, 623–631.(*215*)

Fogel, S., and P. S. Sypherd (1968). Chemical basis for hetergeneity of ribosomal proteins. *Proc. Nat. Acad. Sci. USA*, **59**, 1329–1336.(*159*)

Forchhammer, J., and N. O. Kjelgaard (1968). Regulation of *m*RNA synthesis in *E. coli*. *J. Mol. Biol.*, **37**, 245–256.(*277*)

Forget, B. G., and S. M. Weissman (1969). The nucleotide sequence of ribosomal 5S RNA from KB cells. *J. Biol. Chem.*, **244**, 3148–3165.(*157*)

Fox, M. S. (1960). Fate of transforming DNA following fixation by transformable bacteria. II. *Nature*, **187**, 1004–1006.(*45*)

Fox, M. S. (1962). The fate of transforming DNA following fixation by transformable bacteria. III. *Proc. Nat. Acad. Sci. USA*, **48**, 1043–1048.(*45*)
Fox, M. S. (1966). On the mechanism of integration of transforming DNA. *J. Gen. Physiol.*, **49** part **2**, 183–196.(*46*)
Fox, M. S., and M. K. Allen (1964). On the mechanism of DNA integration in *Pneumococcal* transformation. *Proc. Nat. Acad. Sci. USA*, **52**, 412–419.(*45*)
Fox, M. S., and R. D. Hotchkiss (1957). Initiation of bacterial transformation. *Nature*, **179**, 1322.(*44*)
Frankel, F. R. (1968). DNA replication after T4 infection. *Cold Spring Harbor Symp. Quant. Biol.*, **33**, 485–494.(*346*)
Fresco, J. R., A. Adams, R. Ascione, D. Henley, and T. Lindahl (1966). Tertiary structures in *t*RNAs. *Cold Spring Harbor Symp. Quant. Biol.*, **31**, 527–538.(*194, 195, 196*)
Friedman, H., P. Lu, and A. Rich (1969). Ribosomal subunits produced by cold-sensitive initiation of protein synthesis. *Nature*, **223**, 909–913.(*135*)
Friedman, S. M., R. Berezney, and I. B. Weinstein (1968). Fidelity in protein synthesis. The role of the ribosome. *J. Biol. Chem.*, **243**, 5044–5055.(*218*)
Friefelder, D. (1968). Studies with *E. coli* sex factors. *Cold Spring Harbor Symp. Quant. Biol.*, **33**, 425–434.(*63*)
Friesen, J. D. (1968). A study of the relationship between polyribosomes and *m*RNA in *E. coli. J. Mol. Biol.*, **32**, 183–200.(*290*)
Fuchs, E., R. L. Millette, W. Zillig, and G. Walter (1967). Influence of salts on RNA synthesis by DNA-dependent RNA polymerase from *E. coli. Eur. J. Biochem.*, **3**, 183–193.(*147, 148*)
Fuller, W., and A. Hodgson (1967). Conformation of the anticodon loop in *t*RNA. *Nature*, **215**, 817–821.(*192, 193*)
Furano, A. V., D. F. Bradley, and L. G. Childers (1966). The conformation of the RNA in ribosomes. Dye stacking studies. *Biochemistry*, **5**, 3044–3056. (*161, 162*)
Gallant, J., and M. Cashel (1967). On the mechanism of amino acid control of RNA biosynthesis. *J. Mol. Biol.*, **25**, 545–553.(*291*)
Galper, J. B., and J. E. Darnell (1969). The presence of *N*-formyl methionyl-*t*RNA in Hela cell mitochondria. *Biochem. Biophys. Res. Commun.*, **34**, 205–215.(*123*)
Gamow, G. (1954). Possible relationship between DNA and protein structures. *Nature*, **173**, 318.(*72*)
Ganesan, A. K., and K. C. Smith (1968). Recovery of recombination-deficient mutants of *E. coli* K12 from ultraviolet irradiation. *Cold Spring Harbor Symp. Quant. Biol.*, **33**, 235–242.(*364*)
Garen, A., and H. Echols (1962a). Properties of two regulating genes for alkaline phosphatase. *J. Bacteriol.*, **83**, 297–300.(*252*)
Garen, A., and H. Echols (1962b). Genetic control of induction of alkaline phosphatase synthesis in *E. coli. Proc. Nat. Acad. Sci. USA*, **48**, 1398–1402. (*252*)
Garen, A., and S. Garen (1963). Genetic evidence on the nature of the repressor for alkaline phosphatase in *E. coli. J. Mol. Biol.*, **6**, 433–438.(*231*)
Garen, A., S. Garen, and R. C. Wilhelm (1965). Suppressor genes for nonsense mutations. I. The su_1, su_2, and su_3 genes of *E. coli. J. Mol. Biol.*, **13**, 167–178. (*211*)

Garen, A., and N. Otsuji (1964). Isolation of a protein specified by a regulator gene. *J. Mol. Biol.*, **8**, 841–852.(*252*)

Garen, A., and O. Siddiqui (1962). Suppression of mutations in the alkaline phosphatase structural cistron of *E. coli. Proc. Nat. Acad. Sci. USA*, **48**, 1121–1126.(*211*)

Gavrilova, L. P., D. A. Ivanov, and A. S. Spirin (1966). Studies on the structure of ribosomes. III. Stepwise unfolding of the 50S particles without loss of ribosomal protein. *J. Mol. Biol.*, **16**, 473–489.(*160*, *161*)

Gefter, M. L., A. Becker, and J. Hurwitz (1967). The enzymatic repair of DNA. I. Formation of circular λ DNA. *Proc. Nat. Acad. Sci. USA*, **58**, 240–247.(*318*)

Gefter, M. L., and R. L. Russell (1969). Role of modifications in tyrosine *t*RNA: a modified base affecting ribosomal binding. *J. Mol. Biol.*, **39**, 145–158.(*202*)

Gellert, M. (1967). Formation of covalent circles of lambda DNA by *E. coli* extracts. *Proc. Nat. Acad. Sci. USA*, **57**, 148–155.(*318*)

Gellert, M. W., J. W. Little, C. K. Oshinsky, and S. B. Zimmerman (1968). Joining of DNA strands by DNA ligase of *E. coli. Cold Spring Harbor Symp. Quant. Biol.*, **33**, 21–26.(*319*)

Georgiev, G. P. (1967). The nucleus. In D. B. Roodyn (Ed.), *Enzyme Cytology*, Academic Press, London. pp. 27–102.(*97*, *302*)

Ghosh, H. P., and H. G. Khorana (1967). Studies on polynucleotides. LXXXIV. On the role of ribosome subunits in protein synthesis. *Proc. Nat. Acad. Sci. USA*, **58**, 2455–2467.(*124*)

Ghosh, H. P., D. Soll, and H. G. Khorana (1967). Studies on polynucleotides. LXVII. Initiation of protein synthesis *in vitro* as studied by using polyribonucleotides with repeating nucleotide sequences as messengers. *J. Mol. Biol.*, **25**, 275–298.(*121*, *122*)

Gilbert, W. (1963). Polypeptide synthesis in *E. coli*. I. Ribosomes and the active complex. *J. Mol. Biol.*, **6**, 374–388.(*104*)

Gilbert, W., and D. Dressler (1968). DNA replication: the rolling circle model. *Cold Spring Harbor Symp. Quant. Biol.*, **33**, 473–484.(*343*, *344*, *345*)

Gilbert, W., and B. Muller-Hill (1966). Isolation of the lac repressor. *Proc. Nat. Acad. Sci. USA*, **56**, 1891–1898.(*231*)

Gilbert, W., and B. Muller-Hill (1967). The lac operator is DNA. *Proc. Nat. Acad. Sci. USA*, **58**, 2415–2421.(*231*)

Giles, N. H., M. E. Case, W. H. Partridge, and S. I. Ahmed (1967). A gene cluster in *N. crassa* coding for an aggregate of five aromatic synthetic enzymes. *Proc. Nat. Acad. Sci.*, **58**, 1453–1460.(*260*)

Gillespie, D., and D. Spiegelman (1965). A quantitative assay for DNA–RNA hybrids with DNA immobilised on a membrane. *J. Mol. Biol.*, **12**, 829–842. (*41*)

Glover, S. W., and C. Colson (1969). Genetics of host-controlled restriction and modification in *E. coli. Genet. Res.*, **13**, 227–240.(*325*)

Gold, M., M. Gefter, R. Hausman, and J. Hurwitz (1966). Methylation of DNA. *J. Gen. Physiol.*, **49**, 5–28.(*321*)

Goodgal, S. H. (1961). Studies on transformation of *Hemophilus influenzae*. IV. Linked and unlinked transformations. *J. Gen. Physiol.*, **45**, 205–228.(*46*)

Goodman, H. M., J. Abelson, A. Landy, S. Brenner, and J. D. Smith (1968). *Amber* suppression: a nucleotide change in the anticodon of a tyrosine *t*RNA. *Nature*, **217**, 1019–1024.(*186*, *214*)

Goodman, H., and A. Rich (1962). Formation of a DNA–*t*RNA hybrid and its relation to the origin, evolution, and degeneracy of *t*RNA. *Proc. Nat. Acad. Sci. USA*, **48**, 2101–2109.(*170*)

Gordon, J. (1967). Interaction of guanosine 5′ triphosphate with a supernatant fraction from *E. coli* and aminoacyl-*t*RNA. *Proc. Nat. Acad. Sci. USA*, **58**, 1574–1578.(*110*)

Gordon, J. (1968). A stepwise reaction yielding a complex between a supernatant fraction from *E. coli*, guanosine-5′-triphosphate, and aminoacyl-*t*RNA. *Proc. Nat. Acad. Sci. USA*, **59**, 179–183.(*110*)

Gorini, L., G. A. Jacoby, and L. Breckenridge (1966). Ribosomal ambiguity. *Cold Spring Harbor Symp. Quant. Biol.*, **31**, 657–665.(*217*)

Gorini, L., and E. Kataja (1964). Streptomycin induced over-suppression in *E. coli*. *Proc. Nat. Acad. Sci. USA*, **51**, 995–1001.(*215*)

Gorini, L., and E. Kataja (1965). Suppression activated by streptomycin and related antibiotics in *E. coli* strains. *Biochem. Biophys. Res. Commun.*, **18**, 656–663. (*215*)

Gould, H. J. (1967). The nature of high molecular weight fragments of *r*RNA. *J. Mol. Biol.*, **29**, 307–314.(*155*)

Greenberg, H., and S. Penman (1966). Methylation and processing of *r*RNA in Hela cells. *J. Mol. Biol.*, **21**, 527–535.(*174*, *176*)

Griffith, F. (1928). Significance of *Pneumococcal* types. *J. Hyg.*, **27**, 113–159.(*23*)

Grodzicker, T., and D. Zipser (1968). A mutation which creates a new site for the re-initiation of polypeptide synthesis in the *z* gene of the *lac* operon of *E. coli*. *J. Mol. Biol.*, **38**, 305–314.(*266*)

Gros, F., H. Hiatt, W. Gilbert, C. G. Kurland, R. W. Risebrough, and J. D. Watson (1961). Unstable RNA revealed by pulse labelling of *E. coli*. *Nature*, **190**, 581–585.(*98*)

Gross, J. D., and L. Caro (1966). DNA transfer in bacterial conjugation. *J. Mol. Biol.*, **16**, 269–284.(*65*)

Gross, J. D., D. Karamata, and P. G. Hempstead (1968). Temperature-sensitive mutants of *B. subtilis* defective in DNA synthesis. *Cold Spring Harbor Symp. Quant. Biol.*, **33**, 307–312.(*330*, *338*)

Grossman, L., J. Kaplan, S. Kushner, and I. Mahler (1968). Enzymes involved in the early stages of repair of ultraviolet irradiated DNA. *Cold Spring Harbor Symp. Quant. Biol.*, **33**, 229–234.(*357*)

Gurney, T. J., and M. S. Fox (1968). Physical and genetic hybrids formed in bacterial transformation. *J. Mol. Biol.*, **32**, 83–100.(*46*)

Gussin, G. N., M. R. Capecchi, J. M. Adams, J. E. Argetsinger, J. Tooze, K. Weber, and J. D. Watson (1966). Protein synthesis directed by RNA phage messengers. *Cold Spring Harbor Symp. Quant. Biol.*, **31**, 257–272.(*119*)

Guthrie, C., and M. Nomura (1968). Initiation of protein synthesis: a critical test of the 30S subunit model. *Nature*, **219**, 232–236.(*124*)

Guttman, B., and A. Novick (1963). An *m*RNA for β-galactosidase in *E. coli*. *Cold Spring Harbor Symp. Quant. Biol.*, **28**, 373–374.(*229*)

Haenni, A., and J. Lucas-Lenard (1968). Stepwise synthesis of a tripeptide. *Proc. Nat. Acad. Sci. USA*, **61**, 1363–1369.(*116*)

Hall, B. D., and S. Spiegelman (1961). Sequence complementarity of T2 DNA and T2 specific RNA. *Proc. Nat. Acad. Sci. USA*, **47**, 137–146.(*38*, *98*)

Hall, Z. W., and I. R. Lehman (1968). An *in vitro* transversion by a mutationally altered T4 induced DNA polymerase. *J. Mol. Biol.*, **36**, 321–334.(*304*)

Hanawalt, P. C. (1967). Normal replication of DNA after repair replication in bacteria. *Nature*, **214**, 269–270.(*355*)

Hanawalt, P. C. (1968). Cellular recovery from photochemical damage. *Photo physiol.*, **4**, 203–251.(*353, 355, 356, 359, 361*)

Hanawalt, P. C., and R. H. Haynes (1965). Repair replication of DNA in bacteria: irrelevance of chemical nature of base defect. *Biochem. Biophys. Res. Commun.*, **19**, 464–467.(*357*)

Hanawalt, P. C., and R. H. Haynes (1967). The repair of DNA. *Sci. Amer.*, **216**, part 2, 36–43.(*356*)

Hanawalt, P. C., O. Maaløe, D. J. Cummings, and M. Schaechter (1961). The normal DNA replication cycle. *J. Mol. Biol.*, **3**, 156–165.(*327*)

Hanawalt, P. C., D. E. Pettijohn, E. C. Pauling, C. F. Brunk, D. W. Smith, L. C. Kanner and J. L. Couch (1968). Repair replication of *DNA in vivo*. *Cold Spring Harbor Symp. Quant. Biol.*, **33**, 187–194.(*355, 359*)

Hayashi, M. (1965). A DNA–RNA complex as an intermediate of *in vitro* genetic transcription. *Proc. Nat. Acad. Sci. USA*, **54**, 1736–1743.(*147*)

Hayashi, M., M. N. Hayashi, and S. Spiegelman (1963). Restriction of *in vivo* transcription to one of the complementary DNA strands. *Proc. Nat. Acad. Sci. USA*, **50**, 664–672.(*99, 143*)

Hayashi, M., M. N. Hayashi, and S. Spiegelman (1964). DNA circularity and the mechanism of strand selection in the generation of genetic messages. *Proc. Nat. Acad. Sci. USA*, **51**, 351–361.(*144*)

Hayashi, M. N., and M. Hayashi (1968). The stability of native DNA–RNA complexes during *in vivo* øX174 transcription. *Proc. Nat. Acad. Sci.*, **61**, 1107–1115.(*147*)

Hayes, W. (1952a). Recombination in *Bact. coli* K12: unidirectional transfer of genetic material. *Nature*, **169**, 118–119.(*62*)

Hayes, W. (1952b). Genetic recombination in *Bact. coli* K12: analysis of the stimulating effect of ultraviolet light. *Nature*, **169**, 1017–1018.(*62*)

Hayes, W. (1968a). Trends and methods in virus research. *Symp. Soc. Gen. Mic.*, **18**, 1–14.(*51, 54*)

Hayes, W. (1968b). *The Genetics of Bacteria and their Viruses*, 2nd Ed., Blackwell Scientific Publications, Oxford.(*50, 65, 375*)

Hayward, R. S., G. L. Elicieri, and S. B. Weiss (1966). RNA sulfur activity in extracts from *E. coli*. *Cold Spring Harbor Symp. Quant. Biol.*, **31**, 459–464. (*184*)

Hecht, L. I., M. L. Stephenson, and P. C. Zamecnik (1958). Dependence of amino acid binding to *t*RNA on cytidine triphosphate. *Biochim. Biophys. Acta*, **29**, 460–461.(*100*)

Hecht, L. I., M. L. Stephenson, and P. C. Zamecnik (1959). Binding of amino acids to the end group of a *t*RNA. *Proc. Nat. Acad. Sci. USA*, **45**, 505–518. (*100*)

Hecht, L. I., P. C. Zamecnik, M. L. Stephenson, and J. F. Scott (1958). Nucleoside triphosphates as precursors of RNA end groups in a mammalian system. *J. Biol. Chem.*, **233**, 954–963.(*100*)

Hecht, N. B., M. Bleyman, and C. R. Woese (1968). The formation of 5S *r*RNA in B. subtilis by post transcriptional modification. *Proc. Nat. Acad. Sci. USA*, **59**, 1278–1283.(*177*)

Helmstetter, C. E., and S. Cooper (1968). DNA synthesis during the division cycle of rapidly growing *E. coli* B/r. *J. Mol. Biol.*, **31**, 507–518.(*331*)

Helmstetter, C. E., and O. Pierucci (1968). Cell division during inhibition of DNA synthesis in *E. coli*. *J. Bacteriol.*, **95**, 1627–1633.(*338*)

Helmstetter, C. E., S. Cooper, O. Pierucci, and E. Revelas (1968). On the bacterial life sequence. *Cold Spring Harbor Symp. Quant. Biol.*, **33**, 809–822. (*349*)

Henshaw, E. C. (1968). *m*RNA in rat liver polyribosomes. Evidence that it exists as ribonucleoprotein particles. *J. Mol. Biol.*, **36**, 401–412.(*151*)

Hershey, A. D., and M. Chase (1951). Genetic recombination and heterozygosis in bacteriophage. *Cold Spring Harbor Symp. Quant. Biol.*, **16**, 471–479.(*23*, *373*)

Hershey, A. D., and M. Chase (1952). Independent functions of viral protein and nucleic acid in growth of bateriophage. *J. Gen. Physiol.*, **36**, 39–56.(*23*)

Hershey, J. W. B., K. F. Dewey, and R. E. Thach (1969). Purification and properties of initiation factor f_1. *Nature*, **222**, 944–947.(*130*)

Hershey, J. W. B., and R. E. Monro (1966). A competitive inhibitor of the GTP reaction in protein synthesis. *J. Mol. Biol.*, **18**, 68–76.(*114*)

Hershey, J. W. B., and R. E. Thach (1967). Role of GTP in the initiation of peptide synthesis. I. Synthesis of formyl-methionine puromycin. *Proc. Nat. Acad. Sci. USA*, **57**, 759–766.(*128*, *129*)

Hirt, B. (1969). Replicating molecules of polyoma virus DNA. *J. Mol. Biol.*, **40**, 141–144.(*297*)

Hoagland, M. B., E. B. Keller, and P. C. Zamecnik (1956). Enzymatic carboxyl activation of amino acids. *J. Biol. Chem.*, **218**, 345–357.(*100*)

Hoagland, M. B., M. L. Stephenson, J. E. Scott, L. I. Hecht, and P. C. Zamecnik (1958). A soluble RNA intermediate in protein synthesis. *J. Biol. Chem.*, **231,** 241–257.(*100*)

Hoagland, M. B., P. C. Zamecnik, N. Sharon, F. Lipmann, M. P. Stulberg, and P. D. Boyer (1957). Oxygen transfer to AMP in the enzymatic synthesis of the hydroxamate of tryptophan. *Biochim. Biophys. Acta*, **26**, 215–217.(*101*)

Holley, R. W. (1968). Experimental approaches to the determination of the nucleotide sequences of large oligonucleotides and small nucleic acids. *Prog. Nuc. Acid Res.*, **8**, 37–48.(*184*)

Holley, R. W., J. Apgar, G. A. Everett, J. T. Madison, H. Marquise, S. H. Merrill, J. R. Penswick, and A. Zamir (1965). Structure of a ribonucleic acid. *Science*, **147**, 1462–1465.(*185*, *187*, *188*)

Holliday, R. (1964). A mechanism for gene conversion in fungi. *Genet. Res.*, **5**, 282–304.(*368*, *369*)

Holliday, R. (1966). Radiosensitive mutants of *Ustilago maydis*. *Mut. Res.*, **2**, 557–559.(*360*)

Horikoshi, K., and R. H. Doi (1967). Synthesis of *N*-formylmethionyl-*t*RNAby *B. subtilis* extracts. *Arch. Biochem. Biophys.*, **122**, 685–693.(*122*)

Horikoshi, K., and R. H. Doi (1968). The NH_2 terminal residues of *B. subtilis* proteins. *J. Biol. Chem.*, **243**, 2381–2384.(*122*)

Hosokawa, K., R. K. Fujimura, and M. Nomura (1966). Reconstitution of functionally active ribosomes from inactive sub-particles and proteins. *Proc. Nat. Acad. Sci. USA*, **55**, 190–204.(*164*)

Howard-Flanders, P., R. P. Boyce, and L. Theriot (1966). Three loci in *E. coli* K12 that control the excision of pyrimidine dimers and certain other mutagen products from DNA. *Genetics*, **53**, 1119–1136.(*361*)

Howard-Flanders, P., W. D. Rupp, B. M. Wilkins, and R. S. Cole (1968). DNA replication and recombination after UV irradiation. *Cold Spring Harbor Symp. Quant. Biol.*, **33**, 195–208.(*362*, *364*, *365*)

Howard-Flanders, P., and L. Theriot (1966). Mutants of *E. coli* defective in DNA repair and in genetic recombination. *Genetics*, **53**, 1137–1150.(*378*)

Huberman, J. A., and G. Attardi (1967). Studies of fractionated Hela cell metaphase chromosomes. I. The chromosomal distribution of DNA complementary to 28S and 18S *r*RNA and the cytoplasmic *m*RNA. *J. Mol. Biol.*, **29**, 487–505.(*171*)

Huberman, J. A., and D. A. Riggs (1968). On the mechanism of DNA replication in mammalian chromosomes. *J. Mol. Biol.*, **32**, 327–341.(*302*, *303*)

Hunt, T., T. Hunter, and A. Monro (1969). Control of hemoglobin synthesis: rate of translation of the *m*RNAs for the α and β chains. *J. Mol. Biol.*, **43**, 123–134.(*293*)

Ibuki, F., and K. Moldave (1968). The effect of GTP, other nucleotides, and aminoacyl-*t*RNA on the activity of transferase-I and on its binding to ribosomes. *J. Biol. Chem.*, **243**, 44–50.(*110*)

Ihler, G. and W. D. Rupp (1969). Strand specific transfer of donor DNA during conjugation in *E. coli. Proc. Nat. Acad. Sci. USA*, **63**, 138–143.(*66*)

Imamoto, F. (1968a). On the initiation of transcription of the tryptophan operon in *E. coli. Proc. Nat. Acad. Sci. USA*, **60**, 305–312.(*275*)

Imamoto, F. (1968b). Immediate cessation of transcription of the operator region of the tryptophan operon of *E. coli. Nature*, **220**, 31–35.(*233*)

Imamoto, F. (1969). Intragenic initiations of transcription of the tryptophan operon following dinitrophenol treatment with tryptophan. *J. Mol. Biol.*, **43**, 51–70.(*282*)

Imamoto, F., and J. Ito (1968). Simultaneous initiation of transcription and translation at internal sites in the tryptophan operon of *E. coli. Nature*, **220**, 27–31.(*282*)

Imamoto, F., J. Ito, and C. Yanofsky (1966). Polarity in the tryptophan operon of *E. coli. Cold Spring Harbor Symp. Quant. Biol.*, **31**, 235–250.(*264–6*, *271–2*)

Imamoto, F., N. Morikawa, and K. Sato (1965). On the transcription of the tryptophan operon. III. Multicistronic *m*RNA and polarity for transcription. *J. Mol. Biol.*, **13**, 169–182.(*229*, *271*)

Imamoto, F., N. Morikawa, K. Sato, S. Mishima, T. Nishimura, and A. Matsushiro (1965). On the transcription of the tryptophan operon. II. Production of the specific *m*RNA. *J. Mol. Biol.*, **13**, 157–168(*229*, *271*)

Imamoto, F., and C. Yanofsky (1967a). Transcription of the tryptophan operon in polarity mutants of *E. coli*. I. Characterisation of the tryptophan *m*RNA of polar mutants. *J. Mol. Biol.*, **28**, 1–23.(*271*)

Imamoto, F., and C. Yanofsky (1967b). Transcription of the tryptophan operon in polarity mutants of *E. coli*. II. Evidence for normal production of *trp-m*RNA molecules and for premature termination of transcription. *J. Mol. Biol.*, **28**, 23–35.(*271*)

Imura, N., G. B. Weiss, and R. W. Chambers (1969). Reconstitution of alanine acceptor activity from fragments of yeast $t\mathrm{RNA}_{\mathrm{II}}^{\mathrm{ala}}$. *Nature*, **222**, 1147–1148. (*201*)

Ingram, V. M. (1957). Gene mutations in human hemoglobin: the chemical difference between normal and sickle cell hemoglobin. *Nature*, **180**, 326–328. (*20, 73*)

Inselberg, J. (1968). Physical evidence for the integration of prophage P1 into the *E. coli* chromosome. *J. Mol. Biol.*, **31**, 553–560.(*56*)

Ippen, K., J. K. Miller, J. Scaife, and J. Beckwith (1968). New controlling element in the *lac* operon of *E. coli*. *Nature*, **217**, 825–826.(*233*)

Ishikura, H., and S. Nishimura (1968). Fractionation of serine *t*RNAs from *E. coli* and their coding properties. *Biochim. Biophys. Acta*, **155**, 72–81.(*207*)

Ito, J., and F. Imamoto (1968). Sequential derepression and repression of the tryptophan operon in *E. coli*. *Nature*, **220**, 441–444.(*278*)

Itoh, T., E. Otaka, and S. Osawa (1968). Release of ribosomal proteins from *E. coli* ribosomes with high concentrations of lithium chloride. *J. Mol. Biol.*, **33**, 109–122.(*164*)

Jacob, F., S. Brenner, and F. Cuzin (1963). On the regulation of DNA replication in bacteria. *Cold Spring Harbor Symp. Quant. Biol.*, **28**, 329–348.(*66, 347–8*)

Jacob, F., and J. Monod (1961). Genetic regulatory mechanisms in the synthesis of protein. *J. Mol. Biol.*, **3**, 318–356.(*221–232*)

Jacob, F., A. Ryter, and F. Cuzin (1966). On the association between DNA and the membrane in bacteria. *Proc. Roy. Soc. Ser. B.*, **164**, 267–278.(*346*)

Jacob, F., A. Ullman, and J. Monod (1964). Le promoteur, élément génétique nécessaire à l'éxpression d'un operon. *C.R. Acad. Sci., Paris*, **258**, 3125–3128. (*232*)

Jacob, F., and E. L. Wollman (1961). *Sexuality and the Genetics of Bacteria*, Academic Press, New York.(*64*)

Jacoby, G. A., and L. Gorini (1967). Genetics of control of the arginine pathway in *E. coli* B and K. *J. Mol. Biol.*, **24**, 41–50.(*254*)

Jacoby, G. A., and L. Gorini (1969). A unitary account of the repression mechanism of arginine biosynthesis in *E. coli*. I. The genetic evidence. *J. Mol. Biol.*, **39**, 73–87.(*256, 257*)

Jacquet, M., and A. Kepes (1969). The step sensitive to catabolite repression and its reversal by 3′-5′ cyclic AMP during induced synthesis of β-galactosidase in *E. coli*. *Biochem. Biophys. Res. Commun.*, **36**, 84–92.(*238*)

Jeanteur, Ph., F. Amaldi, and G. Attardi (1968). Partial sequence analysis of ribosomal RNA from Hela cells: II evidence for sequences of non-ribosomal type in 45S and 32S *r*RNA precursors. *J. Mol. Biol.*, **33**, 757–776.(*176*)

Jeanteur, Ph., and G. Attardi (1968). Hybridisation competition experiments with 18S and 28S *r*RNA and purified 45S and 32S *r*RNA precursors in Hela cells. *Fed. Eur. Biochem. Soc. Abstracts, Madrid*, 226.(*176*)

Jones, D. S., S. Nishimura, H. G. Khorana (1966). *In vitro* synthesis of copolypeptides. *J. Mol. Biol.*, **16**, 454–472.(*78, 79*)

Jones, O. W., M. Dieckmann, and P. Berg (1968). Ribosome induced dissociation of RNA from an RNA polymerase–DNA–RNA complex. *J. Mol. Biol.*, **31**, 177–189.(*150*)

Josse, J., A. D. Kaiser, and A. Kornberg (1961). Enzymatic synthesis of DNA. VIII. Frequencies of nearest neighbour base sequences in DNA. *J. Biol. Chem.*, **236**, 864–875.(*30*)

Jovin, T. M., P. T. Englund, and L. L. Bertsch (1969). Enzymatic synthesis of DNA. XXVI. Physical and chemical studies of a homogeneous DNA polymerase. *J. Biol. Chem.*, **244**, 2996–3008.(*306*)

Jovin, T. M., P. T. Englund, and A. Kornberg (1969). Enzymatic synthesis of DNA. XXVII. Chemical modifications of DNA polymerase. *J. Biol. Chem.*, **244**, 3009–3018.(*307*)

Kaempfer, R. (1968). Ribosome subunit exchange during protein synthesis. *Proc. Nat. Acad. Sci. USA*, **61**, 106–113.(*134*)

Kaempfer, R. (1969). Ribosome subunit exchange in the cytoplasm of a eucaryote. *Nature*, **222**, 950–953.(*125*)

Kaempfer, R., and M. Meselson (1968). Permanent association of 5S RNA molecules with 50S ribosomal subunits in growing bacteria. *J. Mol. Biol.*, **34**, 703–708.(*156*)

Kaempfer, R., M. Meselson, and H. J. Raskas (1968). Cyclic dissociation into stable subunits and reformation of ribosomes during bacterial growth. *J. Mol. Biol.*, **31**, 277–289.(*134*)

Kaji, H., and A. Kaji (1965). Specific binding of *t*RNA to ribosomes: effect of streptomycin. *Proc. Nat. Acad. Sci. USA*, **54**, 213–218.(*216*)

Kaplan, J. C., S. R. Kushner, L. Grossman (1969). Enzymatic repair of DNA, I. Purification of two enzymes involved in the excision of thymine dimers from u.v.-irradiated DNA. *Proc. Nat. Acad. Sci.*, **63**, 144–151.(*357*)

Karlstrom, O., and L. Gorini (1969). A unitary account of the repression mechanism of arginine biosynthesis in *E. coli*. II. Application to the physiological evidence. *J. Mol. Biol.*, **39**, 89–94.(*256*)

Kazazian, H. H. Jr., and M. L. Friedman (1968). The characterisation of separated α and β chain polyribosomes in rabbit reticulocytes. *J. Biol. Chem.*, **243**, 6646–6650.(*293*)

Kelly, R. B., M. R. Atkinson, J. A. Huberman, and A. Kornberg (1969). Excision of thymine dimers and other mismatched sequences by DNA polymerase of *E. coli*. *Nature*, **224**, 495–501.(*360*)

Khorana, H. G., H. Buchi, H. Ghosh, N. Gupta, T. M. Jacob, H. Kossel, R. Morgan, S. A. Narang, E. Ohtsuka, and R. D. Wells (1966). Polynucleotide synthesis and theg enetic code. *Cold Spring Harbor Symp. Quant. Biol.*, **31**, 39–49.(*80, 121, 122*)

Kiho, Y., and A. Rich (1965). A polycistronic messenger RNA associated with β-galactosidase induction. *Proc. Nat. Acad. Sci.*, **54**, 1751–1758.(*230, 271*)

Kim, S. H., and A. Rich (1968). Single crystals of *t*RNA: an X-ray diffraction study. *Science*, **162**, 1381–1384.(*198*)

King, J. L., and T. H. Jukes (1969). Non-Darwinian evolution. *Science*, **164**, 788–797.(*87, 88*)

Kingdon, H. S., L. Webster, and E. Davie (1958). Enzymatic formation of adenyl tryptophan: isolation and identification. *Proc. Nat. Acad. Sci. USA*, **44**, 757–765.(*101*)

Knight, E. Jr., and J. E. Darnell (1967). Distribution of 5S RNA in Hela cells. *J. Mol. Biol.*, **28**, 491–502.(*156, 177*)

Kohler, R. E., E. Z. Ron, and B. D. Davis (1968). Significance of the free 70S ribosomes in *E. coli* extracts. *J. Mol. Biol.*, **36**, 71–82.(*135, 136*)

Kolakofsky, D., K. Dewey, and R. E. Thach (1969). Purification and properties of initiation factor f_2. *Nature*, **223**, 694–697.(*130*)

Kornberg, A. (1966). The biosynthesis of DNA. In V. V. Koningsberger and L. Bosch (Eds.), *Regulation of Nucleic Acid and Protein Synthesis* 22–38, Elsevier, Amsterdam.(*302, 314*)

Kornberg, A. (1969). Active center of DNA polymerase. *Science*, **163**, 1410–1418.(*305, 306, 307, 310, 311, 314, 317*)

Kossel, H., A. R. Morgan, and H. G. Khorana (1967). Studies on polynucleotides. LXXIII. Synthesis *in vitro* of polypeptides containing repeating tetrapeptide sequences dependent upon DNA-like polymers containing repeating tetranucleotide sequences; direction of reading of messenger RNA. *J. Mol. Biol.*, **26**, 449–475.(*80*)

Kozinski, A. W. (1968). Molecular recombination in the ligase-negative T4 *amber* mutants. *Cold Spring Harbor Symp. Quant. Biol.*, **33**, 375–392.(*376*)

Kozinski, A. W. (1969). Unbiased participation of T4 phage DNA strands in replication. *Biochem. Biophys. Res. Commun.*, **35**, 294–299.(*346*)

Kozinski, A. W., and P. B. Kozinski (1969). Covalent repair of molecular recombinants in the ligase-negative *amber* mutant of T4 bacteriophage. *J. Virol.*, **3**, 85–88.(*376*)

Kozinski, A. W., P. B. Kozinski, and R. James (1967). Molecular recombination in T4 bacteriophage DNA. I. Tertiary structure of early replicative and recombining DNA. *J. Virol.*, **1**, 758–770.(*377*)

Kurland, C. G., and O. Maaløe (1962). Regulation of ribosomal and *t*RNA synthesis. *J. Mol. Biol.*, **4**, 193–210.(*290*)

Kurland, C. G., M. Nomura, and J. D. Watson (1962). The physical properties of the chloromycetin particles. *J. Mol. Biol.*, **4**, 388–394.(*179*)

Lacey, J. C. Jr., and K. M. Pruitt (1969). Origin of the genetic code. *Nature*, **223**, 799–804.(*89*)

Lacks, S. (1962). Molecular fate of DNA in genetic transformation of *Pneumococcus*. *J. Mol. Biol.*, **5**, 119–131.(*45*)

Lagerkvist, U., L. Rymo, and J. Waldenstrom (1966). Structure and function of *t*RNA. II. Enzyme-substrate complexes with valyl *t*RNA synthetases from yeast. *J. Biol. Chem.*, **241**, 5391–5400.(*102*)

Lake, J. A., and W. W. Beeman (1968). On the conformation of yeast *t*RNA. *J. Mol. Biol.*, **31**, 115–125.(*196, 198, 200*)

Landy, A., J. Abelson, H. M. Goodman, and J. D. Smith (1967). Specific hybridisation of tyrosine *t*RNAs with DNA from a transducing bacteriophage ø80 carrying the *amber* suppressor gene su_{III}. *J. Mol. Biol.*, **29**, 457–471.(*214*)

Lanni, F. (1964). The biological coding problem. *Adv. Gen.*, **12**, 2–143.(*71*)

Lark, C. (1968a). Studies on the *in vivo* methylation of DNA in *E. coli* $15T^-$. *J. Mol. Biol.*, **31**, 389–400.(*322*)

Lark, C. (1968b). Effect of methionine analogues, ethionine and norleucine on DNA synthesis in *E. coli* 15T⁻. *J. Mol. Biol.*, **31**, 401–414.(*322*, *325*)

Lark, K. G. (1966). Regulation of chromosome replication and segregation in bacteria. *Bact. Rev.*, **30**, 1–32.(*337*, *338*, *347*)

Lark, K. G. (1969). *Ann. Rev. Biochem.*, **38**, 569–607.(*301*, *329*)

Lark, K. G., T. Repko, and E. J. Hoffman (1963). The effect of amino acid deprivation on subsequent DNA replication. *Biochim. Biophys. Acta*, **76**, 9–24.(*297*, *300*, *327*)

Lark, K. G., and H. Renger (1969). Initiation of DNA replication in *E. coli* 15T⁻: chronological dissection of three physiological processes required for initiation. *J. Mol. Biol.*, **42**, 221–236.(*329*)

Last, J. A., W. M. Stanley Jr., M. Salas, M. B. Hille, A. J. Wahba, and S. Ochoa (1967). Translation of the genetic message. IV. UAA as a chain termination codon. *Proc, Nat. Acad. Sci. USA*, **57**, 1062–1067.(*132*)

Lavallé, R., and G. de Hauwer (1968). Messenger RNA synthesis during amino acid starvation in *E. coli. J. Mol. Biol.*, **37**, 269–288.(*291*)

Laycock, D. G., and J. A. Hunt (1969). Synthesis of rabbit globin by a bacterial cell free system. *Nature*, **221**, 1118–1122.(*86*, *124*)

Leboy, P. S., E. C. Cox, and J. G. Flaks (1964). The chromosomal site specifying a ribosomal protein in *E. coli. Proc. Nat. Acad. Sci. USA*, **52**, 1367–1374.(*160*)

Leder, P., and M. N. Nau (1967). Initiation of protein synthesis. III. Factor-GTP-codon dependent binding of fmet-*t*RNA to ribosomes. *Proc. Nat. Acad. Sci. USA*, **58**, 774–781.(*122*, *123*)

Leder, P., L. E. Skogerson, and M. N. Nau (1969). Translocation of *m*RNA codons. I. The preparation and characterisation of a homogeneous enzyme. *Proc. Nat. Acad. Sci. USA*, **62**, 454–460.(*117*)

Leder, P., L. E. Skogerson, and D. J. Roufa (1969). Translocation of *m*RNA codons. II. Properties of an anti-translocase antibody. *Proc. Nat. Acad. Sci. USA*, **62**, 928–933.(*117*)

Lederberg, J., and E. L. Tatum (1946). Novel genotypes in mixed cultures of biochemical mutants of bacteria. *Cold Spring Harbor Symp. Quant. Biol.*, **11**, 113.(*62*)

Leive, L., and V. Kollin (1967). Synthesis, utilisation and degradation of *lactose* operon *m*RNA in *E. coli. J. Mol. Biol.*, **24**, 247–259.(*230*, *277*)

Lengyel, P., and D. Soll (1969). Mechanism of protein synthesis. *Bact. Rev.*, **33**, 264–301.(*106*)

Lerman, M. I., A. S. Spirin, L. P. Gavrilova, and V. F. Golov (1966). Studies on the structure of ribosomes. II. Stepwise dissociation of protein from ribosomes by caesium chloride and the re-assembly of ribosome-like particles. *J. Mol. Biol.*, **15**, 268–281.(*164*)

Levinthal, C., and N. Visconti (1953). Growth and recombination in bacterial viruses. *Genetics*, **38**, 500–511.(*51*)

Lewin, S. (1967a). Some aspects of hydration and stability of the native states of DNA. *J. Theor. Biol.*, **17**, 181–212.(*33*)

Lewin, S. (1967b). The conformation stabilising potential of water in biopolymers and its detection by deuteration. *Studia Biophysica*, **4**, 29–44.(*33*)

Lewin, S., and D. S. Pepper (1965). Variation of the melting temperature of calf-thymus DNA with pH and type of buffer. *Arch. Biochem. Biophys.*, **109**, 192–194.(*35*)

Lewis, E. B. (1945). The relation of repeats to position effects in *D. melanogaster*. *Genetics*, **30**, 137–166.(*18*)

Lieb, M. (1969). Allosteric properties of the lambda repressor. *J. Mol. Biol.*, **39**, 379–382.(*231*)

Lipmann, F. (1963). Messenger ribonucleic acid. *Prog. Nuc. Acid Res.*, **1**, 135–163.(*97*)

Lipmann, F. (1969). Polypeptide chain elongation in protein biosynthesis. *Science*, **164**, 1024–1031.(*110, 118*)

Lipsett, M. N. (1966). Disulfide bonds in *t*RNA. *Cold Spring Harbor Symp. Quant. Biol.*, **31**, 449–456.(*193*)

Littlefield, J. W., E. B. Keller, J. Gross, and P. C. Zamecnik (1955). Studies on cytoplasmic ribonucleoprotein particles from the liver of the rat. *J. Biol. Chem.*, **217**, 111–123.(*96*)

Lodish, H. F. (1968a). Polar effects of an *amber* mutation in f_2 bacteriophage. *J. Mol. Biol.*, **32**, 47–58.(*286*)

Lodish, H. F. (1968b). Independent translation of the genes of bacteriophage f_2 RNA. *J. Mol. Biol.*, **32**, 681–686.(*288*)

Lodish, H. F. (1968c). Bacteriophage f_2 RNA: control of translation and gene order. *Nature*, **220**, 345–350.(*288*)

Loftfield, R. B. (1963). The frequency of errors in protein biosynthesis. *Biochem. J.*, **89**, 82-92.(*102*)

Lucas-Lenard, J., and A–L. Haenni (1968). Requirement of GTP for ribosomal binding of aminoacyl-*t*RNA. *Proc. Nat. Acad. Sci. USA*, **59**, 554–561.(*114*)

Lucas-Lenard, J., and A–L. Haenni (1969). Release of *t*RNA during peptide chain elongation. *Proc. Nat. Acad. Sci. USA*, **63**, 93–97.(*117*)

Lucas-Lenard, J., and F. Lipmann (1967). Initiation of polyphenylalanine synthesis by *N*-acetylphenylalanyl-*t*RNA. *Proc. Nat. Acad. Sci. USA*, **57**, 1050–1057.(*110, 127*)

Luria, S. E., and J. Darnell Jr. (1968). *General Virology*. John Wiley and Sons, Chichester.(*55*)

Luzzato, L., D. Apirion, and D. Schlessinger (1968). Mechanism of action of streptomycin in *E. coli*: interruption of the ribosome cycle at the initiation of protein synthesis. *Proc. Nat. Acad. Sci. USA*, **60**, 873–881.(*219*)

Luzzato, L., D. Apirion, D. Schlessinger (1969). Polyribosome depletion and blockage of the ribosome cycle by streptomycin in *E. coli*. *J. Mol. Biol.*, **42**, 315–336.(*219*)

Luzzato, L., D. Schlessinger, and D. Apirion (1968). *E. coli:* high resistance or dependence on streptomycin produced by the same allele. *Science*, **161**, 478–479.(*216*)

Maaløe, O. and P. C. Hanawalt (1961). Thymine deficiency and the normal DNA replication cycle. *J. Mol. Biol.*, **3**, 144–155.(*326, 328*)

Maas, W. K., and A. J. Clark (1964). Studies on the mechanism of repression of arginine biosynthesis in *E. coli*. II. Dominance of repressibility in diploids. *J. Mol. Biol.*, **8**, 365–370.(*254*)

Maas, W. K., R. Maas, J. M. Wiame, and N. Glansdorff (1964). Studies on the mechanism of repression of arginine biosynthesis in *E. coli*. I. Dominance of repressibility in zygotes. *J. Mol. Biol.*, **8**, 359–364.(*254*)

MacHattie, L. A., D. A. Ritchie, C. A. Thomas Jr., and C. C. Richardson (1967). Terminal repetition in permuted T2 bacteriophage DNA molecules. *J. Mol. Biol.*, **23**, 355–364.(*54*)

Mackay, A. L. (1967). Optimization of the genetic code. *Nature*, **216**, 159–160.(*87*)

Maden, B. E. H. (1968). Ribosome formation in animal cells. *Nature*, **219**, 685–688.(*173*)

Madison, J. T. (1968). Primary structure of *t*RNA. *Am. Rev. Biochem.*, **37**, 131–148.(*185*)

Maitra, U., S. N. Cohen, and J. Hurwitz (1966). Specificity of initiation and synthesis of RNA from DNA templates. *Cold Spring Harbor Symp. Quant. Biol.*, **31**, 113–122.(*139*)

Maitra, U., and J. Hurwitz (1967). The role of DNA in RNA synthesis. XIII. Modified purification procedure and additional properties of RNA polymerase from *E. coli* W. *J. Biol. Chem.*, **242**, 4897–4907.(*139*)

Maitra, U., Y. Nakata, and J. Hurwitz (1967). The role of DNA in RNA synthesis. XIV. A study of the initiation of RNA synthesis. *J. Biol. Chem.*, **242**, 4908–4918.(*139*)

Mangiarotti, G. (1969). Role of the initiation factor in the response of 70S ribosomes to R17 RNA. *Nature*, **222**, 947–950.(*131*)

Mangiarotti, G., D. Apirion, D. Schlessinger, and L. Silengo (1968). Biosynthetic precursors of 30S and 50S ribosomal particles in *E. coli*. *Biochemistry*, **7**, 456–471.(*178*, *181*, *182*)

Mangiarotti, G., and D. Schlessinger (1966). Polyribosome metabolism in *E. coli*. I. Extraction of polyribosomes and ribosomal subunits from fragile growing *E. coli*. *J. Mol. Biol.*, **20**, 123–143.(*134*)

Mangiarotti, G., and D. Schlessinger (1967). Polyribosome metabolism in *E. coli*. II. Formation and lifetime of *m*RNA molecules, ribosome subunit couples, and polyribosomes. *J. Mol. Biol.*, **29**, 355–418.(*134*)

Marcker, K. A., B. F. C. Clark, and J. S. Anderson (1966). *N*-formyl methionyl-*t*RNA and its relation to protein synthesis. *Cold Spring Harbor Symp. Quant. Biol.*, **31**, 279–286.(*120*)

Marcker, K. A., and F. Sanger (1964). *N*-formyl-methionyl-*t*RNA. *J. Mol. Biol.*, **8**, 835–840.(*119*)

Marmur, J., and P. Doty (1959). Heterogeniety in DNA. I. Dependence on composition of the configurational stability of DNAs. *Nature*, **183**, 1427–1428.(*36*, *37*)

Marmur, J., and P. Doty (1961). Thermal renaturation of DNAs. *J. Mol. Biol.*, **3**, 585–594.(*36*)

Marmur, J., and C. M. Greenspan (1963). Transcription *in vivo* of DNA from bacteriophage SP8. *Science*, **142**, 387–389.(*143*)

Marmur, J., R. Rownd, and C. L. Schildkraut (1963). Denaturation and renaturation of DNA. *Prog. Nuc. Acid Res.*, **1**, 232–300.(*33*, *35*)

Marshall, R., and M. Nirenberg (1969). RNA codons recognised by *t*RNA from amphibian embryos and adults. *Develop. Biol.*, **19**, 1–11.(*206*)

Martin, R. G. (1963). The one operon—one messenger theory of transcription. *Cold Spring Harbor Symp. Quant. Biol.*, **28**, 357–361.(*229*)

Martin, R. G., H. G. Whitfield Jr., D. B. Berkowitz, and M. J. Voll (1966a). A molecular model of the phenomenon of polarity. *Cold Spring Harbor Symp. Quant. Biol.*, **31**, 215–220.(*266*, *267*)

Martin, R. G., D. F. Silbert, D. W. E. Smith, and H. J. Whitfield Jr. (1966b). Polarity in the histidine operon. *J. Mol. Biol.*, **21**, 357–369.(*266, 267*)

Martin, R. G., and N. Talal (1968). Translation and polarity in the histidine operon. IV. Relation of polarity to map position in *his*-C. *J. Mol. Biol.*, **36**, 219–230.(*268*)

McCarthy, B. J., and E. T. Bolton (1964). Interaction of complementary RNA and DNA. *J. Mol. Biol.*, **8**, 184–200.(*143*)

McConkey, E. H., and J. W. Hopkins (1964). The relationship of the nucleolus to the synthesis of ribosomal RNA in Hela cells. *Proc. Nat. Acad. Sci. USA*, **51**, 1197–1204.(*171*)

McConkey, E. H., and J. W. Hopkins (1969). Molecular weights of some Hela ribosomal RNAs. *J. Mol. Biol.*, **30**, 545–550.(*177*)

McIlreavy, D. J., and J. E. M. Midgley (1967). The chemical structure of bacterial ribosomal RNA. I. Terminal nucleotide sequences of *E. coli r*RNA. *Biochim. Biophys. Acta*, **142**, 47–64.(*154*)

Mendel, G. (1865). Reprinted in English as *Experiments in Plant Hybridisation* J. A. Peters (Ed.), in *Classical Papers in Genetics*, Prentice-Hall, Englewood Cliffs, U.S.A.(*1*)

Meselson, M. (1964). One the mechanism of genetic recombination between DNA molecules. *J. Mol. Biol.*, **9**, 734–745.(*375*)

Meselson, M., M. Nomura, S. Brenner, C. Davern, and D. Schlessinger (1964). Conservation of ribosomes during bacterial growth, *J. Mol. Biol.*, **9**, 696–711. (*163*)

Meselson, M., and F. W. Stahl (1958). The replication of DNA in *E. coli*. *Proc. Nat. Acad. Sci. USA*, **44**, 671–682.(*295*)

Meselson, M., F. W. Stahl, and J. Vinograd (1957). Equilibrium sedimentation of macromolecules in density gradients. *Proc. Nat. Acad. Sci. USA*, **43**, 581–588. (*37*)

Meselson, M., and J. J. Weigle (1961). Chromosome breakage accompanying genetic recombination in bacteriophage. *Proc. Nat. Acad. Sci. USA*, **47**, 857–868.(*375*)

Meselson, M., and R. Yuan (1968). DNA restriction enzyme from *E. coli*. *Nature*, **217**, 1110–1114.(*325*)

Michels, C. A., and D. Zipser (1969). Mapping of polypeptide reinitiation sites within the β-galactosidase structural gene. *J. Mol. Biol.*, **41**, 341–348.(*270*)

Midgley, J. E. M. (1962). The nucleotide base composition of RNA from several microbial species. *Biochim. Biophys. Acta*, **61**, 513–525.(*156*)

Midgley, J. E. M., and D. J. McIlreavy (1967a). The chemical structure of bacterial *r*RNA. II. Growth conditions and polynucleotide distribution in *E. coli r*RNA. *Biochim. Biophys. Acta*, **142**, 345–354.(*154*)

Midgley, J. E. M., and D. J. McIlreavy (1967b). The isonicotinyl hydrazones of *E. coli* ribosomal nucleic acids. *Biochim. Biophys. Acta*, **145**, 512–514.(*154*)

Miller, J. H., J. Beckwith, and B. Muller-Hill (1968). Direction of transcription of a regulatory gene in *E. coli*. *Nature*, **220**, 1287–1289.(*233*)

Miller, O. L. Jr., and B. R. Beatty (1969a). Visualisation of nucleolar genes. *Science*, **164**, 955–957.(*152*)

Miller, O. L. Jr., and B. R. Beatty (1969b). Extrachomosomal nucleolar genes in amphibian oocytes. *Genetics*, **61**, 133–143.(*174*)

Miller, O. L. Jr., and B. R. Beatty (1969c). Portrait of a gene. *J. Cell. Physiol.* **74**, *Suppl.* **1**, 225–232.(*152*)

Milman, G., J. Goldstein, E. Scolnick, and T. Caskey (1969). Peptide chain termination. III. Stimultion of *in vitro* termination. *Proc. Nat. Acad. Sci. USA*, **63**, 183–190.(*133*)

Mirzabekov, A. D., D. Grunberger, A. Holy, A. A. Bayev, and F. Sorm (1967). Recognition of synonym codons by purified $t\mathrm{RNA}_{\mathrm{val}}$ fractions and fragments from $t\mathrm{RNA}_{\mathrm{val}}$. *Biochim. Biophys. Acta*, **145**, 845–847.(*207*)

Milanesi, G., E. N. Brody, E. P. Geideschek (1969). Sequence of the *in vitro* transcription of T4 DNA. *Nature*, **221**, 1014–1016.(*144*)

Miura, K. (1962). The nucleotide composition of ribonucleic acids of soluble and particulate fractions in several species of bacteria. *Biochim. Biophys. Acta*, **55**, 62–70.(*156*)

Monod, J., and G. Cohen-Bazire (1953). L'effet d'inhibition specifique dans la biosynthese de la tryptophane-demase chez *Aerobacter aerogenes*. *C.R. Acad. Sci. Paris*, **236**, 530–532.(*221*)

Monro, R. E. (1969). Protein synthesis: uncoupling of polymerisation from template control. *Nature*, **223**, 903–905.(*116*)

Moore, P. B., R. R. Traut, H. Noller, P. Pearson, and H. Delius (1968). Ribosomal proteins of *E. coli*. II. Proteins from the 30S subunit. *J. Mol. Biol.*, **31**, 441–462.(*159*)

Moore, R. L., and B. J. McCarthy (1967). Comparative study of ribosomal RNA cistrons in Enterobacteria and Myxobacteria. *J. Bacteriol.*, **94**, 1066–1074.(*156*)

Morell, P., and J. Marmur (1968). Association of 5S RNA to 50S ribosomal subunits of *E. coli* and *B. subtilis*. *Biochemistry*, **7**, 1141–1152.(*156*)

Morgan, T. H. (1911). Random segregation versus coupling in mendelian inheritance. *Science*, **34**, 384.(*12*)

Morikawa, N., and F. Imamoto (1969). On the degradation of *m*RNA for the tryptophan operon in *E. coli*. *Nature*, **223**, 37–40.(*279*)

Morris, D. W., and J. A. De Moss (1966). Polysome transitions and the regulation of RNA synthesis in *E. coli*. *Proc. Nat. Acad. Sci.*, **56**, 262–268.(*290*)

Morris, D. W., and N. O. Kjelgaard (1968). Evidence for the *non* co-ordinate regulation of RNA synthesis in *stringent* strains of *E. coli*. *J. Mol. Biol.*, **31**, 145–148.(*291*)

Morse, D. E., R. F. Baker, and C. Yanofsky (1968). Translation of the tryptophan messenger RNA of *E. coli*. *Proc. Nat. Acad. Sci. USA*, **60**, 1428–1435. (*274*, *278*)

Morse, D. E., R. Mosteller, R. F. Baker, and C. Yanofsky (1969). Direction of *in vivo* degradation of tyrptophan *m*RNA—a correction. *Nature*, **223**, 40–43. (*279*)

Morse, D. E., and C. Yanofsky (1968). The internal low efficiency promotor of the tryptophan operon of *E. coli*. *J. Mol. Biol.*, **38**, 447–452.(*282*)

Morse, D. E., and C. Yanofsky (1969). Polarity and the degradation of *m*RNA. *Nature*, **224**, 329–331.(*273*)

Muench, K. H., and P. A. Safille (1968). *t*RNAs in *E. coli:* multiplicity and variation. *Biochemistry*, **7**, 2799–2808.(*209*)

Muller-Hill, B. (1966). Suppressible regulator constitutive mutants of the *lac* system in *E. coli*. *J. Mol. Biol.*, **15**, 374–375.(*231*)

Nagata, T. (1963). Sequential replication of *E. coli* DNA. *Cold Spring Harbor Symp. Quant. Biol.*, **28**, 55–57.(*300*)

Nagata, T., and M. Meselson (1968). Periodic replication of DNA in steadily growing *E. coli:* the localised origin of replication. *Cold Spring Harbor Symp. Quant. Biol.*, **33**, 553–558.(*297*)

Nakada, D. (1967a). Functional diversity of ribosomes formed *in vivo* from *relaxed particles. Biochim. Biophys. Acta*, **145**, 664–670.(*180*)

Nakada, D. (1967b). Proteins of ribosomes formed from *relaxed particles. J. Mol. Biol.*, **29**, 473–485.(*180, 181*)

Nakada, D., I. A. C. Anderson, and B. Magasanik (1964). Fate of the ribosomal RNA produced by a *relaxed* mutant of *E. coli. J. Mol. Biol.*, **9**, 472–488.(*180*)

Nakada, D., and B. Magasanik (1964). The roles of inducer and catabolite repressor in the synthesis of β-galactosidase by *E. coli. J. Mol. Biol.*, **8**, 105–127.(*221, 229, 230*)

Nakada, D., and J. Unowsky (1966). *In vitro* formation of functional ribosomes from the *relaxed particles. Proc. Nat. Acad. Sci. USA*, **56**, 659–663.(*180*)

Nakai, S., and S. Matsumoto (1967). Two types of radiation sensitive mutants in yeast. *Mut. Res.*, **4**, 129–136.(*360*)

Nathans, D. (1964). Puromycin inhibition of protein synthesis: incorporation of puromycin into peptide chains. *Proc. Nat. Acad. Sci. USA*, **51**, 585–592.(*107*)

Nathans, D., M. P. Oeschger, S. K. Polmar, and K. Eggen (1969). Regulation of protein synthesis directed by coliphage MS2 RNA. I. Phage protein and RNA synthesis in cells infected with suppressible mutants. *J. Mol. Biol.*, **39**, 279–292.(*287*)

Naughton, M. A., and H. M. Dintzis (1962). Sequential biosynthesis of the peptide chains of hemoglobin. *Proc. Nat. Acad. Sci. USA*, **48**, 1822–1830.(*93*)

Neidhardt, F. C. (1966). Roles of amino acid activating enzymes in cellular physiology. *Bact. Rev.*, **30**, 701–719.(*290*)

Nelson, J. A., S. C. Ristow, and R. W. Holley (1967). Studies on the secondary structure of yeast alanine *t*RNA: reaction with *N*-bromosuccinimide and with nitrous acid. *Biochim. Biophys. Acta*, **149**, 590–593.(*193*)

Newman, J., and P. C. Hanawalt (1968). Intermediates in T4 DNA replication in a T4 ligase deficient strain. *Cold Spring Harbor Symp. Quant. Biol.*, **33**, 145–150.(*321*)

Newton, A. (1969). Re-initiation of polypeptide synthesis and polarity in the *lac* operon of *E. coli. J. Mol. Biol.*, **41**, 329–340.(*269*)

Newton, A., J. R. Beckwith, D. Zipser, and S. Brenner (1965). Nonsense mutations and polarity in the *lac* operon of *E. coli. J. Mol. Biol.*, **14**, 290–295. (*265*)

Nierlich, D. P. (1968). Amino acid control over RNA synthesis: a re-evaluation *Proc. Nat. Acad. Sci. USA*, **60**, 1345–1352.(*291*)

Ninio, J. (1969). Personal communication.(*198*)

Ninio, J., A. Favre, and M. Yaniv (1969). Molecular model for *t*RNA. *Nature*, **223**, 1333–1335.(*198, 199, 200*)

Nirenberg, M., and P. Leder (1964). The effect of trinucleotides upon the binding of *t*RNA to ribosomes. *Science*, **145**, 1399.(*80*)

Nirenberg, M., and J. H. Matthaei (1961). The dependence of cell-free protein synthesis in *E. coli* upon naturally occuring or synthetic polyribonucleotides. *Proc. Nat. Acad. Sci. USA*, **47**, 1588–1602.(*77*)

Nirenberg, M., and coworkers (1966). The RNA code and protein synthesis. *Cold Spring Harbor Symp. Quant. Biol.*, **31**, 11–24.(*80*)

Nishimura, S., D. S. Jones, and H. G. Khorana (1965). Studies on polynucleotides. XLVIII. The *in vitro* synthesis of a co-polypeptide containing two amino acids in alternating sequence dependent upon a DNA-like polymer containing two nucleotides in alternating sequence. *J. Mol. Biol.*, **13**, 302–324.

Nishimura, S., D. S. Jones, E. Ohtsuka, H. Hayatsu, T. M. Jacob, and H. G. Khorana (1965). Studies on polynucleotides. XLVII. The *in vitro* synthesis of homopeptides as directed by a ribopolynucleotide containing a repeating trinucleotide sequence. New codon sequences for lysine, glutamic acid, and arginine. *J. Mol. Biol.*, **13**, 283–301.(*78, 80*)

Nishimura, S., and I. B. Weinstein (1969). Fractionation of rat liver *t*RNA. Isolation of tyrosine, valine, serine, and phenylalanine *t*RNAs and their coding properties. *Biochemistry*, **8**, 832–842.(*207*)

Nishizuka, Y., and F. Lipmann (1966). The inter-relationship between GTP and amino acid polymerisation. *Arch. Biochem. Biophys.*, **116**, 344–351.(*107, 118*)

Nomura, M., C. V. Lowry, and C. Guthrie (1967). The initiation of protein synthesis: joining of the 50S ribosomal subunit to the initiation complex. *Proc. Nat. Acad. Sci. USA*, **58**, 1487–1493.(*124*)

Nomura, M., and P. Traub (1968). Structure and function of *E. coli* ribosomes. III. Stoichiometry and rate of reconstitution of ribosomes from subribosomal particles and split proteins. *J. Mol. Biol.*, **34**, 609.(*164*)

Nomura, M., P. Traub, and H. Bechmann (1968). Hybrid 30S ribosomal particles reconstituted from components of different bacterial origins. *Nature*, **219**, 793–799.(*169*)

Novack, R. L. (1967). Deoxyribonuclease resistance of DNA–RNA polymerase complexes. *Biochim. Biophys. Acta*, **149**, 593–596.(*146*)

Novelli, G. D. (1967). Amino acid activation for protein synthesis. *Ann. Rev. Biochem.*, **36**, 449–484.(*100*)

Nygaard, A. P., and B. D. Hall (1963). A method for the detection of RNA–DNA complexes. *Biochem. Biophys. Res. Commun.*, **12**, 98–104.(*41*)

Nygaard, A. P., and B. D. Hall (1964). Formation and properties of RNA–DNA complexes. *J. Mol. Biol.*, **9**, 125–142.(*41*)

Ofengand, J., and C. Henes (1969). Specific inhibition of aminoacyl-*t*RNA binding to ribosomes by TΨCG tetranucleotide. *Fed. Proc.*, **28**, 350.(*202*)

Ohlsson, B. M., P. F. Strigini, and J. R. Beckwith (1968). Allelic *amber* and *ochre* suppressors. *J. Mol. Biol.*, **36**, 209–218.(*212*)

Ohta, T., S. Sarkar, R. E. Thach (1967). The role of GTP in the initiation of peptide synthesis. III. Binding of formyl-methionyl-*t*RNA to ribosomes. *Proc. Nat. Acad. Sci. USA*, **58**, 1638–1644.(*129*)

Ohta, T., I. Shimada, K. Imahori (1967). Conformational change of tyrosyl-*t*RNA synthetase induced by its specific *t*RNA. *J. Mol. Biol.*, **26**, 519–524. (*203*)

Ohta, T., and R. E. Thach (1968). Binding of formyl methionyl-*t*RNA and aminoacyl-*t*RNA to ribosomes. *Nature*, **219**, 238–242.(*128*)

Oishi, M., and N. Sueoka (1965). Location of genetic loci of ribosomal RNA on *B. subtilis* chromosomes. *Proc. Nat. Acad. Sci. USA*, **54**, 483–491.(*170, 171*)

Okazaki, R., T. Okazaki, K. Sakabe, K. Sugimoto, R. Kainuma, R. Sugino, and N. Zwatsuki (1968). *In vivo* mechanism of DNA chain growth. *Cold Spring Harbor Symp. Quant. Biol.*, **33**, 129–144.(*313, 315, 321*)

Olivera, B. M., Z. W. Hall, and I. R. Lehman (1968). Enzymatic joining of polynucleotides. V. A DNA-adenylate intermediate in the polynucleotide joining reaction. *Proc. Nat. Acad. Sci. USA*, **61**, 237–244.(*319, 320*)

Olivera, B. M., Z. W. Hall, Y. Anraku, J. R. Chien, and I. R. Lehman (1968). On the mechanism of the polynucelotide joining reaction. *Cold Spring Harbor Symp. Quant. Biol.*, **33**, 27–36.(*319*)

Ono, Y., A. Skoultchi, A. Klein, and P. Lengyel (1968). Peptide chain elongation: discrimination against the initiator *t*RNA by microbial amino acid polymerisation factors. *Nature*, **220**, 1304–1307.(*125*)

Ono, Y., A. Skoultchi, J. Waterson, and P. Lengyel (1969a). Peptide chain elongation: GTP cleavage catalysed by factors binding aminoacyl-*t*RNA to the ribosome. *Nature*, **222**, 645–648.(*114*)

Ono, Y., A. Skoultchi, J. Waterson, and P. Lengyel (1969b). Stoichiometry of aminoacyl-*t*RNA binding and GTP cleavage during chain elongation and translocation. *Nature*, **223**, 697–701.(*115*)

Orgel, L. E. (1968). Evolution of the genetic apparatus. *J. Mol. Biol.*, **38**, 381–394. (*91*)

Osawa, S. (1968). Ribosome formation and structure. *Ann. Rev. Biochem.*, **37**, 109–130.(*178, 181, 182*)

Osawa, S., E. Otaka, T. Itoh, and T. Fukui (1969). Biosynthesis of 50S ribosomal subunit in *E. coli*. *J. Mol. Biol.*, **40**, 321–352.(*181*)

Otaka, E., T. Itoh, and S. Osawa (1968). Ribosomal proteins of bacterial cells: strain and species specificity. *J. Mol. Biol.*, **33**, 93–108.(*159, 160*)

Ozaki, M., S. Mizushima, and M. Nomura (1969). Identification and functional characterisation of the protein controlled by the streptomycin-resistant locus in *E. coli*. *Nature*, **222**, 333–339.(*215, 218, 219*)

Painter, R. B., and A. W. Schaeffer (1969). Rate of synthesis along replicons of different kinds of mammalian cells. *J. Mol. Biol.*, **45**, 467–480.(*314*)

Parsons, J. T., and K. S. McCarty (1968). Rapidly labeled *m*RNA-protein complex of rat liver nuclei. *J. Biol. Chem.*, **243**, 5377–5385.(*151*)

Pastan, I., and R. L. Perlman (1968). The role of the lac promotor locus in the regulation of β-galactosidase synthesis by cyclic 3′-5′ AMP. *Proc. Nat. Acad. Sci. USA*, **61**, 1336–1342.(*238*)

Pastan, I., and R. L. Perlman (1969). Stimulation of tryptophanase synthesis in *E. coli* by cyclic 3′-5′ AMP. *J. Biol. Chem.*, **244**, 2226–2232.(*238*)

Pato, M. L., and D. A. Glaser (1968). The origin and direction of replication of the chromosome of *E. coli* B/r. *Proc. Nat. Acad. Sci. USA*, **60**, 1268–1274.(*301*)

Perlman, R. L., B. De Crombrugghe, and I. Pastan (1969). Cyclic AMP regulates catabolite and transient repression in *E. coli*. *Nature*, **223**, 810–812.(*238*)

Perry, R. P. (1967). The nucleolus and the synthesis of ribosomes. *Prog. Nuc. Acid Res.*, **6**, 219–257.(*171, 173*)

Perry, R. P., and D. E. Kelly (1968). Messenger RNA—protein complexes and newly synthesised ribosomal subunits: analysis of free particles and components of polyribosomes. *J. Mol. Biol.*, **35**, 37–60.(*151*)

Person, S., and M. Osborn (1968). The conversion of *amber* suppressors to *ochre* suppressors. *Proc. Nat. Acad. Sci. USA*, **60**, 1030–1038.(*133*)

Pestka, S. (1968). Studies on the formation of *t*RNA-ribosome complexes. V. On the function of a soluble transfer factor in protein synthesis. *Proc. Nat. Acad. Sci. USA*, **61**, 726–733.(*116*)

Pestka, S., R. Marshall, and M. Nirenberg (1965). RNA codewords and protein synthesis. V. Effect of streptomycin on the formation of ribosome-*t*RNA complexes. *Proc. Nat. Acad. Sci. USA*, **53**, 639–646.(*216*)

Pettijohn, D., and P. C. Hanawalt (1964). Evidence for repair replication of ultraviolet damaged DNA in bacteria. *J. Mol. Biol.*, **9**, 395–410.(*354*)

Pettijohn, D., and T. Kamiya (1967). Interaction of RNA polymerase with polyoma DNA. *J. Mol. Biol.*, **29**, 275–296.(*140*)

Phillips, L. A., B. Hotham-Iglewski, and R. M. Franklin (1969). Polyribosomes of *E. coli*. I. Effects of monovalent cations on the distribution of polysomes, ribosomes, and ribosomal subunits. *J. Mol. Biol.*, **40**, 279–288.(*135*)

Pierucci, O. (1969). Regulation of cell division in *E. coli*. *Biophys. J.*, **9**, 90–112.(*349*)

Pirrotta, V., and M. Ptashne (1969). Isolation of the 434 phage repressor. *Nature*, **222**, 541–544.(*231*)

Preiss, J., P. Berg, E. J. Ofengand, F. H. Bergmann and M. Dieckmann (1959). The chemical nature of the RNA-amino acid compound formed by amino acid activating enzymes. *Proc. Nat. Acad. Sci. USA*, **45**, 319–328.(*100*)

Pritchard, R. H., P. T. Barth, and J. Collins (1969). Control of DNA synthesis in bacteria. *Symp. Soc. Gen. Mic.*, **19**, 263–298.(*349*)

Ptashne, M. (1967a). Isolation of the λ phage repressor. *Proc. Nat. Acad. Sci. USA*, **57**, 306–312.(*231*)

Ptashne, M. (1967b). Specific binding of the λ phage repressor to λ DNA. *Nature*, **214**, 232–234.(*232*)

Revel, M., J. C. Lelong, G. Brawerman, and F. Gros (1968a). Function of three protein factors and ribosomal subunits in the initiation of protein synthesis in *E. coli*. *Nature*, **219**, 1016–1020.(*129*)

Revel, M., M. Herzberg, A. Becarevic, and F. Gros (1968b). Role of a protein factor in the functional binding of ribosomes to natural *m*RNA. *J. Mol. Biol.*, **33**, 231–249.(*129, 151*)

Reznikoff, W. S., J. H. Miller, J. G. Scaife, and J. R. Beckwith (1969). A mechanism for repressor action. *J. Mol. Biol.*, **43**, 201–214.(*233*)

Rich, A., E. Eikenberry, and L. Malkin (1966). Experiments on hemoglobin polypeptide chain initiation and on the shielding action of the ribosome. *Cold Spring Harbor Symp. Quant. Biol.*, **31**, 303–310.(*124*)

Rich, A., J. R. Warner, and H. M. Goodman (1963). The structure and function of polyribosomes. *Cold Spring Harbor Symp. Quant. Biol.*, **28**, 269–285.(*104*)

Richardson, C. C., Y. Masamune, T. R. Live, A. Jacquemin-Sablon, B. Weiss, and G. Fareed (1968). Studies on the joining of DNA by polynucleotide ligase of phage T4. *Cold Spring Harbor Symp. Quant. Biol.*, **33**, 151–164.(*319, 372*)

Richardson, J. P. (1966). The binding of RNA polymerase to DNA. *J. Mol. Biol.*, **21**, 83–114.(*139*)

Richardson, J. P. (1969). RNA polymerase and the control of RNA synethsis. *Prog. Nuc. Acid Res.*, **9**, 75–116.(*138, 140*)

Ritchie, D. A., C. A. Thomas Jr., L. A. MacHattie and P. C. Wensink (1967). Terminal repetition in non-permuted T3 and T7 bacteriophage DNA molecules. *J. Mol. Biol.*, **23**, 365–376.(*54*)

Ritossa, F. M., and S. Spiegelman (1965). Localisation of DNA complementary to *r*RNA in the nucleolus organiser region of *D. melanogaster*. *Proc. Nat. Acad. Sci. USA*, **53**, 737–745.(*170, 172*)

Roberts, J. J., A. R. Crathorn, and T. P. Brent (1968). Repair of alkylated DNA in mammalian cells. *Nature*, **218**, 970–972.(*360*)

Roblin, R. (1968). Nucleotides adjacent to the 5′ terminus of bacteriophage R17 RNA. *J. Mol. Biol.*, **36**, 125–136.(*139*)

Ron, E. Z., R. E. Kohler, B. D. Davis (1968). Magnesium ion dependence of free and polysomal ribosomes from *E. coli*. *J. Mol. Biol.*, **36**, 83–90.(*135*)

Rosenberg, B. H., and L. F. Cavalieri (1968). Shear sensitivity of the *E. coli* genome: multiple membrane attachment points of the *E. coli* DNA. *Cold Spring Harbor Symp. Quant. Biol.*, **33**, 65–72.(*341*)

Rosset, R., and L. Gorini (1969). A ribosomal ambiguity mutation. *J. Mol. Biol.*, **39**, 95–112.(*215, 217*)

Rosset, R., and R. Monier (1963). A propos de la presence d'acide ribonucleique de faible poids moleculaire dans les ribosomes d'*E. coli*. *Biochim. Biophys. Acta*, **68**, 653–655.(*156*)

Roth, J. R., and B. N. Ames (1966). Histidine regulatory mutants in *S. typhimurium*. II. Histidine regulatory mutants have altered *t*RNA synthetase. *J. Mol. Biol.*, **22**, 325–334.(*247*)

Roth, J. R., D. R. Anton, and P. E. Hartman (1966). Histidine regulatory mutants in *S. typhimurium*. I. Isolation and general properties. *J. Mol. Biol.*, **22**, 305–323.(*246*)

Roth, J. R., D. F. Silbert, G. R. Fink, M. J. Voll, D. Anton, P. E. Hartman, and B. N. Ames (1966). Transfer RNA and control of the histidine operon. *Cold Spring Harbor Symp. Quant. Biol.*, **31**, 383–392.(*245, 246, 247, 292*)

Rudland, P. S., W. A. Whybrow, K. A. Marcker, and B. F. C. Clark (1969). Recognition of bacterial initiator *t*RNA by initiation factors. *Nature*, **222**, 750–753.(*125*)

Rupp, W. D., and P. Howard-Flanders (1968). Discontinuities in the DNA synthesised in an excision defective strain of *E. coli* following ultraviolet irradiation. *J. Mol. Biol.*, **31**, 291–304.(*362, 364*)

Ryter, A., Y. Hirota, and F. Jacob (1968). DNA-membrane complex and nuclear segregation in bacteria. *Cold Spring Harbor Symp. Quant. Biol.*, **33**, 669–676.(*340, 346, 348*)

Sadgopal, A. (1968). The genetic code after the excitement. *Adv. Genet.*, **14**, 326–404.(*77*)

Salas, M., M. J. Miller, A. J. Wahba, and S. Ochoa (1967). Translation of the genetic message. V. Effect of Mg^{2+} and formylation of methionine in protein synthesis. *Proc. Nat. Acad. Sci. USA*, **57**, 1865–1869.(*129*)

Salas, M., M. A. Smith, W. M. Stanley Jr., A. J. Wahba, and S. Ochoa (1965). Direction of reading of the genetic message. *J. Biol. Chem.*, **240**, 3988–3995. (*105*)

Sambrook, J. F., D. P. Fan, and S. Brenner (1967). A strong suppressor specific for UGA. *Nature*, **214**, 452–453.(*84, 86*)

Sanger, F., and G. G. Brownlee (1967). Fractionation of radioactive nucleotides. In D. Shugar (Ed.), *Genetic Elements*. Academic Press, London, pp. 303–314.(*185*)

Sanger, F., G. G. Brownlee, and B. G. Barrell (1965). A two dimensional fractionation procedure for radioactive nucleotides. *J. Mol. Biol.*, **13**, 373–398.(*185, 186*)

Sarabhai, A., and S. Brenner (1967). A mutant which reinitiates the polypeptide chain after chain termination. *J. Mol. Biol.*, **27**, 145–162.(*270*)

Sarabhai, A., A. O. W. Stretton, and S. Brenner (1964). Co-linearity of the gene with the polypeptide chain. *Nature*, **201**, 13–17.(*71, 82*)

Sarkar, S., and K. Moldave (1968). Characterisation of the RNA synthesised during amino acid degradation of a stringent auxotroph of *E. coli*. *J. Mol. Biol.*, **33**, 313–224.(*291*)

Scaife, J., and J. R. Beckwith (1966). Mutational alteration of the maximal level of *lac* operon expression. *Cold Spring Harbor Symp. Quant. Biol.*, **31**, 403–408. (*233*)

Schandl, E. K., and J. H. Taylor (1969). Early events in the replication and integration of DNA into mammalian chromosomes. *Biochem. Biophys. Res. Commun.*, **34**, 291–300.(*314*)

Schildkraut, C. L., J. Marmur, and P. Doty (1961). The formation of hybrid DNA molecules and their use in the studies of DNA homologies. *J. Mol. Biol.*, **3**, 595–617.(*38*)

Schildkraut, C. L., C. C. Richardson, and A. Kornberg (1964). Enzymic synthesis of DNA. XVII. Some unusual physical properties of the product primed by native DNA templates. *J. Mol. Biol.*, **9**, 24–45.(*316*)

Schlessinger, D., G. Mangiarotti, and D. Apirion (1967). The formation and stabilisation of 30S and 50S ribosome couples in *E. coli*. *Proc. Nat. Acad. Sci. USA*, **58**, 1782–1789.(*134, 135*)

Schlessinger, S., and B. Magasanik (1964). Effects of α-methyl histidine on the control of histidine synthesis. *J. Mol. Biol.*, **9**, 670–682.(*247*)

Schlief, R. (1968). Origin of the chloramphenicol particle protein. *J. Mol. Biol.*, **37**, 119–130.(*179*)

Schneider, J. A., S. Raeburn, and E. S. Maxwell (1968). Translocase activity in the aminoacyl transferase-II fraction from rat liver. *Biochem. Biophys. Res. Commun.*, **33**, 177–181.(*110*)

Schulman, H. M., and D. M. Bonner (1962). A naturally occuring DNA–RNA complex from *N. crassa*. *Proc. Nat. Acad. Sci. USA*, **48**, 53–63.(*146*)

Schulman, L. H., and R. W. Chambers (1968). Transfer RNA. II. A structural basis for alanine acceptor activity. *Proc. Nat. Acad. Sci. USA*, **61**, 308–315. (*200*)

Scolnick, E., R. Tompkins, T. Caskey, and M. Nirenberg (1968). Release factors differing in specificity for terminator codons. *Proc. Nat. Acad. Sci. USA*, **61**, 768–774.(*133*)

Scott, J. F., R. Monier, M. Aubert, and M. Reynier (1968). Some properties of 5S RNA from *E. coli*. *Biochem. Biophys. Res. Commun.*, **33**, 794–800.(*157*)

Sesnowitz-Horn, S., and A. Adelberg (1968). Proflavin treatment of *E. coli* generation of frameshift mutants. *Cold Spring Harbor Symp. Quant. Biol.*, **33**, 393–402.(*75*)

Setlow, J. K., and R. B. Setlow (1963). Nature of the photoreactivable ultraviolet lesion in DNA. *Nature*, **197**, 560–562.(*350*)

Setlow, R. B. (1968). The photochemistry, photobiology, and repair of polynucleotides. *Prog. Nuc. Acid Res.*, **8**, 257–295.(*357, 362*)

Setlow, R. B., and W. L. Carrier (1964). The disappearance of thymine dimers from DNA: an error correcting mechanism. *Proc. Nat. Acad. Sci. USA*, **51**, 226–231.(*354*)

Setlow, R. B., P. A. Swenson, and W. L. Carrier (1963). Thymine dimers and inhibition of DNA synthesis by UV irradiation of cells. *Science*, **142**, 1464–1465.(*354*)

Shapiro, J., L. MacHattie, L. Eron, G. Ihler, K. Ippen, and J. Beckwith (1969). Isolation of pure *lac* operon DNA. *Nature*, **224**, 768–774.(*235*, *236*)

Sheppard, D., and E. Englesberg (1966). Positive control in the L-arabinose gene-enzyme complex of *E. coli* B/r as exhibited with stable merodiploids. *Cold Spring Harbor Symp. Quant. Biol.*, **31**, 345–348.(*249*)

Sheppard, D. E., and E. Englesberg (1967). Further evidence for positive control of the L-arabinose system by gene *ara*-C. *J. Mol. Biol.*, **25**, 443–454. (*249*)

Shih, A-Y., J. Eisenstadt, and P. Lengyel (1966). On the relationship between RNA synthesis and peptide chain initiation in *E. coli*. *Proc. Nat. Acad. Sci. USA*, **56**, 1599–1605.(*290*)

Shin, D. H., and K. Moldave (1966). Effect of ribosomes on the biosynthesis of RNA *in vitro*. *J. Mol. Biol.*, **21**, 231–245.(*150*, *151*)

Siddiqui, M. A. Q., and N. Hosokawa (1968). Role of 5S *r*RNA in polypeptide synthesis. II. Dissociation of 5S *r*RNA from 50S ribosomes in *E. coli*. *Biochem. Biophys. Res. Commun.*, **32**, 1–8.(*156*)

Signer, E. R. (1968). Lysogeny: the integration problem. *Ann. Rev. Microbiol.*, **22**, 451–488.(*55*)

Silbert, D. F., G. R. Fink, and B .M. Ames (1966). Histidine regulatory mutants in *S. typhimurium*. III. A class of regulatory mutants deficient in *t*RNA for histidine. *J. Mol. Biol.*, **22**, 335–347.(*247*)

Sinsheimer, R. L. (1968). The replication of viral DNA. *Sym. Soc. Gen. Mic.*, **18**, 101–124.(*51*)

Slayter, H. S., J. R. Warner, A. Rich, and C. E. Hall (1963). The visualisation of polyribosome structure. *J. Mol. Biol.*, **7**, 652–657.(*104*)

Smith, D. A., A. M. Martinez, R. L. Ratliffe, D. L. Williams, and F. N. Hayes (1967). Template induced dissociation of RNA polymerase. *Biochemistry*, **6**, 3057–3063.(*138*)

Smith, D. W., and P. C. Hanawalt (1967). Properties of the growing point region in the bacterial chromosome. *Biochim. Biophys. Acta*, **149**, 519–531. (*340*)

Smith, I., D. Dubnau, P. Morell, and J. Marmur (1968). Chromosomal location of DNA base sequences complementary to *t*RNA and to 5S, 16S, and 23S *r*RNA in *B. subtilis*. *J. Mol. Biol.*, **33**, 123–140.(*171*)

Smith, J. D., J. N. Abelson, B. F. C. Clark, H. M. Goodman, and S. Brenner (1966). Studies on *amber* suppressor *t*RNA. *Cold Spring Harbor Symp. Quant. Biol.*, **31**, 479–486.(*213*)

Snow, R. (1967). Mutants of yeast sensitive to ultraviolet light. *J. Bacteriol.*, **94**, 571–575.(*360*)

Soeiro, R., M. H. Vaughan, J. E. Darnell (1968). The effect of puromycin on intra-nuclear steps in ribosome biosynthesis. *J. Cell Biol.*, **36**, 91–101.(*177*)

Soll, D., J. D. Cherayil, and R. M. Bock (1967). Studies on polynucleotides. LXXV. Specificity of *t*RNA for codon recognition as studied by ribosomal binding techniques. *J. Mol. Biol.*, **29**, 97–112.(*206, 207*)

Soll, D., and V. L. Rajh Bhandary (1967). Studies on polynucleotides. LXXVI. Specificity of *t*RNA for codon recognition as studied by amino acid incorporation. *J. Mol. Biol.*, **29**, 113–124.(*207*)

Sonneborn, T. M. (1965). Degeneracy of the genetic code: extent, nature, and genetic implications. In V. Bryson and H. J. Vogel (Eds.), *Evolving Genes and Proteins*, Academic Press, London 377–397.(*90*)

Spahr, P. F., M. Farber, and R. F. Gesteland (1969). Binding site on R17 RNA for coat protein. *Nature*, **222**, 455–459.(*288*)

Spahr, P. F., and R. F. Gesteland (1968). Specific leavage of bacteriophage R17 RNA by an endonuclease isolated from *E. coli* MRE 600. *Proc. Nat. Acad. Sci. USA*, **59**, 876–883.(*288*)

Speyer, J. F. (1963). Synthetic polynucleotides and the amino acid code. *Cold Spring Harbor Symp. Quant. Biol.*, **28**, 559–567.(*78*)

Speyer, J. F. (1965). Mutagenic DNA polymerase. *Biochem. Biophys. Res. Commun.*, **21**, 6–8.(*305*)

Spirin, A. S., and N. V. Belitsina (1966). Biological activity of the reassembled ribosome like particles. *J. Mol. Biol.*, **15**, 282–283(*164*.)

Srinivasan, P. R., and E. Borek (1966). Enzymic alteration of macromolecular structure. *Prog. Nuc. Acid Res.*, **5**, 157–190.(*321*)

Staehelin, T., and M. Meselson (1966a). *In vitro* recovery of ribosomes and of synthetic activity from synthetically inactive ribosomal subunits. *J. Mol. Biol.*, **16**, 245–249.(*164*)

Staehelin, T., and M. Meselson (1966b). Determination of streptomycin sensitivity by a subunit of the 30S ribosome of *E. coli*. *J. Mol. Biol.*, **19**, 207–210.(*214*)

Staehelin, M., N. Rogg, B. C. Baguley, T. Ginsberg, and W. Wehrli (1968). Structure of a mammalian serine *t*RNA. *Nature*, **219**, 1363–1365.(*186, 207*)

Staehelin, M., M. Rouge, and W. Wehrli (1968). The isolation and some structural aspects of three serine *t*RNAs from rat liver. Abstract 132 *Fed. Eur. Biochem. Soc. Madrid*.(*207*)

Stahl, F. W. (1967). Circular genetic maps. *J. Cell. Physiol.*, **70**, pt **II**, 1–12.(*53*)

Stanley, W. M. Jr., M. Salas, A. J. Wahba, and S. Ochoa (1966). Translation of the genetic message: factors involved in the initiation of protein synthesis. *Proc. Nat. Acad. Sci. USA*, **56**, 290–295.(*129*)

Stead, N. W., and O. W. Jones Jr. (1967). The binding of RNA polymerase to DNA: stabilisation by nucleoside triphosphates. *Biochim. Biophys. Acta*, **145**, 679–695.(*138, 139*)

Stent, G. S. (1958). Mating in the reproduction of bacterial viruses. *Adv. Virus Res.*, **5**, 95–149.(*144, 145*)

Stent, G. S. (1964). The operon: on its third anniversary. *Science*, **144**, 816–820.(*275, 292*)

Stent, G. S., and S. Brenner (1961). A genetic locus for the regulation of RNA synthesis. *Proc. Nat. Acad. Sci. USA*, **47**, 2005–2014.(*290*)

Stern, C. (1931). Zytologisch-genetische Untersuchungen als Beweise fur die Morgansche Theories des Faktorenaustauchs. *Biol. Zentrlb.*, **51**, 547–587.(*14*)

Stern, R., L. E. Zutra, U. Z. Littauer (1969). Fractionation of *t*RNA on a methylated albumin—silicic acid column. II. Changes in elution profiles following modification of *t*RNA. *Biochemistry*, **8**, 313–321.(*198*)

Streisinger, G., R. S. Edgar, G. H. Denhardt (1964). Chromosome structure in phage T4. I. Circularity of the linkage map. *Proc. Nat. Acad. Sci. USA*, **51**, 775–779.(*53*)

Streisinger, G., J. Emrich, and M. M. Stahl (1967). Chromosome structure in phage T4. III. Terminal redundancy and length determination. *Proc. Nat. Acad. Sci. USA*, **57**, 292–295.(*55*)

Streisinger, G., Y. Okada, J. Emrich, J. Newton, A. Tsugita, E. Terzhagi, and M. Inouye (1966). Frameshift mutations and the genetic code. *Cold Spring Harbor Symp. Quant. Biol.*, **31**, 77–84.(*75*, *77*, *82*)

Stretton, A. O. W. (1965). The genetic code. *Brit. Med. Bull.*, **21**, No. **3**, 229–235.(*85*)

Stretton, A. O. W., and S. Brenner (1965). Molecular consequences of the *amber* mutation and its suppression. *J. Mol. Biol.*, **12**, 456–465.(*212*)

Stubbs, J. D., and B. D. Hall (1968a). Level of tryptophan *m*RNA in *E. coli. J. Mol. Biol.*, **37**, 289–302.(*291*)

Stubbs, J. D., and B. D. Hall (1968b). Effects of amino acid starvation upon constitutive tryptophan *m*RNA synthesis. *J. Mol. Biol.*, **37**, 303–312.(*291*)

Stulberg, M. P., and K. R. Isham (1967). Studies on the locus of the enzyme recognition site in phenylalanine *t*RNA. *Proc. Nat. Acad. Sci. USA*, **57**, 1310–1315.(*201*)

Sturtevant, A. H. (1913). The linear arrangement of six sex-linked factors in *Drosophila* as shown by their mode of association. *J. Exp. Zoo.*, **14**, 43–59.(*14*)

Subak-Sharpe, H. (1968). Virus-induced changes in translation mechanisms. *Symp. Soc. Gen. Mic.*, **18**, 47–66.(*50*)

Subramamian, A. R., E. Z. Ron, and B. D. Davis (1968). A factor required for ribosome dissociation in *E. coli. Proc. Nat. Acad. Sci. USA*, **61**, 761–767.(*136*)

Sueoka, N., and W. G. Quinn (1968). Membrane attachment of the chromosome replication origin in *B. subtilis. Cold Spring Harbor Symp. Quant. Biol.*, **33**, 695–706.(*341*, *342*)

Sueoka, N., and H. Yoshikawa (1963). Regulation of chromosome replication in *B. subtilis. Cold Spring Harbor Symp. Quant. Biol.*, **28**, 43–54.(*300*)

Sugiura, M., and M. Takanami (1967). Analysis of the 5′ terminal nucleotide sequence of RNAs. II. Comparison of the 5′ terminal nucleotide sequences of *r*RNAs from different organisms. *Proc. Nat. Acad. Sci. USA*, **58**, 1595–1602. (*155–6*)

Sundarajan, T. A., and R. E. Thach (1966). Role of the formyl-methionine codon AUG in phasing translation of synthetic *m*RNA. *J. Mol. Biol.* **19**, 74–90.(*121*)

Sundharadas, G., J. R. Katze, D. Soll, W. Konigsberg, and P. Lengyel (1968). On the recognition of serine *t*RNAs specific for unrelated codons by the same seryl-*t*RNA synthetase. *Proc. Nat. Acad. Sci. USA*, **61**, 693–700.(*202*)

Sypherd, P. (1968). Ribosome development and the methylation of *r*RNA. *J. Bacteriol.*, **95**, 1844–1850.(*182*)

Szybalski, W., H. Kubinski, and P. Sheldrick (1966). Pyrimidine clusters on the transcribing strand of DNA and their possible role in the initiation of RNA synthesis. *Cold Spring Harbor Symp. Quant. Biol.*, **31**, 123–127.(*140*)

Takagi, Y., M. Sekiguchi, S. Okubo, H. Nakayama, K. Shimada, S. Yasuda, T. Mishimoto, and H. Yoshihara (1968). Nucleases specific for ultraviolet-light irradiated DNA and their possible role in dark repair. *Cold Spring Harbor Symp. Quant. Biol.*, **33**, 219–228.(*357*)

Takeda, M., and Webster, R. E. (1968). Protein chain initiation and deformylation in *B. subtilis* homogenates. *Proc. Nat. Acad. Sci. USA*, **60**, 1487–1494. (*120*, *122*)

Taylor, J. H. (1964). Regulation of DNA replication and variegation type position effects. In R. J. C. Harris (Ed.), *Cytogenetics of Cells in Culture*, Academic Press, London, pp. 175–190.(*302*)

Terzhagi, E., Y. Okada, G. Streisinger, J. Emrich, M. Inouye, and A. Tsugita (1966). Change of a sequence of amino acids in phage T4 lysozyme by acridine-induced mutations. *Proc. Nat. Acad. Sci. USA*, **56**, 500–507.(*210*)

Thach, R., T. A. Sundarajan, K. F. Dewey, J. C. Brown, and P. Doty (1966). Translation of synthetic *m*RNA. *Cold Spring Harbor Symp. Quant. Biol.*, **31**, 85–98.(*122*)

Thiebe, R., and H. G. Zachau (1968a). A special modification next to the anticodon of phenylalanine *t*RNA. *Eur. J. Biochem.*, **5**, 546–555.(*201*)

Thiebe, R., and H. G. Zachau (1968b). The role of the anticodon region in homologous and heterologous charging of $t\text{RNA}_{\text{phe}}$. *Biochem. Biophys. Res. Commun.*, **33**, 260–265.(*201*)

Thomas, C. A. Jr. (1963). The arrangement of nucleotide sequences in T2 and T5 DNA molecules. *Cold Spring Harbor Symp. Quant. Biol.*, **28**, 395–396.(*53*)

Thomas, C. A. Jr. (1966). Recombination of DNA molecules. *Prog. Nuc. Acid Res.*, **5**, 315–337.(*367*)

Thomas, C. A. Jr., and L. A. MacHattie (1964). Circular T2 DNA molecules. *Proc. Nat. Acad. Sci. USA*, **52**, 1297–1301.(*53*)

Thomas, R. (1955). Recherches sur la cinetique des transformations bacteriennes. *Biochim. Biophys. Acta*, **18**, 487.(*44*)

Thomas, R. (1968). Lysogeny. *Symp. Soc. Gen. Mic.*, **18**, 313–342.(*55*, *58*)

Tissières, A., J. D. Watson, D. Schlessinger, and B. R. Hollingworth (1959). Ribonucleoprotein particles from *E. coli*. *J. Mol. Biol.*, **1**, 221–233.(*96*)

Tomizawa, J-I., and N. Anraku (1965). Molecular mechanisms of genetic recombination of bacteriophage. IV. Absence of polynucleotide interruption in DNA of T4 and λ phage particles, with special reference to heterozygosis. *J. Mol. Biol.*, **11**, 509–527.(*376*)

Tongur, V. S., N. S. Wladytchenskaya, and W. M. Kotchkina (1968). A natural RNA–DNA complex in bacterial cells. *J. Mol. Biol.*, **33**, 451–464.(*147*)

Traub, P., and M. Nomura (1968a). Structure and function of *E. coli* ribosomes. I. Partial fractionation of the functionally active ribosomal proteins and reconstitution of the artificial subribosomal particles. *J. Mol. Biol.*, **34**, 575–594.(*166*, *168*)

Traub, P., and M. Nomura (1968b). Structure and function of *E. coli* ribosomes. V. Reconstitution of functionally active 30S ribosomal particles from RNA and proteins. *Proc. Nat. Acad. Sci. USA*, **59**, 777–784.(*164*, *166*)

Traub, P., and M. Nomura (1968c). Streptomycin resistance mutation in *E. coli:* altered ribosomal protein. *Science*, **160**, 198–199.(*214*)

Traub, P., and M. Nomura (1969). Structure and function of *E. coli* ribosomes. VI. Mechanism of assembly of 30S ribosomes studied *in vitro*. *J. Mol. Biol.*, **40**, 391–414.(*165*)

Traub, P., M. Nomura, and L. Tu (1966). Physical and functional heterogeneity of ribosomal proteins. *J. Mol. Biol.*, **19**, 215–218.(*159*)

Traub, P., D. Soll, and M. Nomura (1968). Structure and function of *E. coli* ribosomes. II. Translational fidelity and efficiency in protein synthesis of a protein deficient subribosomal particle. *J. Mol. Biol.*, **34**, 595–608.(*167*)

Traut, R. R. (1966). Acrylamide gel electrophoresis of radioactive ribosomal protein. *J. Mol. Biol.*, **21**, 571–576.(*159*)

Traut, R. R., and R. E. Monro (1964). The puromycin reaction and its relation to protein synthesis. *J. Mol. Biol.*, **10**, 63–72.(*107, 115*)

Traut, R. R., P. B. Moore, H. Delius, H. Noller, and A. Tissières (1967). Ribosomal proteins of *E. coli*. I. Demonstration of different primary structures. *Proc. Nat. Acad. Sci.*, **57**, 1294–1301.(*159*)

Travers, A. A. (1969). Bacteriophage sigma factor for RNA polymerase. *Nature*, **223**, 1107–1111.(*262*)

Travers, A. A., and R. R. Burgess (1969). Cyclic re-use of the RNA polymerase sigma factor. *Nature*, **222**, 537–540.(*138, 139, 141, 142, 143*)

Tremblay, G. Y., M. J. Daniels, and M. Schaechter (1969). Isolation of a cell membrane–DNA–nascent RNA complex from bacteria. *J. Mol. Biol.*, **40**, 65–76.(*340*)

Van de Putte, P., H. Zwenck, and A. Rorsch (1966). Properties of four mutants of *E. coli* defective in genetic recombination. *Mut. Res.*, **3**, 381–392.(*378*)

Van den Bos, R. C., and R. J. Planta (1968). Structural comparison of the two ribosomal RNA components of yeast. *Abstracts, Fed. Eur. Biochem. Soc., Madrid.* 191.(*155*)

Vaughan, M. H., J. R. Warner, and J. E. Darnell (1967). Ribosomal precursor particles in the Hela cell nucleus. *J. Mol. Biol.*, **25**, 235–251.(*171*)

Vasquez, D. (1966). Mode of action of chloramphenicol and related antibiotics. *Symp. Soc. Gen. Mic.*, **16**, 169–189.(*179*)

Venetianer, P., M. A. Berberich, and R. F. Goldberger (1968). Studies on the size of the messenger RNA transcribed from the histidine operon during simultaneous and sequential derepression. *Biochim. Biophys. Acta*, **166**, 124–133.(*284*)

Vermeulan, C. W., and K. C. Atwood (1965). The proportion of DNA complementary to *r*RNA in *D. melanogaster*. *Biochim. Biophys. Res. Commun.*, **19**, 221–226.(*171*)

Vielmetter, W., W. Messer, A. Schutte (1968). Growth, direction, and segregation of the *E. coli* chromosome. *Cold Spring Harbor Symp. Quant. Biol.*, **33**, 585–598.(*301*)

Visconti, N., and M. Delbruck (1953). The mechanism of genetic recombination in phage. *Genetics*, **38**, 5–33.(*57*)

Volkin, E., and L. Astrachan (1957). RNA metabolism in T2-infected *E. coli*. In W. D. McElroy, and B. Glass (Eds.), *The Chemical Basis of Heredity*, John Hopkins Press, Baltimore, U.S.A. pp. 686–695.(*98*)

Wagner, E. K., S. Penman, and V. M. Ingram (1967). Methylation patterns of Hela cell *r*RNA and its nucleolar precursors. *J. Mol. Biol.*, **29**, 371–387.(*176*)

Wahba, A. J., R. Mazumder, K. Iwasaki, Y–B. Chae, M. J. Miller, M. A. G. Sillero, and S. Ochoa (1968). Role of ribosome factors in polypeptide chain initiation. *Abstracts, Fed. Eur. Biochem. Soc. Madrid.* 9.(*130*)

Walker, P. M. B. (1969). The specificity of molecular hybridisation in relation to studies on higher organisms. *Prog. Nuc. Acid Res.*, **9**, 301–327.(*39, 41*)

Wallace, H., and M. L. Birnsteil (1966). Ribosomal cistrons and the nucleolar organiser. *Biochim. Biophys. Acta*, **114**, 296–310.(*172*)

Waller, J. P. (1963). The NH_2-terminal residues of the proteins from cell-free extracts of *E. coli. J. Mol. Biol.*, **7**, 483–496.(*119*)

Waller, J. P. (1964). Fractionation of the ribosomal protein from *E. coli. J. Mol. Biol.*, **10**, 319–336.(*159*)

Walter, G., W. Zillig, P. Palm, and E. Fuchs (1967). Initiation of DNA dependent RNA synthesis and the effect of heparin on RNA polymerase. *Eur. J. Biochem.*, **3**, 194–201.(*149*)

Warner, J. R. (1966). The assembly of ribosomes in Hela cells. *J. Mol. Biol.*, **19**, 383–398.(*178*)

Warner, J. R., M. Girard, H. Latham, and J. E. Darnell (1966). Ribosome formation in Hela cells in the absence of protein synthesis. *J. Mol. Biol.*, **19**, 373–382.(*177*)

Warner, J. R., P. M. Knopf, and A. Rich (1963). A multiple ribosome structure in protein synthesis. *Proc. Nat. Acad. Sci. USA*, **49**, 122–129.(*104*)

Warner, J. R., and A. Rich (1964). The number of *t*RNA molecules on reticulocyte polyribosomes. *Proc. Nat. Acad. Sci. USA*, **51**, 1134–1141.(*104*)

Watson, J. D., and F. H. C. Crick (1953a). A structure for DNA. *Nature*, **171**, 736–738.(*28*)

Watson, J. D., and F. H. C. Crick (1953b). Genetic implications of the structure of DNA. *Nature*, **171**, 964–967.(*28*)

Webster, R. E., D. L. Engelhardt, and N. D. Zinder (1965). *In vitro* protein synthesis: chain initiation. *Proc. Nat. Acad. Sci. USA*, **55**, 155–161.(*119*)

Webster, R. E., and N. D. Zinder (1969). Fate of the message-ribosome complex upon translation of termination signals. *J. Mol. Biol.*, **42**, 425–440.(*286*)

Weigert, M. G., and A. Garen (1965a). Amino acid substitutions resulting from suppression of nonsense mutations. I. Serine insertion by the su_1 suppressor gene. *J. Mol. Biol.*, **12**, 448–455.(*211*)

Weigert, M. G., and A. Garen (1965b). Base composition of nonsense codons in *E. coli. Nature*, **206**, 992–994.(*211*)

Weigert, M. G., E. Lanka, and A. Garen (1965). Amino acid substitutions resulting from suppression of nonsense mutations. II. Glutamine insertion by su_2 gene. Tyrosine insertion by su_3 gene. *J. Mol. Biol.*, **14**, 522–527.(*211*)

Weinstein, I. B., S. M. Friedman, and M. Ochoa Jr. (1966). Fidelity during translation of the genetic code. *Cold Spring Harbor Symp. Quant. Biol.*, **31**, 671–682.(*218*)

Werner, R. (1968). Initiation and propagation of growing points in DNA of phage T4. *Cold Spring Harbor Symp. Quant. Biol.*, **33**, 501–508.(*346*)

Whitehouse, H. L. K. (1963). A theory of crossing over by means of hybrid DNA. *Nature*, **199**, 1034–1040.(*368*)

Whitehouse, H. L. K. (1966). An operator model of crossing over. *Nature*, **211**, 708–713.(*373*)

Whitehouse, H. L. K. (1969). *The Mechanism of Heredity*, 2nd Edition, Edward Arnold, London.(*371*)

Whitehouse, H. L. K., and P. J. Hastings (1965). The analysis of genetic recombination on the polarom hybrid DNA model. *Genet. Res.*, **6**, 27–92.(*371*)

Wilkins, M. F. H. (1963). Molecular configuration of nucleic acids. *Science*, **140**, 941–950.(*28*)

Wilkins, M. F. H., A. R. Stokes, and H. R. Wilson (1953). Molecular structure of deoxypentose nucleic acids. *Nature*, **171**, 738–740.(*28*)

Willems, M., E. Wagner, R. Laing, and S. Penman (1968). Base composition of ribosomal RNA precursors in the Hela cell nucleolus: further evidence of non-conservative processing. *J. Mol. Biol.*, **32**, 211–220.(*176*)

Williamson, A. R., and B. A. Askonas (1967). Biosynthesis of immunoglobulins: the separate classes of polyribosomes synthesising heavy and light chains. *J. Mol. Biol.*, **23**, 201–216.(*293*)

Williamson, R., and G. G. Brownlee (1969). The sequence of 5S *r*RNA from two mouse cell lines. *Febs Lett.*, **3**, 306–308.(*157*)

Willson, C., D. Perrin, M. Cohn, F. Jacob, and J. Monod (1964). Non-inducible mutants of the regulator gene in the lactose system of *E. coli*. *J. Mol. Biol.*, **8**, 582–592.(*225*)

Witkin, E. M. (1967). Mutation-proof and mutation-prone modes of survival in derivatives of *E. coli* B differing in sensitivity to ultraviolet light. *Brookhaven Symp. Biol.*, **20**, 17–55.(*361, 363*)

Wittman, H. G. (1963). Tobacco mosaic virus mutants and the genetic coding problem. *Cold Spring Harbor Symp. Quant. Biol.*, **28**, 589–595.(*73*)

Woese, C. R. (1965). On the evolution of the genetic code. *Proc. Nat. Acad. Sci. USA*, **54**, 1546–1552.(*90*)

Woese, C. R. (1967). The present status of the genetic code. *Prog. Nuc. Acid Res.*, **7**, 102–172.(*71, 77, 89, 90, 91*)

Woese, C. R. (1968). The fundamental nature of the genetic code: prebiotic interactions between polynucleotides and polyamino acids of their derivatives. *Proc. Nat. Acad. Sci. USA*, **59**, 110–117.(*89, 90*)

Woese, C. R. (1969). Models for the evolution of codon assignments. *J. Mol. Biol.*, **43**, 235–240.(*88, 89*)

Wolf, B. M., M. L. Pato, C. B. Ward, and D. A. Glaser (1968). On the origin and direction of replication of the *E. coli* chromosome. *Cold Spring Harbor Symp. Quant. Biol.*, **33**, 575–584.(*301*)

Yaniv, M., A. Favre, and B. G. Barrell (1969). Evidence for interaction between two non-adjacent nucleotide residues in *t*RNA$_1^{val}$ from *E. coli*. *Nature*, **223**, 1331–1333.(*198*)

Yaniv, M., and F. Gros (1969). Studies on valyl-*t*RNA synthetase and *t*RNA^{val} from *E. coli*. II. Interaction between valyl-*t*RNA synthetase and valine acceptor *t*RNA. *J. Mol. Biol.*, **44**, 17–30.(*203, 205*)

Yanofsky, C. (1963). Amino acid replacements associated with mutation and recombination in the A gene and their relation to *in vitro* coding data. *Cold Spring Harbor Symp. Quant. Biol.*, **28**, 581–588.(*73*)

Yanofsky, C., B. C. Carlton, J. R. Guest, D. R. Helinski, and U. Henning (1964). On the co-linearity of gene structure and protein structure. *Proc. Nat. Acad. Sci. USA*, **51**, 266–272.(*70*)

Yanofsky, C., G. R. Drapeau, J. R. Guest, and B. C. Carlton (1967). The complete amino acid sequence of the tryptophan synthetase A protein (α-subunit) and its co-linear relationship with the genetic map of the A gene. *Proc. Nat. Acad. Sci. USA*, **57**, 296–298.(*70*)

Yanofsky, C., and J. Ito (1966). Nonsense codons and polarity in the tryptophan operon. *J. Mol. Biol.*, **21**, 313–324.(*266*, *267*, *269*)

Yanofsky, S. A., and S. Spiegelman (1962). The identification of the *r*RNA cistron by sequence complementarity. II. Saturation of and competitive inhibition at the RNA cistron. *Proc. Nat. Acad. Sci. USA*, **48**, 1466–1472.(*170*)

Yarus, M., and P. Berg (1969). Recognition of *t*RNA by isoleucyl-*t*RNA synthetase. Effect of substrates on the dynamics of *t*RNA-enzyme function. *J. Mol. Biol.*, **42**, 171–190.(*203*, *205*)

Yates, R. A., and A. B. Pardee (1957). Control by uracil of formation of enzymes required for orotate synthesis. *J. Biol. Chem.*, **227**, 677–692.(*221*)

Ycas, M. (1969). *The Biological Code*, North-Holland Publishing Co.(*86*, *88*)

Yoshida, K., and S. Osawa (1968). Origin of the protein component of the chloramphenicol particles in *E. coli*. *J. Mol. Biol.*, **33**, 559–570.(*179*, *180*)

Yoshikawa, H. (1967). The initiation of DNA replication in *B. subtilis*. *Proc. Nat. Acad. Sci. USA*, **58**, 312–319.(*342*, *346*)

Yoshikawa-Fukada, M. (1967). The intermediate state of ribosome formation in animal cells in culture. *Biochim. Biophys. Acta*, **145**, 651–663.(*178*)

Zabin, I. (1963). Proteins of the *lactose* system. *Cold Spring Harbor Symp. Quant. Biol.*, **28**, 431–435.(*267*)

Zachau, H. G., G. Acs, and F. Lipmann (1958). Isolation of adenosine amino acid esters from a ribonuclease digest of soluble, liver RNA. *Proc. Nat. Acad. Sci. USA*, **44**, 885–888.(*100*)

Zachau, H. G., D. Dutting, and H. Feldman (1966). Nucleotide sequences of two serine specific *t*RNAs. *Ang. Chem. Eng. Ed.*, **5**, 422–423.(*192*, *209*)

Zamenhof, S., and H. H. Eichoran (1967). Study of microbial evolution through loss of biosynthetic functions: establishment of 'defective' mutants. *Nature*, **216**, 456–458.(*261*)

Zimmerman, E. F. (1968). Secondary methylation of *r*RNA in Hela cells. *Biochemistry*, **7**, 3156–3163.(*176*)

Zimmerman, E. F., and B. W. Holler (1967). Methylation of 45S and *r*RNA precursors in Hela cells. *J. Mol. Biol.*, **23**, 149–161.(*174*)

Zinder, N. D., D. L. Engelhardt, and R. E. Webster (1966). Punctuation in the genetic code. *Cold Spring Harbor Symp. Quant. Biol.*, **31**, 251–256.(*286*)

Zipser, D. (1967). UGA: a third class of suppressible polar mutants. *J. Mol. Biol.*, **29**, 41–45.(*84*)

Zipser, D. (1969). Polar mutations and operon function. *Nature*, **221**, 21–25. (*264*)

Zipser, D., and A. Newton (1967). The influence of deletions on polarity. *J. Mol. Biol.*, **25**, 567–569.(*266*)

Zubay, G. (1962). A theory on the mechanism of RNA synthesis. *Proc. Nat. Acad. Sci. USA*, **48**, 456–461.(*144*)

Author Index

Papers with more than three authors are cited in the text as First Author and coworkers; in this case, although all the authors are given in the references and in this index, their names do not appear at this point in the text.

Subject index

An asterisk (*) indicates that the topic is indexed separately.